D1531859

AUTOMATION TECHNOLOGY AND

HUMAN PERFORMANCE:

CURRENT RESEARCH AND TRENDS

Proceedings of the Third Conference on Automation Technology and Human Performance held in Norfolk, VA, March 25-28, 1998

Executive Committee:

Mark W. Scerbo, *Old Dominion University*
Kristin Krahl, *Old Dominion University*

Scientific Program Committee:

Mark W. Scerbo, Chair, *Old Dominion University*
Raja Parasuraman, *The Catholic University of America*
Mustapha Mouloua, *University of Central Florida*
J. Raymond Comstock, Jr., *NASA Langley Research Center*

Cover Graphic Design by:

Victor Riley, Honeywell

AUTOMATION TECHNOLOGY AND

HUMAN PERFORMANCE:

CURRENT RESEARCH AND TRENDS

Edited by

Mark W. Scerbo, Ph.D.

Department of Psychology
Old Dominion University
Norfolk, VA 23529-02067

Mustapha Mouloua, Ph.D.

Center for Applied Human Factors in Aviation
Department of Psychology
University of Central Florida
Orlando, FL 32816-1780

LAWRENCE ERLBAUM ASSOCIATES, PUBLISHERS
1999   Mahwah, New Jersey   London

Copyright © 1999 by Lawrence Erlbaum Associates, Inc.
All rights reserved. No part of this book may be reproduced
in any form, by photostat, microform, retrieval system, or any
other means, without the prior written permission of
the publisher.

Lawrence Erlbaum Associates, Inc., Publishers
10 Industrial Avenue
Mahwah, New Jersey 07430

Library of Congress Cataloging-in-Publication Data

Automation technology and human performance : current research and
trends / edited by Mark W. Scerbo, Mustapha Mouloua.
        p.   cm.
    "Proceedings of the Third Conference on Automation Technology
and Human Performance held in Norfolk, VA, March 25-28, 1998"--
    Includes bibliographical references and index.
    ISBN 0-8058-3135-5 (alk. paper)
    1. Human-machine systems--Congresses. 2. Automation--Human
factors--Congresses.   I. Scerbo, Mark W. II. Mouloua, Mustapha.
III. Automation Technology and Human Performance Conference (3rd :
1998 : Norfolk, VA)
    TA167.A965 1998
    620.8'2--dc21   98-30954
                                                           CIP

Books published by Lawrence Erlbaum Associates are printed
on acid-free paper, and their bindings are chosen
for strength and durability.

Printed in the United States of America

10 9 8 7 6 5 4 3 2 1

TABLE OF CONTENTS

AIR TRAFFIC CONTROL

ADAPTIVE AUTOMATION

DESIGN AND INTERFACE ISSUES

TECHNOLOGY AND AGING

FOREWORD

Advancements in technology today are occurring at an unprecedented rate. Coupled with these changes is the increasing prevalence of automation. Automation technology can be found in the highly complex and unique systems in the cockpit of a Boeing 777 as well as in the common telephone answering machine. Automation promises to increase the sophistication of machine and systems operation while relieving users of burdensome activities.

Over the last twenty years, human factors researchers and developers have begun to examine more closely the impact that automation technology has on human performance. Indeed, many of the promises of automation have been realized in increased efficiency and in extending human capabilities. On the other hand, automation has fallen short of expectations on issues of safety and workload reduction. Recent research shows that automation alters the roles of operators and machines. As automated systems perform more and varied functions, humans are left with new kinds of responsibilities that often lead to decreases in situation awareness, increased mental workload, poorer monitoring efficiency, and a degraded ability to intervene and exercise manual control when automated systems fail.

In 1994, Mustapha Mouloua and Raja Parasuraman of the Cognitive Science Laboratory at the Catholic University of America organized a meeting to bring together researchers, developers, students, and users from a variety of domains to address human factors issues associated with automation and technology. Since that time, we have seen the successful exploration of Mars by NASA's *Sojourner* and the increasing system failures on the space station, *MIR*. We watched IBM's Deep Blue computer beat World Champion Garry Kasparov in chess and were introduced to Dolly, the first successfully cloned sheep. During this interval, the World Wide Web evolved into a major communication and commerce medium and the standard PC clock speed increased from 66 to 400 MHz. Also, during this time the aviation industry unwittingly provided headlines of catastrophes on almost a weekly basis. There were temporary blackouts of primary air traffic control systems in major metropolitan areas as well as high profile disasters both at home and abroad (e.g., the explosion of TWA Flight 800 outside of Long Island, NY; the collision of a Cessna Citation jet and a single-engine Cessna over an Atlanta suburb; the crash of an American Eagle ATR-72 commuter flight; and crashes of a Singapore operated 737 in Indonesia; a Chinese airbus in Taiwan; and a Russian Antonov-124 military cargo jet into a Siberian suburb. These examples serve to showcase the pinnacle of technological achievement, the horror of technological failures, as well as the rapid acclimation to and complacent acceptance of new technology. Moreover, they underscore the necessity for continued, systematic study of human interaction with automated technology.

The papers in this volume, *Automation Technology and Human Performance: Current Research and Trends* are from the Third Automation Technology and Human Performance Conference held in Norfolk, Virginia, on March 25-28, 1998. They are a representative sample of current experimental and investigative research concerned with the effects of automation and technology on human performance. Topics cover a variety of domains such as aviation, air traffic control, medical systems, and surface transportation, as well as user concerns such as situation awareness, stress and workload, and monitoring and vigilance behavior. Other sections address methodology and design issues. Further, the topics address both theoretical and applied aspects of human interaction with technology.

Like the previous volumes, this book should appeal to students, scientists, and researchers in academia, government, and industry. It should enable them to:

1. Keep current with basic and applied knowledge in several domains where automation technology is implemented.

2. Address training issues and guidelines for the design of intelligent machine systems and hybrid human-machine systems.

3. Discuss future trends in automation research as we approach the 21st century.

Finally, it is our hope that this volume will continue the trend established by its predecessors by making a valuable contribution to the academic, research, and professional communities at large. We are indebted to all of the participants who helped make the Third Conference on Automation Technology and Human Performance such a success and particularly, the authors who contributed to this volume. Their ideas, research, experiences, and dedication to understanding and improving the fit between people and technology have gone a long way to help ensure the effectiveness, efficiency, safety, reliability, and usability of the next generation of human-machine systems.

Mark W. Scerbo
Mustapha Mouloua

Acknowledgments

The editors would like to thank the following people for their help in organizing and running the conference which produced this book: Todd M. Eischeid, Gerry Hadley, Jamie L. LoVerde, Anthony Macera, Lawrence J. Prinzel, III, Victoria S. Schoenfeld, Julie Stark, Kenneth Vehec, Aaron Dudek, and Elaine Justice and Woody Turpin. Further, we owe a special debt of gratitude to the conference coordinator, Kristin Krahl.

We would also like to thank Daniel K. Mayes, Dennis Vincenzi, Jefferson Koonce, and Dennis Peache for their help in preparing this book.

We also wish to acknowledge and thank the following organizations for their help in sponsoring the Third Conference on Automation Technology and Human Performance. Their financial assistance contributed significantly to the success of this conference.

The Department of Psychology at Old Dominion University

The College of Sciences at Old Dominion University

NASA Langley Research Center

The Office of Naval Research

The Tidewater Chapter of the Human Factors and Ergonomics Society

Finally, Mark would like to thank Sue for her understanding. You can have me back now.

Mark W. Scerbo
Mustapha Mouloua
Norfolk, VA, 1998

FEATURED PRESENTATIONS

Automation in Air Traffic Control: The Human Performance Issues

Christopher D. Wickens
University of Illinois at Urbana-Champaign

INTRODUCTION

The North American air traffic control system has been an incredibly safe one, coordinating roughly 60 million operations (takeoffs and landings) per year; and not having been directly associated with a single fatality resulting from mid-air collisions, during that last 40 years. Such a record of safety is attributable to the skills of air traffic controllers, and to the inherent redundancies that are built into the system. At the same time, increasing demands for air travel are projected to place excessive demands on the capacity of the ATC system over the next 20 years, even as the current system appears to be incapable of meeting such demands (Aviation Week & Space Technology, 1998). In large part, such limitations are related to antiquated equipment, (Perry, 1997) as well as to the resolution of current radar systems limiting their ability to precisely identify the 3D location of aircraft. Three general approaches have been proposed to address these needs: **modernization** by which older, less reliable equipment will be replaced by newer equipment (Aviation Week & Space Technology, 1998); **automation**, by which computers will assume responsibility for certain tasks, previously performed by controllers (Wickens et al., 1998); and **free flight**, by which greater authority for maneuvering and traffic separation will be provided to pilots (RTCA, 1995a,b; Planzer & Jenny, 1995).

In late 1994, a panel was convened by the National Research Council, at the request of the House Aviation Subcommittee and the FAA, to examine the second of these issue--air traffic control automation--, and in the third year, the panel was also asked to examine the issue of free flight, from the controller's perspective (Wickens, Mavor, & McGee, 1997; Wickens, Mavor, Parasuraman, & McGee; 1998). The current paper highlights the conclusions of the panel regarding three critical human factors issues in the automation of the air traffic control system. While I provide here my own interpretation of these issues, I also acknowledge the major contributions of the co-editors of the panel reports, Anne Mavor, Raja Parasuraman, and Jim McGee in particular, and the remaining panel members in general[1]. I believe that my words below are quite consistent with the overall views of the panel.

All three of the major conclusions discussed below are anchored in roughly three decades of human performance research in automation, the focus of the present conference, and these conclusions clearly point to the relevance of this research to some of the most important safety issues confronting our society -- the safety of air travel -- even as they also lay out an agenda for more research, particularly addressing the concept of **trust**, and the human response to infrequent events.

Our panel spent three years studying air traffic control systems and learning about their details, benefiting from the expertise of two experienced controllers on the panel. We performed a cognitive task analysis, considering the delicate balancing acts that controllers must carry out between safety and efficiency; between ground and air concerns; and between following procedures to deal with routine situations, and using some creative problem solving to deal with the novel, unexpected events. We examined the cognitive vulnerabilities that should be supported by automation, as well as the cognitive strengths of controllers, that should not be neutralized or replaced by automation (Wickens, Mavor, & McGee, 1997).

---

[1]Charles B. Aalfs, Tora K. Bikson, Marvin S. Cohen, Diane Damos, James Danaher, Robert L. Helmreich, V. David Hopkin, Jerry S. Kidd, Todd R. Laporte, Susan R. McCutchen, Joseph O. Pitts, Thomas B. Sheridan, Paul Stager, Richard B. Stone, Earl. L. Wiener, Christopher D. Wickens, and Laurence R. Young.

We took a close look at ATC automated functionality that exists, is in test, or has been proposed. We coupled these with our understanding of human performance issues of automation that are now beginning to reach the maturity of some solid empirically-based conclusions and principles; and finally, we used as a backdrop the framework of **human-centered automation** discussed by Billings (1996), our definition focused both on the ideal properties of automation as well as the **processes** by which automation should be introduced into the workplace. We used this definition in order to offer a set of formal conclusions and recommendations, three of the most important of which will be the focus of this paper: levels of automation, failure recovery, and free flight. These are discussed in greater detail in Wickens et al. (1998).

LEVELS OF AUTOMATION

Several taxonomies of automation have been proposed by various authors. We choose to use as a framework, a scale of automation **level** proposed by Sheridan (1980), that varies in the degree of constraints placed upon what a human can do, and in the level of authority of what an automation system **will** do (See also Endsley & Kiris, 1995). However, as shown in Figure 1, we found it necessary to expand the concept of "levels" into three different "human-centered" scales of automated functionality, corresponding approximately to the human information processing functions of (1) Information integration to maintain situation awareness (through attention, perception and inference), (2) choosing or deciding upon actions, and (3) executing those actions. Each scale represents a continuum along which computers can do more or less "work," and we explain these continua in more detail below. Our first major conclusion is embodied in a set of recommendations for the appropriate "levels" that should be selected on these three scales. In general, we recommend that high levels of automation be pursued for information integration, as long as these processes are reliable. We recommend only intermediate levels for decision choice, if such decisions involve risk, and we recommend high levels of automation for response execution, as long as automation choice levels are not high. The basis for these recommendations are provided as we "zoom in" to each scale in turn.

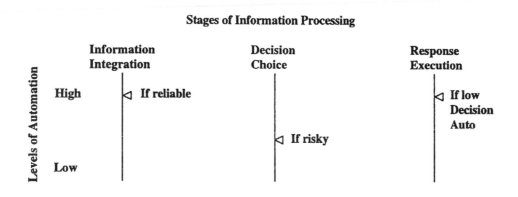

Figure 1. Three scales of automation functionality levels. The designated level on each scale is a recommended upper limit, given the qualifying conditions, as discussed in the text.

Information integration. We consider high levels of automation of information integration to result from the combination of several different features or functions of computer "work" on raw data, prior to presenting the human with a display with which to update situation awareness. One framework for looking at several of these functions is provided across the top row of Figure 2. Thus, computers may serve the critical **attentional** functions of selection and filtering, by **decluttering** displays (Mykityshyn, Kuchar, & Hansman, 1994) That is, by "hiding" or "revealing" certain classes of information. Alternatively, they may serve the attentional filtering function less severely by **highlighting** or **cueing** critical channels of information believed to be most relevant, while leaving less-relevant channels still visible, but "lowlighted" or uncued.

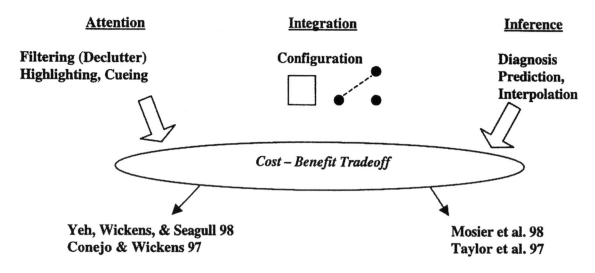

Figure 2. Examples of features of automation integration related to attention, integration and inference. Higher levels of automation will contain more of these features. However, if attentional filtering and inference making are less than perfectly reliable they will induce costs (when wrong), along with benefits (when right).

Computers may also carry out various forms of information integration, for example, by **configuring** separate channels of information graphically in the configural (Bennett & Flach, 1992) or object display (Wickens & Carswell, 1995). Finally, computers can draw **inferences** from raw data -- using relatively lower levels of intelligence in interpolating or predicting data, or using more sophisticated algorithms and heuristics in the process of **diagnosis**.

Across all three functions--attention, integration, and inference--human performance data support the conclusion that higher levels of automation, supporting higher levels of situation awareness are of near universal benefit (MacMillan, Deutsch, & Young, 1997; Parasuraman, Mouloua, Molloy, & Hilburn, 1996), **as long as such automation is reliable**, a qualification we address below. Careful task analysis can insure that highlighting and attentional cueing will guide users to the most relevant information, while low-lighting or backgrounding of less relevant information, will still make that information accessible should it be unexpectedly needed. Both integration and inference can substantially offload human working memory demands. In air traffic control automation systems, both the User Request Evaluation Tool (URET; Brudnicki et al., 1996), and the Controller TRACON Automation System (CTAS; Erzberger et al., 1993) offer positive examples of proposed functions that will integrate information for controllers, and draw inferences about future events (prediction) that can improve decision making, while reducing workload (Harwood, Sanford, & Lee, 1998).

We note, however, that in the case of both attention guidance (through cueing, filtering or highlighting), and inference (through prediction, interpolation or diagnosis), computers (or their software designers) must inherently make assumptions about information that is important for the user to see (attention guidance) or about states that the user is encouraged to believe (inference). These assumptions will be of considerable value when they are correct. But we do not fully understand the potential **costs** that may be incurred when they may be wrong. Thus, the cost benefit analysis shown in Figure 2 is of critical importance. Can automation provide benefits with no costs? Basic attentional work by Posner (1978) and his colleagues (Jonides, 1981), reveals, for example, that attentional cueing can sometimes provide benefits with no costs, and at other times dictates a symmetry or tradeoff between costs and benefits.

In more complex forms of automated cueing and inferencing these costs are beginning to be well illustrated by experimental results. For example, Yeh, Wickens, and Seagull (1998) examined an attention cueing tool for Army threat target recognition. When such a cue was present, subjects clearly detected the cued targets with greater facility than when cueing was not available. However, they were also more likely to MISS non-cued targets of considerably greater priority and danger, visible on the same scene with the cued target. Conejo and Wickens (1997) used target cueing or highlighting in a simulated air-ground targeting task. They found that on the occasions when the cue was **unreliable**, directing attention to something that was not the designated target, pilots were still very likely to choose this "non-target" as the target, despite the fact that the true target, whose position and identity was known to the pilot, was sometimes directly visible on the display.

In inference tasks, Mosier et al. (1998) provided pilots with partially reliable automation diagnostic aids, offering them advice regarding the nature of certain aircraft failures. In a finding echoing that of Conejo and Wickens (1997), the authors found that several pilots trusted (and followed) this advice even when it conflicted with directly visible and contradicting evidence. Taylor et al., (1997) examined a decision aid in an air-to-air combat scenario that would provide predictive advice as to enemy intent. These authors too found evidence that such advice was often followed, even when it conflicted with other visible evidence related to unexpected enemy actions.

What these four sets of experimental results have in common is the finding that humans may follow and believe unreliable automation even as the latter conflicts with evidence visible to their own eyes. This then is a direct manifestation of excessive **trust** in automation (Parasuraman & Riley, 1997); a trust which may induce complacency (Parasuraman, Molloy, & Singh, 1993), an issue we address below. The human factors research that is critical to address this issue, is how to design displays that can intrinsically signal the level of unreliability of the cueing or inference, in order to avoid the syndrome of overtrust by the user (and hence, perhaps avoid an overreaction to mistrust once the unreliability has been noticed). An example of such a design is the likelihood alarm (Sorkin, Kantowitz, & Kantowitz, 1988).

Degree of automation choice: The constraints. Shown in Table 1 is an elaboration of the middle scale in Figure 1: the scale of automation level of decision choice, which ranges from full authority of humans to consider and select all authorities, to mid-levels in which computers can recommend options, to still higher levels at which computers fully decide (Sheridan, 1980; Wickens et al., 1998). As with the scale of information integration automation, a key feature with this scale is the reduction of operator workload that is realized at higher levels, along with the possibility that decisions might be made better by computers. Yet research that is both relatively old, in the domain of autopilots and manual control (Ephrath & Young, 1981; Wickens & Kessel, 1981), as well as that of more recent vintage (Parasuraman, Molloy, & Singh, 1993; Endsley & Kiris, 1995; see Parasuraman & Riley, 1997), has addressed the concerns that result when such automation at its higher levels fails.

If the decisions made are those with little risk (e.g., the decision to implement one display viewpoint over another; or the decision to hand off control of an aircraft from one controller to another), then high levels of automation may suffer only minor consequences even if there are failures. However, if decisions have high risks associated with them (e.g., recommending a flight maneuver in heavy traffic), then three sources of human performance deficiencies at high automation levels mitigate against the use of those higher levels: (1) humans become less likely to **detect** failures in the automation itself, or in the processes controlled by the automation, if they were only passively observing the automation, (2) humans lose some awareness of the state of the processes controlled by the automation, and (3) humans will eventually lose skills in performing the decisions manually, if they are made only by automation. The first and second of these phenomena are often attributed to a cognitive state of **complacency** (Parasuraman, Molloy, & Singh,1993), in which operators are not forced to consider the nature of the decision, since it is made by automation, and hence, may direct attention to other tasks. Collectively, all three phenomena may be captured by what I have referred to as a syndrome of "Out of the Loop Unfamiliarity" or **OOTLUF**.

Table 1.
Levels of Automation (From Wickens, Mavor, Parasuraman, and McGee, 1998).
Scale of Levels of Automation of Decision and Control Action

HIGH 10. The computer decides everything and acts autonomously, ignoring the human
9. informs the human only if it, the computer, decides to
8. informs the human only if asked, or
7. executes automatically, then necessarily informs the human, and
6. allows the human a restricted time to veto before automatic execution, or
5. executes that suggestion if the human approves, or
4. suggests one alternative, and
3. narrows the selection down to a few, or
2. The computer offers a complete set of decision/action alternatives, or
LOW 1. The computer offers no assistance: the human must take all decisions and actions.

Response execution automation. Many simple automation devices simply replace our own muscles in executing the same task. In air traffic control, an automated handoff will automatically relay responsibility of an aircraft to an adjacent sector, without requiring vocal requests and communications by the two controllers involved. We suggest that, as long as the decision to implement that action has retained some level of human involvement (middle scale of Figure 1), then it is appropriate for automation to execute the action (for example, the choice by the human of a decision option could serve to trigger automatic implementation of the action). But we have concerns if automation levels on both decision and execution scales are high, since in this case, "error trapping" mechanisms have been eliminated from the system; the operator cannot "catch" a risky decision that may be incorrect, if he has no opportunity either to actively approve the choice, nor implement its actions. We note that it is redundancy such as this that has inherently preserved the safety of ATC.

FAILURE RECOVERY SCENARIOS

Section 2 has provided some evidence about how the human-computer air traffic control system COULD break down under high levels of automation -- a failure occurring while the human is complacently supervising an airspace about which he has less than desirable awareness and for which manual control intervention skills may be degraded. As Bainbridge (1983) has pointed out, an irony in such a situation is that the **more** reliable the automation, the **more** susceptible to OOTLUF will the operator be if the failure **does** occur, both because of increased complacency and because of lower skills of manual intervention. Our panel also recognized that the explicit goal of ATC automation is to increase the capacity of the airspace, a goal that will likely be achieved, in part, by reducing the separation between aircraft (i.e., so that more aircraft can land and take off per unit time). These two assumptions about future ATC automation -- a potential invitation for OOTLUF and a reduced separation between aircraft -- meet face to face in the analysis of a failure recovery situation: that is, the time **available** to safely respond to an emergency situation will be diminished (by decreased separation), just as the time **required** to respond (recovery response time) may well be increased by OOTLUF. Figure 3 provides an overview of this tradeoff on the right side of

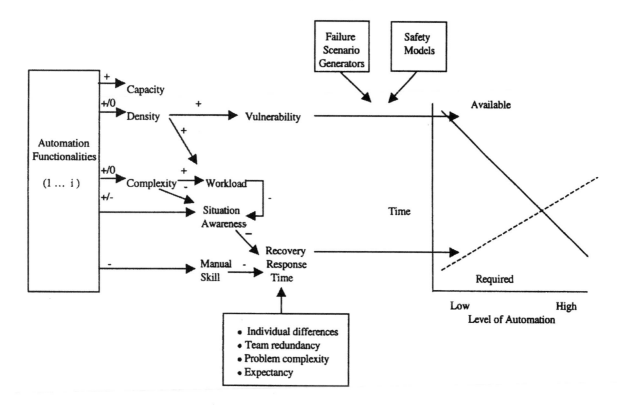

Figure 3. Model of failure recovery in air traffic control. Where two nodes are connected by an arrow, signs (+, -, 0) indicate the direction of effect on the variable depicted in the right node, caused by an increase in the variable depicted in the left node. From Wickens, Mavor, Parasuraman, and McGee (1998).

the figure, as it might be reflected in higher levels of decision automation. That is, at higher levels of automation, the time available diminishes, as the time required to respond appropriately may increase. On the left side of the figure we consider the possible influences of different forms of automated functionalities for air traffic control, like predictors or decision aids. For example, certain functionalities--operating successfully to increase capacity--will increase the density of the airspace, thus directly influencing the "vulnerability" to accident (inherent in the reduced separation). As shown in the bottom center of the figure, these same functionalities may also have effects on controller cognitive processes, mediated by workload, situation awareness, and the manual skill proficiency necessary to assume control should automation fail (e.g., visual prediction should predictor symbology fail, working memory for aircraft position should radar fail).

There are adequate human performance data to be confident that the direction of effects shown in the figure are correct, for example, reduced situation awareness will increase the time to respond appropriately, and higher levels of workload will reduce situation awareness. However, we have very little data to suggest what the **actual numerical estimates** of response time may be, particularly in "worst case" situations of unexpected failures. Those data we do have suggest that these numbers may be disconcertingly

7

high. For example, Beringer (1996) found that the time for well-trained general aviation pilots to notice and diagnose the presence of a serious system failure in an automated aircraft was sometimes in the 30-40 second range. In one of the few naturalistic studies of driver reaction time, Summala (1981) observed response times of well over 4 seconds (by slower responders) to an unexpected event on the roadway ahead. Ozmore and Morrow (1996) observed a few dangerously long response times for pilots on simulated parallel runway approaches to respond to controller directives regarding a dangerous incursion of the parallel flight into the pilots' own path. Research on head-up displays (Wickens, 1997b) has revealed correspondingly long response times (in excess of 15 seconds) for pilots reacting to unexpected obstructions on a landing runway.

The point of concern here is that much of the basic laboratory data on response time to unexpected events may not scale up well to RT to events that are not only infrequent, but are truly unexpected because they may never have been experienced by the operator (or team of operators) in question (Huey & Wickens, 1993). We also emphasize here the parameters of the **distribution** of such response times, rather than the mean, since it is often the worst case or slowest response that will be the weak link in the safety of such a system (Wickens, 1997a).

In conclusion thus, our recommendations regarding the second issue of failure recovery is that a great deal of human performance data on emergency team responses to unexpected events must be collected before models can adequately predict the critical crossover point between time available and time required shown on the right side of Figure 3.

Free flight. Neither the movement toward reducing current ATC constraints on air routes generally assumed to be accomplished by providing pilots with greater authority for maneuvering space nor time allows me to deal extensively with the issues addressed by the panel regarding free flight. Furthermore, in some sense, free flight is not an automation issue (pilots, rather than controllers **or** automation will be making more decisions on routings); although free flight will depend heavily upon much automation technology to improve inter-cockpit communications. However, in brief, we reached the conclusion that whether authority was shifted from the controller to high levels of decision automation, or to pilots, the human performance concerns of OOTLUF for the controller would be similar. That is, when other agents, whether computers or pilots, are making active decisions about flight trajectories, the abilities to note breakdowns in the system, and deal rapidly with them should controller intervention be required, would be compromised. Furthermore, in both cases, if the (free flight or automation) system worked effectively as planned, the reduced separation would increase the vulnerability, as shown in Figure 3.

However, we also concluded that free flight would add at least one other source of cognitive vulnerability, that was NOT necessarily evident in high levels of ground based automation. This is the increased **complexity** of the airspace that would result from pilots flying locally preferred routing (favorable to local weather conditions, and negotiating local traffic conflicts). Such complexity would likely be enhanced considerably above its level in the current airspace (Wyndemere, 1996); or even an airspace designed with considerably greater efficiency than that of today (Wickens et al., 1998). Controllers, like chess masters, or experts in any other highly visual domain, depend fairly heavily upon certain consistent rule-based structures of their visual environment, in order to maintain awareness of the dynamic changes in that environment (Durso et al., in press; Endsley, 1995). If that structure is badly degraded, as would be the case to the extent that pilots are given their own decisions for maneuvering, this fact could render the timeliness of effective intervention of even greater concern than in the case of ground-based automation failure.

In conclusion, I have highlighted here, only three of the major conclusions and recommendations from our Panel. Several others pertain to the human factors of implementing automation in the workplace, as well as to specific automation systems like data link, or collision protection systems in the air, and on the airport surface. What is important about the three I have highlighted however, is their foundation in the fundamental psychological research on human interaction with automation, research which this conference has helped to foster. I trust that the papers that follow will continue to contribute to this exciting and growing research domain.

ACKNOWLEDGMENTS

I acknowledge the financial support provided by the Federal Aviation Administration for the 3-year panel study whose conclusions provide the basis for this paper. Thanks also go to Susan McCutchen of the National Research Council whose extensive administrative skills were responsible for the smooth functioning of the pond and the preparation of its reports.

REFERENCES

Aviation Week and Space Technology (Feb. 2,1998). Answers to the Gridlock (pp. 42-62).

Bainbridge, L. (1983). Ironies of automation. *Automatica, 19*(6), 775-779.

Bennett, K. B., & Flach, J. M. (1992). Graphical displays: Implications for divided attention, focused attention, and problem solving. *Human Factors, 34*(5), 513-533.

Billings, C. (1996). *Toward a human centered approach to automation.* Englewood Cliffs, NJ: Lawrence Erlbaum.

Beringer, D. B. (1996). Automation in general aviation: Responses of pilots to autopilot and pitch trim malfunctions. *Proceedings of the 40ᵗʰ Annual Meeting of the Human Factors & Ergonomics Society* (pp.86-90). Santa Monica, CA: Human Factors Society.

Brudnicki, D. J., McFarland, A. L., & Schultheis, S. M. (1996). *Conflict probe benefits to controllers and users* (MP 96W0000194). McLean, VA: MITRE Corporation.

Conejo, R., & Wickens, C.D. (1997). *The effects of highlighting validity and feature type on air-to-ground target acquisition performance.* University of Illinois Institute of Aviation Technical Report (ARL-97-11/NAWC-ONR-97-1). Savoy, IL: Aviation Res. Lab.

Durso, F. (Ed.). (in press). *Handbook of Applied Cognition.* Sussex, England: Wiley.

Endsley, M. R. (1995). Toward a theory of situation awareness in dynamic systems. *Human Factors, 37*(1), 85-104.

Endsley, M.R., & Kiris, E.O. (1995). The out-of-the-loop performance problem and level of control in automation. *Human Factors, 37*(2), 381-394.

Ephrath, A.R., & Young, L. R. (1981). Monitoring vs. man-in-the-loop detection of aircraft control failures. In J. Rasmussen & W. B. Rouse (Eds.), *Human detection and diagnosis of system failures.* New York: Plenum Press.

Erzberger, H., Davis, T. J., & Green, S. (1993). Design of center-TRACON automation system. *AGARD Conference Proceedings 538: Machine Intelligence in Air Traffic Management* (pp. 11-1/11-12). Seine, France: Advisory Group for Aerospace Research & Development, AGARD.

Harwood, K., Sanford, B. D., & Lee, K. K. (in press, 1998). Developing ATC automation in the field: It pays to get your hands dirty. *The Air Traffic Control Quarterly.*

Hopkin, D. (1995). *Human factors in air traffic control.* London: Taylor & Francis.

Huey, M.B., & Wickens, C. D. (Eds.). (1993). *Workload transition: Implications for individual and team performance.* Washington, D.C.: National Academy Press.

Jonides, J. (1981). Voluntary versus automatic control over the mind's eye movement. In J. B. Long & A. Baddeley (Eds.), *Attention and Performance IX* (pp. 187-203). Hillsdale, NJ: Lawrence Erlbaum.

MacMillan, J., Deutsch, S. E., & Young, M. J. (1997). A comparison of alternatives for automated decision support in a multi-task environment. *Proceedings of the 41ˢᵗ Annual Meeting of the Human Factors & Ergonomics Society* (pp. 190-194). Santa Monica, CA: Human Factors Society.

Mosier, K. L., Skitka, L. J., Heers, S., & Burdick, M. (1998). Automation bias: Decision making and performance in high-tech cockpits. *The International Journal of Aviation Psychology, 8,* 47-63.

Mykityshyn, M. G., Kuchar, J. K., & Hansman, R. J. (1994). Experimental study of electronically based instrument approach plates. *The International Journal of Aviation Psychology, 4*(2), 141-166.

Ozmore, R. E., & Morrow S. L. (1996). *Evaluation of dual simultaneous instrument landing system approaches to runways spaced 3000 feet apart with one localizer offset using a precision runway monitor system* (DOT/FAA/CT-96/2). Atlantic City International Airport: Federal Aviation Administration.

Parasuraman, R. Molloy, R., & Singh, I. L. (1993). Performance consequences of automation-induced complacency. *The International Journal of Aviation Psychology, 3*(1), 1-23.

Parasuraman, R., Mouloua, M., Molloy, R., & Hilburn, B. (1996). Monitoring of automated systems. In R. Parasuraman & M. Mouloua (Eds.), *Automation and human performance: Theory and applications* (pp. 91-115). Mahwah, NJ: Lawrence Erlbaum.

Parasuraman, R., & Riley, V. (1997). Humans and automation: Use, misuse, disuse, abuse. *Human Factors, 39,* 230-253.

Perry, T. S. (1997). In search of the future of air traffic control. *IEEE Spectrum* (August): 19-35.

Planzer, N., & Jenny, M. T. (1995). Managing the evolution to free flight. *Journal of Air Traffic Control* (Jan-Mar), 18-20.

Posner, M. I. (1978). *Chronometric explorations of the mind.* Hillsdale, NJ: Lawrence Erlbaum.

RTCA, (1995a). *Report of the RTCA Board of Directors' Select Committee on Free Flight.* Washington, DC: RTCA, Inc.

RTCA (1995b). *Free flight implementation. RTCA Task Force 3 Report.* Washington, DC: RTCA Inc.

Sheridan, T. B. (1980). Computer control and human alienation. *Technology Review, 10,* 61-73.

Sorkin, R. D., Kantowitz, B. H., & Kantowitz, S. C. (1988). Likelihood alarm displays. *Human Factors, 30,* 445-460.

Summala, H. (1981). Driver/vehicle steering response latencies. *Human Factors, 23,* 683-692.

Taylor, R. M., Finnie, S., & Hoy, C. (1997). Cognitive rigidity: The effects of mission planning and automaton on cognitive control in dynamic situations. *Proceedings of the 9th International Symposium on Aviation Psychology* (pp. 415-421). Columbus, OH: Dept. of Aviation, The Ohio State University.

Wickens, C.D. (1997a). The tradeoff of design for routine and unexpected performance: Implications of situation awareness. In D. J. Garland & M. R. Endsley (Eds.), *Situation awareness analysis and measurement.* Mahwah, NJ: Lawrence Erlbaum.

Wickens, C .D. (1997b). Attentional issues in head-up displays. In D. Harris (Ed.), *Engineering psychology and cognitive ergonomics: Integration of theory and application.* London: Avebury Technical Pub. Co.

Wickens, C .D., & Carswell, C.M. (1995). The proximity compatibility principle: Its psychological foundation and relevance to display design. *Human Factors, 37*(3), 473-494.

Wickens, C. D., & Kessel, C. (1981). Failure detection in dynamic systems. In J. Rasmussen and W. Rouse (Eds.), *Human detection and diagnosis of system failures.* New York: Plenum Press.

Wickens, C. D., Mavor, A. S., & McGee, J. P. (Eds.) (1997). *Flight to the future: Human factors in air traffic control.* Washington, DC: National Academy Press.

Wickens, C .D., Mavor, A., Parasuraman, R., & McGee, J. M. (1998). *The future of air traffic control: Human operators and automation.* Washington, DC: National Academy Press.

Yeh, M., Wickens, C. D., & Seagull, J. (1998). *Effects of frame of reference and viewing condition on attentional issues with helmet mounted displays.* University of Illinois Institute of Aviation Technical Report (ARL-98-1/ARMY-FEDLAB-98-1). Savoy, IL: Avia. Res. Lab.

Wyndemere, (1996). *An evaluation of air traffic control complexity* (Final Report Contract # NAS 2014284). Boulder, CO: Wyndemere.

Automation and Human Purpose:
How Do We Decide What Should Be Automated?

Raymond S. Nickerson
Tufts University

I was very pleased to be asked to participate in this conference, but I had some difficulty deciding on a topic. One possibility that came to mind was the historical roots of automation. In A *history of western technology,* Klemm (1964) points out the importance that the role of changing attitudes toward human rights in the much-maligned middle ages had in the building of a civilization that "no longer, like that of antiquity, rested on the backs of slaves, but to a greater and greater extent derived power from machinery" (p. 56). In classical Greece and Rome, among other ancient cultures, labor was viewed as the province of slaves, and human beings were the primary source of power. Progress during the Middle Ages in tapping the wind, water, and draught animals as alternatives to slaves as power sources was made in the context of promotion by the church of the idea that all people are created free. The building of sailing ships that could be used without oarsmen and the use of water wheels to drive mills of various sorts are only two of many technological innovations that appeared during this time. It would be fascinating to consider the historical roots of automation, especially with a view to attempting to understand some of the more influential psychological and cultural factors involved, but I decided not to try this.

Another possible topic that struck me as particularly interesting was automation in literature. In *The fourth discontinuity,* Bruce Mazlish (1993) discusses this, citing such examples as "The Nightingale" of Hans Christian Andersen, Mary Shelley's *Dr. Frankenstein,* the Tiktok of L. Frank Baum's Oz stories, and Karel Kapek's robots in his play *R.U.R.*. More recent well-known fictional treatments of humanoid robots include Isaac Asimov's *I Robot,* and Arthur Clarke's *2001: A Space Odyssey.* There are many other works of fiction in which automation, or some aspect thereof, figures as a prominent theme. Charlotte Bronte's *Shirley,* George Orwell's *1984,* and Kurt Vonnegut's *Player Piano* are among them. In many of these works, automation has received somewhat bad press. It would be interesting to consider whether this is representative of the treatment of automation in literature generally and, if so, why that is the case. But again, I decided not to go down this path.

A number of other topics suggested themselves as possibilities. The one that I decided I would most like to spend some time thinking about was automation and human purpose. It strikes me as not only a very important topic, but one that probably does not get the attention it deserves from technologists. So, for better or worse, that is what I will focus on in this paper. What I propose to do is toss out a few of the questions that consideration of this topic brings to mind. The questions are not rhetorical; I do not come with answers to them. My hope in expressing them is to make the case that they are worthy of continuing thought and reflection.

WHAT SHOULD BE AUTOMATED?

The technology of our age is information technology, by which I mean computer and communication technology, and especially their blending. Progress in this area during the last half of the 20th century has been nothing short of phenomenal, and its importance for automation would be difficult to overstate. In large part because of the rapid development of information technology, the future of automation appears to be unbounded. Many processes -- including especially cognitive or intellective processes -- that one could hardly have imagined being automated even a few decades ago have been automated or now appear to be automatable within the foreseeable future. Given the apparently unlimited possibilities, the question naturally arises: What *should* be automated?

One answer to the question is "everything that can be automated." One climbs the mountain because it is there. One automates because it is possible to do so. This is an answer that might be given by those who see automation primarily as an intellectual challenge, whose main goal is to push the boundaries of what is technologically feasible. We might call this an engineer's, or, more generally, a problem-solver's, view.

I want to tell a story -- apocryphal, I presume, but with a point. During the Reign of Terror following the French Revolution, public executions were commonplace in Paris. At one such event, three men waited their turn to experience the marvelous efficiency of Dr. Guillotin's invention, which had proven to be far more reliable and effective than the sometimes clumsy axemen it had replaced. The first condemned man to mount the scaffold and place his neck beneath the poised blade was a priest. When the executioner pulled the trip cord, the machine malfunctioned and the blade did not descend. Interpreting this as an omen, the crowd insisted that the man be freed, and he was. The next person in line was a politician. The machine again malfunctioned, and again the man was released at the insistence of the crowd.

The third person in line, who happened to be an engineer, observed these proceedings with intense interest, as one might imagine. As he mounted the scaffold to take his turn, he paused long enough to get a better look at the device that, if it worked properly, would presently snuff out his life. "You know," he said, to the official in charge, "I think I see your problem."

The point of this story is, of course, that some problems might better be left unsolved. Some intellectual challenges might be better never met. It makes sense, before attempting to solve a problem, to try to understand the larger implications of doing so. The fact that a problem *can* be solved does not mean that it necessarily *should* be solved. The fact that a process can be automated does not mean that it should be automated.

HOW SHOULD WE DECIDE WHAT SHOULD BE AUTOMATED?

But if technical feasibility is not a sufficient criterion for determining what should be automated -- or if even technical and economic feasibility in combination are not so -- how should we decide what should be automated? What should the considerations in making this type of decision be? Imagine that you were the czar of automation with the power to automate anything you wanted to automate. How would you decide what should be automated?

It is easy to identify several reasons why we *do* automate, or attempt to automate, various processes. Four that come immediately to mind are financial benefit, safety, capability and preference.

Processes are often automated because it is believed that their automation will be financially beneficial. Productivity will be increased, or the benefit-cost ratio of some product will be improved. I say "believed," because there are many examples of financially-motivated automation, or the application of technology more generally, that have not yielded the improvements in productivity or benefit-cost ratios they were expected to yield. Thomas Landauer makes the point forcefully in *The Trouble with Computers.* On the other hand, I have no doubt that gains in productivity or benefit-cost ratios have sometimes been realized, at least as evidenced by conventional measures of these variables.

Sometimes automation has been motivated by a desire to make operations and processes safer. Replacing human operators by robots can remove humans from dangerous work situations. One would prefer, for example, to have robots run mine detectors through mine fields than to have.
humans do so. Or to work in environments that are radioactive, chemically contaminated, or otherwise toxic or potentially injurious to humans. Automation can also increase safety in the sense of making systems -- especially systems that have the potential for failures with catastrophic consequences -- less vulnerable to human error. Systems with the capability to shut down automatically when they get close to a run-away state illustrate the point.

Automation can make it possible to do things that could not be done without it. Effective exploration of the surface of Mars, or other planets, would not be possible, at the present time, without the use of automation. Even assuming a device with the ability to move around the surface could be landed on a planet without the aid of automation, controlling the movement of such a device in real-time from the earth would be highly impractical, given the round-trip communication time of several minutes between the earth and even the nearest of the planets.

There are many processes that can be performed either by humans or by machines that most humans prefer not to perform. Each of us could come up with his or her own list of such processes. My list would certainly

12

include regulating the temperature in my house, and washing clothes and dishes. I suspect that most of us would find it easy to generate a list of things that we now do ourselves that we would gladly pass off to machines, were it feasible to do so.

I think these are all legitimate reasons for automating, provided that they do not conflict with compelling reasons *not* to automate in specific instances.

ARE THERE COMPELLING REASONS NOT TO AUTOMATE?

Job destruction has been a concern associated with automation since at least as early as the 16th century. This concern was epitomized by the machine-demolishing activities of the Luddites in England in the early part of the 19th century. The counter argument to the concern is that automation has typically created more jobs than it has made obsolete. This argument is easier to accept if one is not among those whose jobs have been made obsolete and are unable to accommodate to the change. How automation will affect job opportunities in the future is a debatable question and one can find predictions that range from dire to rosily optimistic.

The possibility of job deskilling -- the replacement of jobs that require skilled workers with those that do not -- is a closely related concern. Automation has had the effect of decreasing the skill requirements of some jobs to the point of making them almost intolerably boring, and it has made other jobs more interesting and satisfying. Again, as we look to the future, it is possible to find a range of predictions regarding what the long-term effects of automation on job-skill requirements will be.

One effect of automation, in some contexts, is depersonalization of human services. Automated phone systems illustrate the point. I do not remember where I recently saw the following quip (or something close to it): "To place an order, press 1; to check a balance, press 2, to leave a message, press 3; if you wish to speak to a human being, move to a small New England hamlet." Automated phone systems usually mean increased efficiency and decreased costs from the service providers' point of view; their advantages from the customers' perspective are not so clear. Unquestionably they decrease human contact. It is a pleasure, is it not, to call a company and be surprised by having the phone answered by a human being.

Automation is sometimes used to create the illusion of familiarity. "Personalized" letters, in which one's name is inserted by computer in appropriate places in the text is a case in point. To my taste, it is a particularly irritating one.

Was there more personal contact when things were less automated? It is easy to think of instances in which this was the case. I am old enough to remember going to the grocery store, giving the grocer a list my mother had written, and having him get the items from the shelves, writing the prices on a paper bag and adding them up by hand. I am not suggesting that that was a better system than the one we have today -- but it was more personal. We knew the grocer and he knew us. My own family has lived in the same town and bought groceries at the same supermarket for almost 40 years. But I don't know the grocer's name; indeed, there is no "grocer" in the old sense of the term.

Will the ubiquitous computer increase or decrease interpersonal interactions in the future? I wonder about this when I see children spending what seem to me to be excessive amounts of time playing computer games, including games, like chess, that, until recently, one normally played against a human opponent.

You may have read about the results of a recent study by researchers at the Johns Hopkins University School of Medicine on the problem of obesity among U.S. children. They found a correlation between obesity and the amount of time spent watching TV, and they concluded from their data that a fourth of all children in the U.S. watch four or more hours of TV per day. There are few, if any, indications in the results of educational testing that U.S. children are benefitting intellectually from this experience. I worry a little that computer technology has the potential of becoming an even more addictive time killer than (much of) TV has proved to be.

Will computer-mediated interperson communication more and more replace face-to-face, or voice-to-voice, interperson communication? And if it does, is this good or bad? Or, perhaps a better question is, what are the good and bad aspects of this likely to be?

My dad was a letter carrier for 40 years. He knew just about everyone in his town, as individuals. I know lots of people by name -- correspond by regular mail and email with several hundred people a year just in the course of editing a journal, but know relatively few of them as individuals. I am not making a value judgment here, but I find this contrast thought-provoking.

I recently attended a talk by a VP of Information Systems at a major U.S. university. In his enthusiasm for expanding the number of university functions that can be automated, he raised the question: How do we structure things so that a student can spend four years at the university and never realize there is a registrar's office on campus? My question is: Why should we want to do that?

Are these compelling reasons not to automate? I doubt it. But they are compelling reasons to attempt to anticipate the probable effects of automating specific tasks and functions and to take the measures necessary to minimize the negative ones. As is true of most complex phenomena, automation can be seen from more than one perspective. It is seen by some people as a means of empowering workers and dignifying work and by others as a cause of deskilling workers and debasing work. It can be seen as increasing the range of possibilities for communication between and among people; it can also be seen as creating impediments to truly personal contact. The potential for all of these scenarios, and many others, is there; which of them will be realized remains to be determined.

WHAT PURPOSES SHOULD AUTOMATION SERVE?

I gave a talk recently at a university colloquium on the topic: "How do we tell what questions are worth asking?" One reason for choosing this topic was the belief that too little attention is given in graduate programs to this question. Why graduate students -- and more seasoned researchers, for that matter -- choose the research problems they do is an interesting question, and one that invites speculation.

I wonder what percentage of psychology departments make the problem of deciding what research to do an explicit focus of graduate training. My suspicion, based only on conversations with a few colleagues, is that it is a small one. If the suspicion is correct, I think this is unfortunate. A course designed to get PhD candidates to think carefully about how to decide what research is worth doing would, in my view, be a very valuable element of their graduate training (Nickerson, in press). The goal of such a course should not be to ensure that all students would necessarily acquire the same set of criteria for selecting research problems, but that they all would have thought deeply about the issue and arrived at some working conclusions, consistent with their own values, about what makes a research question worth asking.

A similar question can be raised about the problem of deciding what processes should be automated. I suspect that less attention is given by technologists to this problem than it deserves -- that the question of whether some process *can* be automated is more likely to motivate thought and discussion than that of whether it *should* be. But "Why?" is a question that should be asked with respect to any effort to automate. Not only is it a legitimate question to ask in any particular instance, I want to argue that not to ask it is a dangerous mistake.

It is much too easy to assume, unthinkingly, that what can be done should be done. Bright and energetic people who want to demonstrate what they can do can easily lose sight of the purposes of their activities, and what were originally intended to be means can become ends in themselves. Management-information systems sometimes outgrow their purposes and become more burdensome than helpful, in my view, for just this reason. If a number *can* be produced by a management information system, someone is very likely to assume that it *should* be produced, and will produce it, whether anyone really needs it or not.

Many possible applications of automation undoubtedly are desirable and beneficial. Equally undoubtedly, some are silly, and others are dangerous. It is important to try to distinguish among these types. And the distinction needs to be made in terms of human purposes and values -- in terms of what the effects of automation, in any particular instance, are likely to be on aspects of life that matter to us. Identifying goals and objectives toward which developers of automation should work, because they are desirable from a humanistic point of view, should be a continuing concern.

HOW SHOULD THE EFFECTS OF AUTOMATION BE ASSESSED?

This is a very difficult question, in part because attempting to answer it is likely to require dealing with tradeoffs that are not easily weighed in the balance. Evaluating technological developments generally is complicated by the fact that what has constituted a solution to one or more problems often has created one or more others.

• Development of industrial mass production techniques have made consumer goods available on a much greater scale than otherwise would be possible. It also has contributed to the depletion of scarce resources, to the pollution of the environment and, in some instances, to the depersonalization of the worker.

• Development and use of pesticides have increased agricultural yields manyfold, but have also contributed to air and water pollution and to the evolution of increasingly hardy strains of pests.

• Development of the commercial nuclear reactor has provided a new source of energy, but at the expense of the prospects of nuclear accidents and the problem of disposal of nuclear wastes.

One could generate a long list of such tradeoffs, all of which represent difficult problems of judgment and decision regarding what the proper balance between the trading variables is. We have clearly gained much from automation, as from technology more generally. But we have also lost something. How can one put this all in proper perspective? Get a balance-sheet view? Assets and liabilities. Benefits and costs. Pros and cons. I raise the question not because I have an answer to propose, but because I believe it deserves thought -- especially by people working on the cutting edge of technological development.

One standard approach to the assessment of any technology is to attempt to measure its effect on productivity. This can be productivity of an individual, a business, an industry, or a nation. I believe that productivity is an important concern, but I have some misgivings about the amount of emphasis that seems to be put on it. Some discussions of it would lead one to believe that it is all that matters.

I have two concerns about this. First is the fact that, in many contexts, how productivity should be measured -- what it means -- is far from clear. This is especially true in white collar work, in service industries -- as distinct from agriculture or manufacturing. How does one measure the productivity of a school teacher, of an artist, of an entertainer, of a scientist? Quantity of output is a very crude yardstick and, in some cases, probably not applicable even in principle.

My other concern is that, even when it is reasonably clear what productivity means, I am not convinced that it is the be-all and end-all that it is often made out to be. According to one point of view, the best hope of increasing the standard of living across the world is to increase the productivity of the global workforce. What makes this relevant in the present context is that increasing automation is seen by many as perhaps the most effective way of raising productivity.

It may be true that raising productivity globally is the best hope of increasing the standard of living worldwide. I confess to believing it strongly at one time. I argued in a book published a few years ago that a major benefit of increased productivity, worldwide, is that "it makes goods and services more widely available to consumers at relatively lower prices. Increased productivity, in the sense of more efficient, less wasteful, use of energy and other resources, is, in the long run, good for everybody; it is the best hope of being the rising tide that can lift all ships" (Nickerson, 1992)

I have had some second thoughts about this, and now wonder if the great emphasis we put on productivity tends to make us insensitive to other considerations that are equally important to general human well-being. I think it is a great tragedy that many people, from all appearances, do not like what they do for work. They do it, not because it is an interesting, challenging, satisfying or, in any way, fulfilling, but because it is necessary to make a living.

What may be lost in the focus on productivity is the fact that work serves purposes other than that of "making a living." The time that one spends at work represents a very significant fraction of one's *life*. To what extent should increased productivity, however measured, be bought at the expense of having people perform tasks that give them no pleasure, no satisfaction except what is represented by the paycheck, no sense of accomplishment, of personal growth or development?

HOW MIGHT AUTOMATION INCREASE THE MEANINGFULNESS OF WORK?

The nature of work has been changing for a long time, and the rate of change has been increasing. Automation is one of the factors that has been effecting this change and is likely to continue to be so. One of my sons-in-law is a medical technologist and, by all appearances, a very good one. He was the head of the medical lab in a major Boston hospital. After 10 years at the job, he began to find it unbearably boring, because, as he explained to me "All the testing is automated." He seriously considered going to school to become a teacher, but finally decided to go to medical school, where he is at the moment.

Is deskilling a serious problem? I don't know. Will the typical jobs of tomorrow be more or less challenging, interesting, and fulfilling than the typical jobs of today? Although one can find predictions of the full range of possibilities, I doubt if anyone knows. A possibility that I see is that there will be some jobs that are very demanding and a great many that will require little skill at all -and that the demand for unskilled workers will grow faster than the demand for those with high levels of skills. Just a hunch.

I feel confident of this. For the physical, mental and spiritual well-being of the worker, it is more important that a job be meaningful and challenging than that it be easy and, perhaps, even safe. People take pride in doing things that are not easy to do. They value skills they have acquired with some effort. They get little satisfaction from doing things that anyone can do. Job simplification has been a primary goal of engineering and industrial psychologists and ergonomists in the past; helping to make jobs meaningful and intrinsically rewarding is a major challenge for the future.

HOW MIGHT AUTOMATION IMPROVE THE EFFECTIVENESS OF HUMAN REASONING, PROBLEM SOLVING, DECISION MAKING?

When I was a child in school, I was taught an algorithm for finding the square root of any number. (I don't know if this is still taught in school.) When small calculators became ubiquitous, this became a relatively useless bit of knowledge. Some people would argue that many of the computational aspects of mathematics are no longer worth learning, because one can so easily do the computations with pocket calculators. (I am not making that argument here.)

Imagine a time when people will have easy interactive access, via natural language, to a storehouse of encyclopedic knowledge. Imagine a question-answering device that would be able to answer most questions of fact of the type that one would now try to answer by consulting an encyclopedia, almanac, or other similar resource. Where is the Gulf of Martaban? Who was Prime Minister of Great Britain in 1938? Who discovered penicillin and what were the circumstances of the discovery? What is the atomic weight of copper? . . .

I find it easy to believe that such systems will be commonplace in the not-greatly-distant future, although I would not want to try to guess exactly when. This speculation raises the following interesting question: When such encyclopedic question-answering capabilities become available, will knowledge acquisition by individuals still be important? I think the answer is yes, for two reasons. First, question-asking is a knowledge-based activity (as is answer understanding). One who is knowledgeable about a particular topic (automobile mechanics, biochemistry) is able to pose questions about the topic that one who is not knowledgeable about it would not know enough to ask, and to understand answers that a person who lacked knowledge of the topic would find opaque. Second, I believe that knowledge is a basic human need. Having information at one's fingertips (in a personal library, in readily accessible databanks) is very desirable and useful, but not nearly as satisfying as having knowledge in one's head.

It seems clear that as a consequence of advancing technology, information of many sorts will be increasingly available to potential users everywhere. The question of interest to me is whether people will become more effective problem solvers and decision makers -- better thinkers in general -- as a consequence. I want to believe that the answer to this question is yes, but only time will tell.

HOW MIGHT AUTOMATION IMPROVE THE
QUALITY OF LIFE, GENERALLY, FOR PEOPLE EVERYWHERE?

One can easily make the case that automation already has greatly improved the quality of our lives. At least it has made them easier in specific respects. We need only remind ourselves of the numerous automata that we use, directly or indirectly, every day: thermostats (in cars, in house heating systems, in refrigerators), governors, flush toilets, sump pumps, automatic transmissions in cars, washing machines and numerous other household appliances, car washes, vending machines, range finders on cameras, . . .

I suspect that few of us would gladly do without most of these devices. We have come to take them so much for granted that were we suddenly to be deprived of them we would feel put upon indeed. Someone has estimated that the energy equivalent of one servant is about 67 kilowatt hours per year, and that, as of around 1960, the average American home used about 3150 kilowatt hours of electrical energy per year, or the energy equivalent of about 47 servants. I do not know how one would update this calculation to take into account the types of services one now gets from computational circuitry, which is built into many household appliances and devices, that consumes little power, but I am sure that any plausible calculation would show very modestly equipped homes to be capable of providing services and comforts that no number of servants -- without the help modern technological conveniences -- could provide.

In raising questions about automation, I do not intend to suggest that things were much better before automation -- and technology more generally -- had progressed to where they are today. For anyone who is inclined to look back nostalgically to the "good old days," I recommend a little book by Otto Bettmann titled *The Good Old Days -- They were Terrible!* Bettmann looks at housing, food, work, crime, health, education, leisure and other topics and reminds us of a side of life of bygone days that we have a tendency to forget or overlook. But the fact that we are better off today than were our predecessors -- I believe it is a fact -- does not mean that the best strategy for the future is to not worry about it and let nature take its course.

In *The Human Use of Human Beings: Cybernetics and Society,* Norbert Wiener (1950) made the following thought provoking observation: "It is a degradation to a human being to chain him to an oar and use him as a source of power; but it is an almost equal degradation to assign him a purely repetitive task in a factory, which demands less than a millionth of his brain capacity. It is simpler to organize a factory or a gallery which uses human beings for a trivial fraction of their worth than it is to provide a world in which they can grow to their full stature" (p. 16). Who can disagree with this sentiment? But what are the characteristics of a world in which people can grow to their full stature? And what is the place -- what are the places -- for automation in such a world?

Automation offers the prospects of great benefits of effective control systems, efficient use of energy, unprecedented access to information and to other people, . . . It also represents the possibility of electronic surveillance, predatory uses of computer networks, depersonalization . . . How do we ensure that the desirable effects outweigh the undesirable ones, and by a large margin? And what is the responsibility of technologists who are advancing the state of the art in that regard?

HOW MIGHT AUTOMATION AFFECT
OUR IDEAS OF WHAT IT MEANS TO BE HUMAN?

Reflecting on automation leads more or less invariably to questions of how people differ (if they do) from machines. This question can be interpreted at a relatively superficial level to mean how do they differ with regard to their capabilities and limitations, or with respect to the functions they can perform. Not long ago psychologists frequently drew up lists of what people can do better than machines and what machines can do better than people. Such lists provided a basis for allocating functions to people and machines. As computers have gotten smarter and people have stayed more or less the same, the list of things that computers can do better than people has gotten longer and that of the things that people can do better than computers shorter. Psychologists tend not to draw up these lists so much anymore

The question can also be asked at a deeper level: Are there really fundamental differences between people and machines? This is not a new question, but the increasing ability of machines to do things that not long ago were considered distinctively human capabilities has given the question new meaning. I have already mentioned Bruce Mazlish in connection with the treatment of automation in literature. The book in which the discussion of Mazlish's to which I referred is *The Fourth Discontinuity: The Co-evolution of Humans and Machines.* In this book, Mazlish, a professor of history at MIT, starts with the idea that we, as human beings, like to see ourselves as unique or special and that our self-perceived special status has been reflected in a variety of beliefs about the universe and our relationship to it. Mazlish argues that three of the more important of these beliefs, which can be described as beliefs in discontinuities, have been shown to be incorrect by science in the past and that a fourth one (to which the title of the book refers) is likely to be shown to be wrong as well.

The first discontinuity, which, thanks to Copernicus, proved not to be a discontinuity after all, concerned our position in the universe. Before Copernicus, the prevailing belief was that the earth was the center of everything. Everything else revolved about it and served its purposes. Now we believe, on compelling evidence, that the earth is an inconspicuous speck in an unimaginably large universe that is inhabited by billions of galaxies much like that to which we belong, each of which has billions of stars much like that that we call the sun. The second discontinuity was the discontinuity between humanity and other forms of life. This was discredited by Charles Darwin, who demonstrated to the satisfaction of the scientific community the continuity of all life forms, from the simplest to the most complex. The third discontinuity was that between reason and instinct. If we are not discontinuously different from animals physically, at least it could be held that we are rational creatures and therefore still unique in this way. But Sigmund Freud, by his own assessment, discredited this discontinuity by showing that we have less conscious control over our thoughts and actions, and are more in the control of processes that occur below the level of consciousness, than we had realized.

Mazlish now is arguing that what we had believed to be a fourth discontinuity -- a discontinuity between people and machines -- is not a true discontinuity either. In Mazlish's words, "We are coming to realize that humans are not as privileged in regard to machines as has been unthinkingly assumed" (p. 3). "To put it bluntly, we are now coming to realize that humans and the machines they create are continuous and that the same conceptual schemes that help explain the workings of the brain also explain the workings of a 'thinking machine.' Human pride and its attendant refusal or hesitation to acknowledge this continuity form a substratum upon which much of the distrust of technology and an industrialized society has been reared" (p. 4).

I am not comfortable with the idea that the discontinuity between people and machines is an imaginary discontinuity. In fact, I dislike it a lot. I suppose that makes me a reactionary, but there it is.

There is a related idea with which I have been becoming increasingly uncomfortable and that is the idea of person-computer partnership. I do not want to think of the computer as a partner, a collaborator, a friend, . . . I want to think of it -- and I want to argue that it *should* be thought of as a tool -- an unusually versatile and powerful tool, but a tool nonetheless.

Just so with automation and technology more generally. They are means to ends, not ends in themselves. I believe it is good to remind ourselves of this perspective from time to time and of the importance of maintaining it; it is easy, especially in the challenge and excitement of creative technological work, to forget to consider the purposes for which technology might be developed or the uses to which it might be put.

RECAP

• Automation, like technology in general, is neither good nor bad, but, again like all technology, it can be put to good and bad uses.

• The more powerful the technology, the greater the potential for both good and bad applications. The potential for automation, especially in view of the extraordinarily rapid advances in information technology over the last few decades, appears to be unbounded.

- The most difficult challenges relative to automation in the future will not be technological but philosophical and ethical. They will have to do with human purposes and human values. The most important goal is not that of determining what *can* be automated, but that of deciding what *should* be automated.

"As engineering technique becomes more and more able to achieve human purposes, it must become more and more accustomed to formulate human purposes. In the past, a partial and inadequate view of human purpose has been relatively innocuous only because it has been accompanied by technical limitations that made it difficult for us to perform operations involving a careful evaluation of human purpose. This is only one of the many places where human impotence has hitherto shielded us from the full destructive impact of human folly." Norbert Wiener (1964, p. 64) made this observation thirty-five years ago, but it is as timely now as it was then.

I fear, in retrospect, that my comments could be perceived as more pessimistic than I intended. In fact, I look to the future with considerable optimism, in large measure because of the potential that automation and other aspects of technology have, if used wisely, to enhance the quality of life everywhere. If the optimism is somewhat guarded, it is because I would like to be more confident than I am that the proviso will be realized. In my view, it is imperative that human wisdom in the applications of technology increase apace with increases in the power of technology, and its potential for good and ill. We need a perspective that is reflective and that gives priority to considerations of value and purpose. We need, somehow, to maintain the ability to distinguish between dog and tail, and to tell which is wagging which.

REFERENCES

Bettmann, O. L. (1974). *The good old days -- They were terrible!* New York: Random House.

Klemm, F. (1964). A *history of western technology.* (Translated by D. W. Singer.) Cambridge, MA: MIT Press. (Original published in German in 1954.)

Landauer, T. K. (1995) *The trouble with computers: Usefulness, usability and productivity.* Cambridge, MA., MIT Press.

Mazlish, B. (1993). *The fourth discontinuity: The co-evolution of humans and machines.* New Haven: Yale University Press.

Nickerson, R. S. (1992). *Looking ahead: Human factors challenges in a changing world.* Hillsdale, NJ: Erlbaum.

Nickerson, R. S. (In press). Basic versus applied research. In R. J. Sternberg (Ed.), *The concept of cognition.* Cambridge, MA: MIT Press.

Wiener, N. (1950). *The human use of human beings: Cybernetics and society.* Boston, MA: Houghton Mifflin.

Wiener, N. (1964). *God and Golem, Inc.: A comment on certain points where cybernetics impinges on religion.* Cambridge, MA: MIT Press.

PERSPECTIVES ON HUMAN

INTERACTION WITH

AUTOMATION AND TECHNOLOGY

The *R.M.S. Titanic*: A Siren of Technology

Mark W. Scerbo
Old Dominion University

The Third Conference on Automation Technology and Human Performance was held in Norfolk, Virginia, part of the greater Hampton Roads region. This area is home to one of the largest natural harbors in the world. It is no accident that Hampton Roads is the East Coast headquarters for the U.S. Navy.

Because of the importance of the maritime industry to this area, I thought it would be appropriate to open the meeting with the story of a ship. It is quite a magnificent ship. In fact, they recently made a movie about this ship. Of course, I am referring to the *Titanic*. The movie, of the same name won 11 academy awards including one for Best Picture. Moreover, the gross receipts for this film total over $500,000,000 making *The Titanic* the highest grossing movie in history.

What is it about this film that has brought people into theaters in record numbers? Well, it has many of the important elements of a successful film. It has spectacular special effects. It has a handsome and popular leading man, Leonardo DiCaprio, and a beautiful leading woman, Kate Winslet. But most important, it has a great story. In true Hollywood fashion, it is a film that tells a story of love and one of seduction, but I am not referring to the affair between the two fictional characters nor am I necessarily referring to the ship itself.

PRELUDE TO A DISASTER

The *R.M.S. Titanic* was the grandest ship of her time. She was the second in a series of three sister ships envisioned by the White Star Line. She measured 882 ft 9 in from bow to stern and weighed 46,329 gross registered tons. In fact, the *Titanic* was the largest moving man-made object built at that point in history. She had as many as 9 decks, room for 885 crew and 2,435 passengers. She was built to compete with the ships of Cunard Line, not in terms of speed, but in luxury and safety, and she was the most luxurious ship of her time. She had Turkish baths, an indoor swimming pool, a regulation size squash court, a gymnasium with all the latest electronic exercise equipment, four electronic passenger elevators, and a specially-designed crane to load automobiles. Moreover, her interior was decorated to rival the finest European hotels. First class suites each had unique architectural designs (Lynch & Marschall, 1992). The two largest suites measured 50 ft. in length and included a private veranda. Fares for first-class passage ranged from $2,000 to $40,000 in today's dollars. Perhaps her most distinguishing feature was the grand staircase of polished oak, covered by a wrought iron and glass dome to allowing the circular stairway to be illuminated with natural light.

Most everyone knows the fateful story of the *Titanic*. She struck an iceberg in the North Atlantic and sank. But those of us in the human factors profession are schooled in accident investigation and critical incident analysis. We understand that there are many precipitating factors that contribute to a disaster such as this. According to Reason (1990), human errors made in the system design process or in interactions with a system form "holes in planes of unsafe acts". Under most circumstances, one or several errors can occur without serious consequence. On rare occasions, however, when enough "holes" occur in the right sequence, relatively minor errors can produce catastrophic consequences. Some have argued that the sinking of the *Titanic* represents this very type of catastrophe. Although this is partly true, I contend that the *real* story behind the *Titanic* is one of seduction.

It is easy to see how people could succumb to her. The *Titanic* was the crowning technological achievement of the early 1900s. This was a period of the industrial revolution where rapid expansion in technology promised to solve all of life's problems. The *Titanic* was to be the most luxurious and safest liner in the world. She was "unsinkable". She had 15 traverse bulkheads that were "water tight." In an emergency, the doors could be electronically shut in an instant. Well, the press (notably, *Shipbuilder* magazine) claimed that she was practically unsinkable. Alexander Carlisle, the ship's designer, knew that she could withstand a hull breech of any two of the 16 consecutive compartments or the first four compartments. If these first four compartments flooded, she would still stay afloat. But if more than the first four flooded, the bow of the ship would sink below the water line and water would flow over the tops of the bulkheads which were *not* water tight at the ceiling and did not extend up through all decks of the ship. Under these conditions, water would flow into and fill successive compartments like that of an ice cube tray until the ship was heavy enough to sink. Alexander Carlisle knew that. Thomas Andrews, Managing Director of the Harland & Wolff shipyard which built the *Titanic* knew that. J. Bruce Ismay, President of While Star line which owned and operated the *Titanic* knew it as well, but no one believed it was necessary to correct the stories in the press.

Any British ship built in the early 1900s weighing over 10,000 tons was required to carry lifeboats with enough room for 980 people. This code had been enforced since 1894, but was woefully inadequate for ships the size of *Titanic* being built in the early 1900s. The *Titanic* carried 20 lifeboats. There were 14 boats each with a capacity of 65, two emergency cutters which could each hold 35, and 4 collapsible boats each with a capacity of 49. All total, the *Titanic* had lifeboat capacity for 1,176 which met and exceeded the 1894 requirement, but fell far short of 3,295 needed for a full ship and crew (Eaton & Haas, 1996). Original plans, however, had included another 12 boats, but these were eliminated because J. Bruce Ismay wished to use more space along the midsection of the ship for scenic verandas for the first class passengers. Moreover, there was a general feeling that the presence of lifeboats on a ship like the *Titanic* was a public relations gesture to convey safety rather than a necessity (Peltier, 1994a).

DISASTER STRIKES

The *Titanic* was put through her sea trials on April 2, 1912 which included starting, stopping, and turning at various speeds (Eaton & Haas, 1996). At a high speed of 20 knots the engines were thrust into full reverse. It took the ship took 850 yds (approximately half a mile) and about 37 seconds to come to a complete stop.

The *Titanic* was commanded by Captain Edward J. Smith, a veteran of the White Star Line. He was transferred from the sister ship, the *R.M.S. Olympic*, to take *Titanic* on her maiden voyage. It was rumored that he had plans to retire after this trip. Captain Smith had an outstanding record with White Star and a respected reputation among crew and passengers alike. Many first class passengers believed Smith to be a safe captain and would only travel with him at the helm. For that matter, Smith, himself claimed to have had an uneventful career at sea and had not been a party to any wrecks or threatening situations. (Perhaps Smith was merely modest, because records indicate he was in command of the *Germanic* when it capsized and the *Olympic* when it collided with the *Hawke*, see Eaton & Haas, 1996).

Most of the crew had never served on a ship like *Titanic*. Only 41 officers and senior crewmen as well as 78 coal stokers were aboard a week before picking up passengers in Southampton. The more than 900 crew and staff began boarding from April 6 until April 10, when the first passengers came on board. Some of the staff, stewards, bell-boys, and attendants signed on looking for any kind of work due to a lengthy coal strike in Southampton. Many of the crew and staff did not know one another and received no formal training regarding their duties and responsibilities. In fact, Second Officer Charles Lightoller stated that it took him 14 days to become proficient at navigating within the ship (Peltier, 1994a).

On April 10, the *Titanic* left Southampton and headed toward Cherbourg, France and then Queenstown, Ireland to pick up additional passengers. In the early afternoon on April 11, the *Titanic* departed from Queenstown and made its way into the North Atlantic shipping lanes toward its destination, New York. Aboard, were approximately 2,235 passengers and crew members.[1]

[1] The chronology of events described below is taken largely from Eaton and Haas (1996).

April had been an unusually warm month in 1912, a year that had seen one of the mildest winters in nearly 30 years. Consequently, there was more ice than usual drifting into the shipping lanes of the North Atlantic. Other ships traveling along the same route had reported seeing ice in this area and the *Niagara* was damaged by ice on April 11. Between April 11 and 14, at least 19 other ships encountered ice which lay directly ahead in *Titanic*'s course.

While at sea, it is thought that J. Bruce Ismay encouraged the Captain to make better time. He was interested in beating the transatlantic time of the *Olympic* and hoped to reach New York a half day ahead of schedule.

On Sunday, April 14, the *Titanic* was heading west at approximately 21 knots. At about 9:00 am she received a wireless message from the *Caronia* about ice. This message was taken immediately to the bridge. The Captain read the message and then posted it for the other officers to see. At 10:30 am, the Captain led a church service in the first class dining room. Afterward, Captain Smith dispensed with the standard lifeboat drill (there was also no daily inspection of the ship that Sunday). At 1:42 pm, the *Titanic* received another report about ice originating from the *Athinai*. This message was taken to the bridge, and then to the dining room and given to the Captain who was having lunch with J. Bruce Ismay.

That evening the temperature began to drop and it became quite cold very quickly. At about 7:00 pm, an iceberg watch was ordered. Around 7:30 pm, a message from the *Californian* to the *Antillian* was intercepted. It indicated ice in an area less than 20 miles north of *Titanic*'s track. This message was not delivered to the Captain who was the guest of honor at a dinner party in the *à la carte* restaurant. Shortly before 9:00 pm, the Captain returned to the bridge and then headed to his cabin at 9:30 pm.

At 9:40 pm, junior Marconi operator, Harold Bride, received a message from the *Mesaba* about having sighted many icebergs, heavy pack ice, and field ice in an area which lay directly ahead in *Titanic*'s path. Harold Bride and senior operator, John Phillips, were actually employees of the Marconi company, not White Star. They were besieged with personal (and profitable) message traffic from paying passengers to loved ones ashore. The message from the *Mesaba* did not make it to the bridge.

At 10:00 pm, Reginald Lee and Frederick Fleet took their positions in the lookout cage. It was a clear, but dark moonless night. The temperature was about 32 degrees. Although there were binoculars in the cage at the beginning of the trip, they were not in the lookout cage during their shift.

Around 11:00 pm, the *Titanic* received a warning from the *Californian* about 20 miles away that they had stopped in a field of ice. Phillips, irritated by the interruption told Operator Evans, of the *Californian*, to "Shut up". Shortly thereafter, Evans removed his headset and retired for the evening without rewinding the mechanically powered detector on the Marconi radio. Thus, he would be unable to receive any further communications that evening.

At 11:40 pm, *Titanic* was moving at about 22 knots. Frederick Fleet spotted an iceberg and shouted "Iceberg, Right ahead" and rang the bell. Fleet and Lee braced themselves for the impending collision. The ship was about 500 yds (or less than 37 secs) away from impact.

First Officer Murdoch issued the order, hard-a-starboard, to turn the ship to port. At the same time he ordered the engines thrust into full reverse. Unfortunately, given the design of the *Titanic* this action disrupted the water flowing past the rudder making the ship harder to steer. The end result was that rather than fully turning to port, the ship continued to "glide" forward for a brief interval and eventually grazed the iceberg on the starboard side. The iceberg opened holes in the hull plating below the water line damaging the first five compartments and a portion of the sixth.[2] The water tight bulkhead doors were closed immediately, but water began to pour into the first six consecutive compartments of the ship.

To most everyone's surprise, the impact was not all that noticeable. Those on the lower decks on the

---

[2] Originally, it was thought that the iceberg produced a 300 ft. gash in the hull. More recent analyses suggest that the steel plating used on the *Titanic* was particularly brittle in cold water and was held in place by rivets that could snap under low levels of stress. It is thought that the when the iceberg struck the ship, it fractured the hull plating, popped rivets, and separated the plating from the ship producing a series of holes totaling only 12 sq. ft. (see Foecke, 1998).

starboard side reported hearing a grinding noise and feeling a rumbling vibration. On the upper decks on the port side, the impact went largely unnoticed. Captain Smith, however, did notice and rushed from his quarters to the bridge.

First Officer Murdoch reported that the Titanic had struck an iceberg and that the bulkheads had been shut. Captain Smith sent Fourth Officer Boxhall to inspect the damage, but Boxhall only did a cursory inspection and did not look below F deck. He reported back to the Captain that he saw no damage. Smith then sent Boxhall to find the ship's carpenter who met up with Boxhall along the way and informed him that the mail room on G deck was beginning to flood.

Captain Smith then sent for Thomas Andrews and together they made their own inspection of the damage. Andrews informed the Captain that they had no more than 2 hours before the ship would founder.

Throughout the ship, many of the passengers (and crew for that matter) did not understand the gravity of the situation. It was business as usual. Some of the third class passengers played soccer with pieces of ice that had shaken loose and fallen on board after the collision. Some passengers in the first class smoking room passed around pieces of the ice as token souvenirs and made jokes about getting ice for their drinks. Others wandered out of their rooms to see what all the talk was about. Many of the crew did not realize that the ship had struck an iceberg and informed the passengers that *Titanic* had stopped as a precaution. Rumor and misinformation ran rampant because the *Titanic* had no P.A. or alarm system to allow communication throughout the ship.

At 12:05 am, Captain Smith issued an order to muster the crew and uncover the lifeboats. Shortly thereafter, he asked Boxhall to determine the ship's position. Boxhall calculated a position based upon his *estimates* of the ship's speed and stellar observations he had taken earlier. (The *Titanic* actually continued moving at half speed for a brief period after the impact and Murdoch issued a "hard-a-port" command so that the stern would move out of the way of the ice. Moreover, there was a southerly current of 1 knot in the route that *Titanic* was traveling, all of which diminished the accuracy of Boxhall's estimated position.) He wrote down his estimate and gave it to the Captain who took it to wireless operator Phillips and ordered the call for assistance, CQD – Come Quick, Danger. Interference from other transmissions caused the location to be recorded incorrectly by some ships in the area. (Seventy years later when the wreck of the *Titanic* was located, it was found to be 13.5 miles east-southeast of the estimated position sent in the distress call.)

The wireless operator, Harold Cottam, on the *SS Carpathia*, received *Titanic*'s distress call and was incredulous. He asked if he should alert his captain. He received a second message – "Yes, Come Quick". Cottam informed his Captain, Arthur Rostron, who determined that they were about 58 miles south-east of *Titanic*'s distress position. He ordered the ship to change course and headed toward *Titanic* at top speed (about 17 knots).

At 12:30 am, Lightoller had to suggest to Captain Smith that they begin to get women and children into the lifeboats. The crew began knocking on doors to wake the passengers and have them put on their life vests. Many of the passengers were irritated by the request. They did not perceive any immediate danger. Rather, they believed it was much more dangerous to venture out into the cold, dark Atlantic Ocean in a little wooden boat than to remain in the "safety" and comfort of the warm, well-lit *Titanic* (Peltier, 1994a). In the gymnasium, John Jacob Astor cut open a life vest with his knife to show his wife what was inside (Lord, 1955).

Loading the first lifeboats was a difficult task. Many of the crew had not been trained in the procedures. Moreover, most were unaware that special Welin davits for lowering the boats had been designed for *Titanic*. They had been tested in Belfast with 68 full grown men in the boats. On this night, however, the crew were uncertain as to how much weight they could hold. So, the first lifeboat to be lowered, No. 7, was only ⅓ full. Lightoller claimed that he initially lowered the boats at less than full capacity so that they could pick up additional passengers at an open hatch at the bottom of the gangway. (This never happened, however, because the gangway doors were never opened.) Once they were on the water, Lightoller believed they would return and pick up other passengers (this too did not happen).

Loading the initial lifeboats was complicated further by the activities of the passengers. Some were told that this was merely a precaution until the problems with the ship could be remedied. Many others were reluctant to be separated from loved ones. Some hesitated or changed their minds and climbed back out of the boats. As a rule, the boats were to be loaded with only women and children; however, this was not adhered to strictly on the starboard side

Lightoller was one of the first officers to begin loading passengers into the lifeboats. Initially, he attempted to load passengers into lifeboat No. 4, however he believed it would be safer to lower the boat to Deck A and load the passengers there. So, he sent the passengers down to Deck A. He soon realized it was safe to load passengers from the boat deck and began to load people into lifeboat No. 6. Unfortunately, the group sent down to Deck A found the promenade closed and the windows shut. Because Lightoller needed as many hands as possible on the boat deck, he could not send someone to open the windows. These passengers waited on Deck A for about half an hour before a steward directed them back upstairs to the boat deck. Again, they waited because lifeboat No. 4 was still down at Deck A. It had not been raised back to the boat deck level. Eventually, Lightoller recognized that passengers were still waiting to board lifeboat No. 4 and sent them back down to Deck A where the windows had finally been opened. (Lifeboat No. 4 would finally be lowered 1 ½ hours later with only 32 people aboard.)

At 12:45 am, Captain Smith ordered Boxhall to fire the distress rockets. He then went to the wireless room and asked Phillips which message he was sending. Phillips replied, "CQD". Bride joked that they should try the new international code, SOS, because it might be the last chance they had to use it. Phillips laughed and began sending SOS. It was one of the first times this distress call was ever issued at sea (Heyer, 1995).

Boxhall and others reported seeing the light of another ship nearby, at first with binoculars and then with unaided vision. The ship appeared to be about 5 miles away, close enough that Boxhall could identify the red and green port and starboard lights. Lightoller saw the lights as well and told the passengers he was loading into the lifeboats that it would soon come to their rescue. Captain Smith ordered Boxhall to try and signal the ship with the morse lamp in between firing the rockets. The other ship did not respond and eventually appeared to turn and head away from *Titanic*.

Crew members of the *Californian* which had issued an ice warning earlier that evening reported seeing another vessel about 5 miles away. Captain Lord and Second Officer Stone both saw the ship. Stone believed the ship was a small tramp steamer and attempted to contact this other ship with his morse lamp shortly after midnight, but received no response. After 12:45 am, Stone saw what appeared to be a flash of light in the sky coming from beyond the steamer. He saw several more within the next half hour. He notified his captain, who ordered him to try signaling the ship again. Again, Stone got no response. It then appeared that this other vessel steamed away.

By the time the rockets were being fired, it became obvious to most passengers that the ship was indeed sinking. Still, lifeboats were being lowered only ½ or ⅔ full. Panic soon set in. Firearms were distributed to officers to maintain order. As one of the last boats was being lowered, collapsible boat C, a group of men gathered around and Chief purser McElroy, who was assisting Murdoch, fired his pistol into the air to thwart any attempt at rushing the boat. As the boat was being lowered, Ismay who was on the next deck below, looked around and seeing no one else nearby, jumped into it.

Of the 20 lifeboats on *Titanic*, only 4 went to sea at nearly full capacity. Two of the collapsible boats were never launched. The last lifeboat was lowered at 2:05 am and within 15 minutes the ship foundered. Only 1 boat returned to the wreckage to look for survivors in the water. Fourteen people would be pulled from the sea after the ship went under, but only 7 of those would survive. Another 1,523 lost their lives.

The *Carpethia*, heading toward *Titanic* picked up the first survivors at 4:10 am. The last of the 705 survivors reached the *Carpethia* at 8:10 am. Four days later, at 9:25 pm, the *Carpethia* finally docked in New York.

AFTERMATH

In the weeks that followed the *Titanic* disaster, both the United States and Great Britain held hearings to fix responsibility for the disaster. At the British inquiry the senior surviving officer, Charles Lightoller, stated:

> Of course, we know now the extraordinary combination of circumstances that existed at that time which you would not meet again in 100 years; that they should all have existed just on that particular night shows, of course, that everything was against us. (Lord, 1986, p. 47).

To be sure, there were numerous factors contributing to the *Titanic* disaster. From a design perspective, the bulkheads did not extend high enough above the water line and were not water tight at the ceiling. Two of the collapsible lifeboats were stowed on the roof of the officers' quarters making them virtually inaccessible (and, in fact, these two lifeboats were never launched). From a safety perspective, there were too few life boats for the ship's capacity, there was no warning system for the passengers, and there were no regulations regarding wireless communication. In fact, Lord (1986) suggests that to be successful in the increasingly competitive shipping industry, White Star actually abandoned many of the safety design measures that had been used and validated on other ships. According to Lord, the *Titanic*'s "appearance of safety was mistaken for safety itself." From an operational perspective, the vast majority of the crew received no formal training about the ship or their responsibilities. There were no evacuation plans and clearly no boat assignments for the passengers. Activities in the wireless room and on the bridge were virtually independent of one another. The total picture about ice in that region was not perceived by the officers because only some of the warnings were posted on the bridge and not every officer availed himself of that information. Further, the significance of the warnings was lost on the two Marconi operators who were not experienced sailors. Finally, Captain Smith may have been ill-equipped to handle the situation. He had been in command of the largest ship ever built for less than 2 weeks, was pushing her at nearly top speed (which would severely compromise the ability to deal with a sudden, unexpected event), and though he had received several warnings about ice, failed to appreciate the seriousness of the operating conditions (Hancock, 1996).

Lightoller's comment about the rare combination of circumstances is wholly consistent with James Reason's ideas about human error and large-scale disasters. Any one of the contributing factors to the *Titanic* disaster would be unlikely to pose a serious threat in and of itself, but in the proper combination or sequence could be catastrophic. In truth, many of the problems that compounded the disaster were common practice throughout the shipping industry in that era. For instance, none of the other grand luxury liners operating at that time carried enough lifeboats for all passengers and crew, but within a week of the *Titanic*'s sinking British, German, and French liners were outfitted with the necessary lifeboats (Peltier, 1994b).

As a result of the *Titanic* disaster, other regulations were introduced in the shipping industry (Lord, 1955). First, no ship would ever again be referred to as unsinkable. Second, all ships were required to carry lifeboat capacity for all passengers and crew. Third, shipping lanes across the North Atlantic were moved further south to keep free of ice. In addition, the United Sates and Great Britain established the International ice patrol to chart and in some instances, destroy large icebergs. Fourth, ships were soon required to man the wireless station 24 hrs a day.

The sinking of the *Titanic* may be one of the great tragedies of the 20th century, but the real story behind this tragedy is one of seduction. In an era marked by a steady succession of extraordinary technological achievements, the *Titanic* stood apart from the rest. We were infatuated with her beauty. We were awe stricken with her size. We marveled at her amenities. We revered her power. She was our mistress. As Wyn Wade (1986) put it, "For the *Titanic* was the incarnation of man's arrogance in equating size with security; his pride in intellectual (apart from spiritual) mastery; his blindness to the consequences of wasteful extravagance; and superstitious faith in materialism and technology."(p. 323).

But she was a deceitful mistress and she turned our vanity against us. Like a Greek siren, beckoning sailors to certain doom, she filled our ears with a song of self-indulgence so captivating that we chose not to hear the warnings. This siren of technology lulled us into believing we could create something mightier than Mother Nature herself and that we had dominion over the sea. And then, with an icy touch, she hardened a handful of that sea and smacked us back into reality taking our carelessness as well as our pride back with her to the very depths of the ocean.

The sinking of the *Titanic* was a wake-up call for the 20th century. We would learn tragic lessons about the short comings of technology. For a while, we would approach technology with guarded enthusiasm and place supreme emphasis on issues of safety, but the lessons would soon fade into the background of the Siren's song because the Siren of Technology has never been silenced. She could be heard singing at the nuclear mishaps at Three Mile Island and Chernobyl. She could be heard on the space shuttle, *Challenger*. In fact, if one listens carefully, one can hear her singing at every human factors case study and accident we are called upon to investigate.

When James Cameron accepted the academy award for his film, he asked everyone to remember the victims of the *Titanic* tragedy. Likewise, it is imperative that we as human factors researchers, designers, and practitioners remember that the work we do has a direct and genuine impact on peoples' lives. We may take a modicum of comfort in knowing that there was no human factors profession when *Titanic* was built, but that only places a greater burden upon us to avert such disasters now and in the future. For if we do not champion the crusade for safety, efficiency, and usability for the consumers and users of technology then who will?

ACKNOWLEDGMENTS

I would like to thank P.A. Hancock, C.A. Haas, and J.P. Eaton for their inspiration, encouragement, and guidance on this manuscript.

BIBLIOGRAPHY

Eaton, J.P., & Haas, C.A. (1996). *Titanic: destination disaster*. New York: W. W. Norton.

Foecke, T. (1998). *Metallurgy of the RMS Titanic* (Report No. NIST-IR-6118). National Institute of Standards and Technology Gaithersburg, MD.

Hancock, P.A. (1996, May). *Directly perceiving disaster*. Paper presented at the Festschrift for William N. Dember, Cincinnati, OH.

Heyer, P. (1995). *Titanic legacy: Disaster as media event and myth*. Westport, CN: Praeger Publishers.

Lord, W. (1986). *The night lives on*. New York: Avon books.

Lord, W. (1955). *A night to remember*. New York: Holt, Rinehart, & Winston.

Lynch, D., & Marschall, K. (1992). *Titanic: An illustrated history*. Ontario, Canada: Madison Press.

Peltier, M.J. (1994a). *Titanic: Death of a dream*. C. Haffner & D.E. Lusitana (Producers). A&E Productions.

Peltier, M.J. (1994b). *Titanic: The legend lives on*. C. Haffner & D.E. Lusitana (Producers). A&E Productions.

Reason, J. (1990). *Human error*. New York: Cambridge University Press.

Wade, W. C. (1986). *The Titanic: End of a dream*. New York: Penguin Books.

Caveman in the Cockpit? Evolutionary Psychology and Human Factors

Alan F. Stokes[1] and Kirsten Kite
[1]Minds & Machines Laboratory, Rensselaer Polytechnic Institute

Human factors has been defined, perhaps just a little tongue-in-cheek, as the consideration of "fallible humans in interaction with systems designed by fallible humans" (Stokes & Kite, 1994). However, it would be misleading to mistake human evolution's cognitive design specifications for human fallibilities. After all, we do not blame the 747 (or its pilots, or designers) for its relatively poor hovering performance. But human factors and engineering psychology have taken their cues from cognitive psychology, and this, in turn, has shown surprisingly little interest in the systems engineering of humans; that is, what humans were designed to do, and for what ecological contexts (rather than what they *can* do -- like learning lists of nonsense words). Just because the system engineer in this case is a self-organizing Darwinian process acting (still) over time should not blind us to the fact that, at any point in that development, *Homo sapiens* does have an engineering specification, a 'factory fit', as it were. Engineering psychologists, we argue, are therefore the very folk who should be in the vanguard of the new developments in evolutionary psychology and related sciences, and should not merely be pulled along by conventional cognitive psychology in general, and an unreconstructed information-processing (IP) framework in particular.

THE EVOLUTIONARY APPROACH TO COGNITION

For most of its history cognitive psychology has, in essence, wholly disregarded our evolutionary past. The mind has been treated idealistically and prescriptively, as a unitary general-purpose learning mechanism, and, in general, likened to a computer in its information-processing functions. For example, there has been a marked focus on isolated "knowledge in the head," rather than on functionally and ecologically relevant "worlded knowledge" (in Merleau-Ponty's phrase). Reasoning processes that diverge from the traditional precepts of mathematical logic and Bayesian statistics have been labeled as inefficient or biased. Emotion, when it has been considered at all, has tended to be regarded as a "spoiler" that has been tacked on as an afterthought to standard, "cold" models of intellectual functioning (and generally to the detriment of that functioning). Consciousness, the jewel in the cognitive crown, has, until recently, been hidden in the discipline's basement less like a jewel than some mad aunt too embarrassing to deal with. Moreover, traditional cognitive psychology has failed to apply the Darwinian test of evolutionary plausibility to its accounts of human mental processes. Given this, the question of human interaction with automation can be recast. What happens when a conscious and self-monitoring, ecologically-embedded, socially-oriented, biologically-motivated and emotionally 'hot' cognition (as described by evolutionary psychology), interacts with the unconscious, context-insensitive (unworlded), 'cold' processing of machine cognition? The latter, of course, is much closer to architectures described by cognitive psychology in standard IP models, and realized as automation hardware by engineers.

JUST WHO ARE THE HUMANS IN 'HUMAN FACTORS'?

Cognitively, we are hunter-gatherers, the descendants of Pleistocene hunter-gatherers, and the products of a lifestyle and environment that has characterized virtually the entire history of the hominid lineage --several million years. Even in the past 100,000 years of fully modern humans, only the most recent 10,000 have seen any agriculture, and then only in some societies. Some remain hunter-gatherers today. The Pleistocene environment lacked not only most of the technologies that we now take for granted, but also a reliable food supply, writing, permanent settlements, and formalized class divisions. Communication was conducted face-to-face, and with members of one's own band (the modern ubiquity of strangers and contact with other cultures was also unknown). It was difficult to cheat and run, to lose face and start anew. Rejection by the band, or being dispossessed of food, status, social intelligence (gossip), or mates, could mean personal as well as genetic disadvantage -- including death. We have the genes of those who fought and won, the psychological characteristics that contributed to survival and inclusive fitness in that environment. They inform our feelings about and interactions with Disneyland crowds, foreigners, soap-operas, gossip shows, BMWs and Rolex watches. But we also take our Pleistocene cognition into the factory, the computerized office, the launch silo, and as the title suggests, onto the automated flight-deck.

BEHAVIOR AND THE DARWINIAN PLAUSIBILITY TEST

Many psychological phenomena that have traditionally been regarded as "faulty" or maladaptive deserve to be reevaluated in light of evolutionary insights. As a case in point, consider the topic of decision heuristics or "biases" (see, for example, Kahneman, Slovic, & Tversky, 1982). Among other things, such biases are potentially very important in human interaction with Decision Support Systems, the diagnosis of automation failures, etc. This corpus of work has been popularized by Piattelli-Palmerini (1994), in a book called "Inevitable illusions: How mistakes of reason rule our minds." The title is an accurate description of the prevailing view of these biases. Applying the basic Darwinian plausibility test, however, should at the very least stir in us some curiosity. Something is not right here. After all, we are a fairly successful species -- so far. We get by in arctic and desert, in submarine and in orbit. We repair car engines, and defeat invisible pathogens. These are impressive feats for poor deluded minds ruled by the irrational.

But by what mechanism did these putative bugs in our decision software cause our ancestors to outbreed and outsurvive their stodgily rational brethren? How have these bugs been selected for and preserved over human evolution? Or are they some unwanted, nonselected artifact of design (i.e., *spandrels*, discussed later)? If not, we might ask if these biases (a) really exist at all, or (b) are in fact positive in most ecological circumstances (past, perhaps even present).

In fact there is another literature on biases that is rarely cited in human factors papers. For example, Leda Cosmides' famous (1989) research with the Wason selection task provides strong evidence that human reasoning capacities evolved for evolutionary fitness rather than for abstract truth-seeking. Cosmides' findings suggest the existence of a hard-wired and powerful "cheater-detection" algorithm that helps us avoid being taken advantage of in reciprocal social exchanges (which could mean starvation or in fecundity in the Pleistocene), and which appears to override several precepts of traditional, "correct" logic. This may throw light on the case of the demoted union captains flying as first officers to promoted non-union 'scab' junior officers (see Stokes & Kite, 1994, pp. 350-355). Non-communication, rivalry and even fist-fights on the flight deck certainly failed to follow the cool, logical imperatives of modern system safety. But were these pilots being irrational (or feeling cheated), or merely behaving according to another rationality, that of a hot, social cognition rooted in Pleistocene survival imperatives?

Equally interesting observations have been made about human probability judgments. In one oft-cited study, (Kahneman et al., 1982), even Harvard Medical School faculty and students had great difficulty answering questions such as this: "if a disease has a prevalence of 1/1000 and the test for it has a false positive rate of 0.1%, what is the likelihood that a person who has tested positive is actually infected?". However, Gerd Gigerenzer of the Max Planck Institute has suggested that the very concept of single-event probabilities is not one our ancestors would have found useful. The frequency of events in nature (lightning igniting fires, geese arriving in spring) is more likely to have underpinned the evolution of probabilistic reasoning in our forebearers. In a paper entitled "How to make cognitive illusions disappear," he reframed the same type of problem in terms of frequencies, e.g., "if 1 person in 1,000 has a disease, and if 1 of the other 999 healthy individuals also (falsely) tests positive, what is the likelihood that a positive-testing individual is in fact infected?" When the problem was presented this way, people suddenly became much better at judging probabilities (Gigerenzer, 1991). Thus, an evolutionary perspective can also help us avoid misleading methodological errors.

MISCONCEPTIONS ABOUT EVOLUTIONARY THEORY

Environmental mismatch theory (EMT) is a passe platitude: EMT is a concept borrowed from the study of psychopathology, and addresses the issue of (a) the differences between ancestral and current environments, and (b) the relevance of this to modern functioning (Bailey, 1995, cited in Crawford, 1998). Bowlby (1969) pointed out that the range of human environments must now be far more diverse than in ancestral times, and that the speed of this diversification is likely to have proceeded faster than humans can adapt via natural selection. Within psychiatry these mismatches have been suggested as a source of stress and pathology. This idea has spawned its share of Rousseauistic cliches about the happy lives of our more primitive ancestors, and even the occasional mad bomber. However, cogent criticisms have helped refine the theory (see, for example, Crawford, 1998), and the fundamental concept continues to provide a powerful heuristic tool for human factors.

Culture is an alternative framework for explanation. Most of twentieth-century psychology and social thought can be described in terms of what Tooby and Cosmides (1992) call the "Standard Social Science Model." This view represents biology and psychology as separate and opposing domains, and further posits (a) that humans (uniquely among all species) do not have an innate and universal set of evolved psychological characteristics; and (b) that human psychology is instead molded more or less entirely by an arbitrary and self-perpetuating entity known as culture. For many reasons these ideas do not meet the test of Darwinian plausibility (see Tooby & Cosmides for a fuller discussion). Rather than a cause, culture is a result of the human cognitive 'specification' in interaction with the natural and engineered environment over time. Nevertheless, there has sometimes been a tendency to view humans as an almost endlessly malleable system component. This is, in part, what prolongs the "blame then train" mentality familiar to most human factors practitioners.

Evolutionary hypotheses cannot be tested. In addition to the Gigerenzer work discussed above, consider the question of the nature of any stress-related linguistic change in the content of communications between remote teams using telecommunications. Pinker and Bloom (1992) have suggested, an evolved characteristic, like our mating strategies, and not a cultural artifact like music. So, beginning with the circumstances of Pleistocene hunter-gatherers described above, it can be posited that communications under stress could be expected to have been optimized for (a) brevity, (b) face to face exchanges, in (c) shared environmental and cultural contexts. Brevity should lead, inter alia, to fewer subordinate clauses. Since such clauses are often where we find causal, additive, and other relationship-clarifying connectives, we might expect to see fewer under stress. The common environment supposition leads to the expectation of more exophoric/deictic reference. In modern English this might imply more ellipses, more pronouns, especially exophoric pronouns, and fewer substantives. The common culture assumption suggests more linguistic "shorthand," group slang, etc. These changes can be sought in the content of modern telecommunications under stress, where these assumptions do not hold, and message coherence would therefore be reduced (Stokes, Pharmer, & Kite,1997). The point here is that an evolutionary perspective can help generate specific, interesting, testable and falsifiable hypotheses, not that it precludes reference to other mechanisms in the interpretation of results, or that such hypotheses (uniquely in the modern philosophy of science) are amenable to once-and-for-all verification.

Woolly notions like 'human nature' have little relevance to real world human factors. Consider the federal rule that airliner cabins be designed such that they can be evacuated in 90 seconds. For many years (and continuing) this capability was demonstrated in manufacturer and FAA tests, the "passengers" emptied from the cabin in a smooth orderly way, in 90 seconds or less. Analysis shows that regularity and efficiency is assured by waiting, deference and turn-taking -- in short by individuals acting for the welfare of the group in the simulated emergency. But is this descriptive of the human psychological fit, the real output of hot cognition when the prime biological directive "survive!" is invoked? Put another way, are we the descendants of folk who deferred to others and waited their turn in emergencies, or are we blessed with the genes of those who dropped everything and ran like the blazes? When a cabin evacuation experiment was conducted at Cranfield, U.K., the mere chance of collecting about $8 (for the first 30 out) was enough to create bedlam in the cabin-jammed exits, people trapped under seats, etc. (Muir & Marrison, 1991). This is hardly an unexpected outcome of esoteric genetic theory, yet we had to wait for a fatal 737 cabin fire at Manchester Airport before such a study was funded, revealing the 90-second rule for what it is -- a polite fiction rooted more in corporate wishful thinking than in the operational "specs" for *Homo sapiens*.

EVOLUTIONARY THEORY: SOME CAVEATS

It has been said that if you give a small boy a hammer, you will quickly discover that everything needs hammering. The effect is not confined to small boys or hammers, however. It can be observed wherever a new tool becomes available and attracts its enthusiastic adherents, whether that tool be a new interpretive framework or an analytical/statistical technique. Evolutionary theory can be a powerful additional tool in the human factors toolbox -- perhaps even one that will, in the long run, become the single most powerful, encompassing explanatory, predictive and hypothesis generating framework of all. But it should not and will not replace all other models and techniques available to us.

Second, we must be wary of the Panglossian fallacy. In Voltaire's *Candide*, Dr. Pangloss persisted with an absurdly optimistic outlook against all the available evidence, taking the Leibnizian view that "all is for the best, in this the best of all possible worlds." An evolutionary Pangloss holds that if a human feature or characteristic exists, Darwinian sifting will have ensured that it is optimum. Thus English (and other natural languages) must be a perfected means of expression, our five senses the perfected method for apprehending the world, and so on. But not all features observed in a system are design features. Some are epiphenomena: spin-offs or accidental products dubbed *spandrels* by Gould and Lewontin (1979), who give an example from cathedral architecture. (The triangular spaces formed when a number of archways are set side by side provide spaces which seem so ideal for the placement of statuary and bas-reliefs that they appear to have been designed for that purpose.)

Finally, and perhaps, most importantly, we need to avoid what has become known as the "Just So" pitfall. The original Just So Stories are the splendid children's tales written early in the century by Rudyard Kipling (*How The Leopard Got His Spots; How the Elephant Got His Trunk*, etc.). How did the elephant obtain his trunk? From a crocodile's ministrations to an (originally modest) proboscis. Just so. The point is that anyone can devise myriad evolutionary stories of varying plausibility to account for observed characteristics of physiology and cognition. Thus: because of our arboreal banana-seeking past, human vision will be very sensitive to the yellow part of the spectrum (and so it is); salience "bias," or perceptual tunneling will have been selected for, since those with attention that wandered off the threat (saber-toothed tigers, angry rivals, etc), didn't stick around long enough to pass on their genes; over time blood has become red (a universal alarm color), since those without sufficiently alarming blood were too casual about wounds, and thus didn't stick around ... etc. (Interestingly, blood color is a *spandrel* from the oxygen carrying properties of iron compounds such as hemoglobin).

The point here is twofold. On the one hand, we do need to go beyond superficially plausible, glib, or post-hoc "explanatory" accounts set in the (evolutionary) lingo, as it were (but constructed without sympathy for the relevant biology, population genetics and dynamics of inclusive fitness). On the other hand, we need not shrink from making bold, cogent hypotheses based on evolutionary principles for fear of being labeled purveyors of Just So Stories. But those hypotheses should represent the start, not the end of the agenda. They should be testable in principle, at best falsifiable by experiment, but otherwise amenable via convergent evidence from the human factors laboratory, field observation, historical accounts and from

31

theoretical biology, Darwinian anthropology, and other relevant disciplines.

CONCLUSION

Evolutionary psychology provides a powerful framework for generating testable hypotheses about the nature of human interaction with automation, and for checking on the methodological assumptions built into studies. There is strong evidence for the existence of a pan-cultural "human nature" underlying both cross-cultural and individual differences. This can be viewed as a cognitive/emotional specification concerning the functioning of, and the evolved *purposes for* the modules of mind. Thus evolutionary psychology, like human factors, is a field science where ecological validity, the role of motivation, emotion, and actual human objectives and on-the-job performance take center stage. Both evolutionary psychology and human factors are inadequately informed by a laboratory based cognitive psychology with its emphasis on cold cognition and general purpose information processing. It has been said that there are no general purpose problems, only specific ones -- not being cheated, diagnosing an automation problem, finding a mate, etc. Human factors, at least, has a potential mate -- a natural partner in the study of the performance specifications of the 'liveware' component of human-machine systems -- evolutionary psychology.

REFERENCES

Bowlby, J. (1969). *Attachment and loss: Volume 1. Attachment.* New York: Basic Books.

Crawford, C. (1998). Environments and adaptations: Then and now. In C. Crawford & D. L. Krebs (Eds.), *Handbook of evolutionary psychology: Ideas, issues, and applications* (pp. 275-302). Mahwah, NJ: Erlbaum.

Cosmides, L. (1989). The logic of social exchange: Has natural selection shaped how humans reason? Studies with the Wason selection task. *Cognition, 31,* 187-276.

Gigerenzer, G. (1991). How to make cognitive illusions disappear: Beyond heuristics and biases. *European Review of Social Psychology, 2,* 83-115.

Gould, S. J, & Lewontin, R. C. (1979). The spandrels of San Marco and the Panglossian program: A critique of the adaptationist programme. *Proceedings of the Royal Society, 205,* 281-288.

Kahneman, D., Slovic, P., & Tversky, A. (Eds.). (1982). *Judgment under uncertainty: Heuristics and biases.* Cambridge and New York: Cambridge University Press.

Muir, H. C., & Marrison, C. (1991). Human factors in cabin safety. In E. Farmer (Ed.), *Stress and error in aviation* (pp. 135-141). Aldershot, England: Ashgate.

Piattelli-Palmarini, M. (1994). *Inevitable illusions: How mistakes of reason rule our minds.* New York: Wiley.

Pinker, S., & Bloom, P. (1992). Natural language and natural selection. In Barkow, J. H., Cosmides, L., & Tooby, J. (Eds.), *The adapted mind: Evolutionary psychology and the generation of culture* (pp. 451-494). New York and Oxford: Oxford University Press.

Stokes, A., Pharmer, J., & Kite, K. (1997). Stress effects upon performance in distributed teams. *Proceedings of the IEEE Proceedings on Systems, Man, and Cybernetics* (pp. 4171-4176). New York: IEEE.

Stokes, A., & Kite, K. (1994). *Flight stress: Stress, fatigue, and performance in aviation.* Aldershot, England: Ashgate.

Tooby, J., & Cosmides, L. (1992). The psychological foundations of culture. In Barkow, J. H., Cosmides, L., & Tooby, J. (Eds.), *The adapted mind: Evolutionary psychology and the generation of culture* (pp. 19-136). New York and Oxford: Oxford University Press.

Opportunities and Challenges of Automation in the Training of Pilots

Jefferson M. Koonce
Center for Applied Human Factors in Aviation
University of Central Florida

The training of pilots has been a challenge since the earliest days of aviation (Koonce, 1984; Koonce, 1998). In early flight training there were no formal instructor pilots, but only the words of advice from those who designed the air machines and the friendly and skeptical advice from friends and on-lookers. Many pilots were self-taught from books, magazines or their own imagination while constructing their own airplanes. After the First World War, many of the returning aviators engaged in barnstorming which attracted individuals eager to learn how to fly. Flight instruction was not well structured and it often took its toll in accidents and, unfortunately, lives.

TODAY'S PILOT TRAINING

Most of the pilot training today is quite similar to the methods used to train pilots fifty years ago (Koonce, 1984, December). Typically, flight instructors guide their students through the information published in the various applicable FAA Advisory Circulars (AC 61-21A, 61-23C, & 61-65C) and the Federal Aviation Regulations (FARs) to learn about the knowledge areas required for certification. As the student is learning about the knowledge areas, the instructor teaches the students about the basics of flight in the airplane. The knowledge and skills to be developed in the flight instruction are specified in the FARs. After the first solo flight the instruction tends to focus on cross-country flying including en route emergencies or abnormal situations and pilot judgement and decision making. The student must take a multiple-choice examination on the specified knowledge areas with 70% correct being required for a passing grade before being recommended for the practical test for the private pilot certificate. To ensure the standardization of the training and evaluation, the FAA has issued Practical Test Standards (i.e. FAA-S-881-1A) for the instructors and the evaluators to follow.

Beyond these basics, students may also acquire knowledge from a variety of books, audiotapes, video tapes and study guides to develop the knowledge specified in the regulations. Some student gain this knowledge through formal classroom training at aviation related college/university courses or at larger more formal flight training academies. The more formal flight training environments frequently include the use of flight simulators and part-task trainers to aid in developing the students' skills and knowledge. Still, the most common place for giving flight instruction has been and still is the airplane itself, a noisy, vibrating, unstable environment with the ever-present threat of bodily harm if one errs too greatly – not the best training environment.

The preponderance of persons providing flight instruction are relatively new pilots, flight instructors who are in the business, not to become professional flight instructors but, to gain flight hours so that they can apply for jobs as professional pilots with airlines or other commercial ventures. Due to the economics of the situations, highly experienced flight instructors move up to supervisory or managerial positions in flight training academies/schools and generally have little contact with novice flight students. Occasionally one might find a retired professional pilot giving flight instruction at a local airfield because of the love of flying and desire to share their years of experience with persons just starting their years of flying.

INNOVATIONS IN PILOT TRAINING

With the ever-increasing role of computers in our everyday lives, it is natural to expect them to enter the realm of the training of pilots. Personal computers (PCs) are becoming more and more popular in the general aviation industry (Koonce & Bramble, 1998). PCs are used as educational devices to develop individual's declarative knowledge of a specific domain of information, such as weather, aerodynamics, and regulations. These systems are generally electronic page-turners, that is, text on a CD-ROM that one uses a keyboard function to advance from page to page. Many have both sound and dynamic visual illustrations to hold the student's attention and enhance the training effectiveness of the program. Some even present questions regarding the materials covered so the student can check his level of understanding of the materials.

There is a blossoming industry providing aviation instruction via computers, generally sold on CD ROM disks. These cover topics such as the Federal Aviation Regulations (FARs), aviation weather, principles of flight, emergency procedures and decision making, and some even providing sample examinations of the questions that the FAA will have on the written examination that the applicant must pass for a pilot certificate. Typically, these knowledge development systems do not require the learner to demonstrate a specific level of knowledge and understanding on a particular lesson before permitting the learner proceeds to the next lesson. That is, they are not criterion-referenced training systems. Generally, the student is able to "browse" the lesson materials and dwell on any of particular interest and skip other areas. The multiple-choice examination of the knowledge domains required for a pilot certificate or rating can now be taken at certified testing sites via computer with near-immediate feedback regarding performance on the test(s) (FAA, AC 60-23).

Sadlowe (1991) presented a collection of papers focusing on the use of personal computers in the training of instrument pilots. There are several software programs that allow a student to "fly" a variety of types of airplanes in visual as well as instrument condition. The students typically interface with the system through the keyboard and/or mouse and a set of flight controls, sometimes including rudder pedals to control yaw. Instructor pilots in some flight instruction programs to provide instrument flight instruction use such PC-based flight simulators. The instructor will conduct the lesson while the student performs the maneuvers. For post-maneuver critique, the instructor generally tells the student what errors took place and perhaps why and sometimes the PC-based system might provide a ground plot of the student's flight and possibly even a vertical profile view. Many flight instructors do not care for flight training in the simulators, whether they are full-enclosure flight simulators or desktop PC-based training systems. They would rather be in an airplane giving instruction where they earn a greater hourly wage for their time. Also, instructors cannot log the time in the ground-based trainers as flight time in their logbooks, and often their main goal as flight instructor is to accumulate flight hours to secure a pilot position with an airline. Studies (Caro, 1977; Caro, Shelnutt, & Spears 1981) have shown that the instructor pilot might be the greatest factor in the training effectiveness of ground training devices (simulators), and instructors who are not positively motivated in the use of these devices will detract from their training effectiveness.

Many flight training organizations have noted the economic advantages and training effectiveness of PC-based flight training devices and have been utilizing them as training aids to enhance the student's skills. The use of PC-based flight simulators are now so popular that the Federal Aviation Administration (1997) has recently issued an Advisory Circular (AC 61-126) detailing the extent to which a PC-based flight simulator can be used in the training of pilots seeking an instrument rating. This Advisory Circular identifies specific features that such a device must have and be capable of to be an acceptable training device.

Not only must a student know about the aviation arena and how to fly an airplane, but also the student must be capable of communicating in the aviation system. That is, they must have the ability to understand pertinent radio communications as well as know how to conduct two-way communication within the airspace system. Koonce and Weller (1991) demonstrated a multimedia communications simulator that trained the student in listening to and understanding aviation radio communications and being capable of making appropriate radio transmissions during the course of a flight scenario. This system was developed from an earlier computer-based program that taught the student what communications should be made and when, and it also taught the student to understand the messages received by his/her radios. Unfortunately, many of the students in that program who developed good knowledge of the communications still had difficulty in the airplane because they had no experience in actually producing communications in response to messages received or requesting information at the appropriate times. The MARCS taught the students to listen and understand and to product the appropriate communications using a headset with microphone. The student could later compare their actual transmissions to what should have been produced and a digitized record of the student's performance was made available to the flight instructor.

A recent development in computer-based flight training incorporates lessons to develop the student's understanding about flight and instruction on how to actually perform basic flight maneuvers (Benton, Corriveau, Koonce, & Tirre, 1992). The Basic Flight Instruction Tutoring System (BFITS) is a fully automated computer-based criterion-referenced system for the training of novices how to fly an airplane. It provides a basic understanding of the physics of flight, parts of an airplane, how to control an airplane in flight, what the flight display instruments tell the pilot about the state of the airplane in flight, demonstrations of how to perform most of the basic flight maneuvers required for first solo flight, and trains the pilots to perform those maneuvers to a pre-set standard of performance. When the student pilot is performing a particular flight maneuver, BFITS will provide informational hints whenever the student's performance deviates significantly from desired performance. The student must continue to perform a set of maneuvers in a particular lesson until the performance meets the pre-set criteria. The student cannot progress to the next lesson until satisfactory performance on the current lesson has been demonstrated, criterion-referenced training.

BFITS has demonstrated the capability to save a significant amount of flight time in the pre-solo training as well as the total time to the private pilot certificate (Koonce, Moore & Benton, 1995). It also provides the additional benefit of a 38% reduction in landings prior to the first solo flight, saving wear and tear on the airframe. BFITS does not teach about the Federal Aviation Regulations (FARs), weather phenomena, cross-country navigation, and emergency procedures.

THE FUTURE?

With the great amount of flight training material being offered on CD ROMs and the introduction of concepts, such as BFITS, the future of pilot training will become increasingly dependent upon computer-based training systems. While BFITS demonstrated a rather remarkable capability, it was written for a 286 machine (remember those?) with a 30 MB hard drive which limited the amount of material that could be provided and the style of presentation of material. Sound capability and animation was rather limited. With today's computer systems speed, storage and RAM, virtually the entire flight curriculum for a private pilot's certificate could be put on a computer.

An extension of the BFITS has been the Semi-Automated flight Evaluation System (SAFES) (Benton, Corriveau, & Koonce, 1997) in which the performance criteria of the BFITS has been put into a box for an airplane. The SAFES box is powered from the airplane's cigar lighter receptacle, interfaces with the communications system the pitot/static system and the manifold pressure line, and is controlled through a small hand held keyboard with LCD display. When a flight maneuver is selected and the letter "B" is depressed to signal the beginning of a maneuver, SAFES provides the pilot with "hints" as to when a flight parameter is beginning to deviate significantly from desired performance; sort of an instructor in a box. After a flight the SAFES downloads the data of all of the maneuvers performed so that they can be replayed on an instructor's PC and the data trail can be converted to ASCII format for statistical analysis. Some have envisioned the SAFES information being down-linked to the flight instructor's office so the instructor can monitor the student's performance and provide "tailored" feedback beyond that of the SAFES and not have to actually be in the airplane.

It is tempting to say that most all of a pilot's training could be accomplished on the ground via a computer-based automated system and a good bit of it performed by automated systems in the airplane (Koonce, 1997). As one might expect, concerns have been expressed regarding the widespread use of automation in the training of pilots will not be accepted because it will greatly reduce the role of the flight instructors and his/her income.

The work of Herold, Parsons, and Fedor (1997) strongly encourages the use of flight instructors to provide the optimum feedback to the students instead of fixed feedback provided by present systems. However, adaptive computer-based systems could be developed to change the amount and type of feedback provided to the students based upon their individual propensities for feedback and the particular stage of skill development.

With the rapid advancement of the computer technologies and automation, the future is limited only by the imagination of the future human factors and systems designers.

REFERENCES

Benton, C.J., Corriveau, P., Koonce, J.M., & Tirre, W.C. (1992). *Development of the Basic Flight Instruction Tutoring System (BFITS)*. (Technical Paper AL-TP-1991-0060). Brooks Air Force Base, Texas: USAF Armstrong Laboratory.

Benton, C.J., Corriveau, P., & Koonce, J.M. (1993). *Concept Development and Design of a Semi-Automated Flight Evaluation System (SAFES)* (Technical Report AL/HR-TR-1993-0124). Brooks Air Force Base, Texas: USAF Human Resources Directorate, Armstrong Laboratory.

Caro, P.W. (1977). Some Current Problems in Simulator Design, Testing and Use. Pensacola, Florida: Seville Research Corporation.

Caro, P.W., Shelnutt, J.B. & Spears, W.D. (1981). Aircrew Training Devices Utilization, Vol.1. Brooks Air Force Base, Texas: Air Force Systems Command, TR: AFHRL-TR-80-35.

Federal Aviation Administration (FAA AC 60-23)(1994). Conversion *to the Computer Based Airman Knowledge Testing Program* (6/17/94). Washington, D.C.: Superintendent of Documents, U.S. Government Printing Office.

Federal Aviation Administration (FAA AC 61-21A)(1980). *Flight Training Handbook* (Revised 1980 with Errata Sheet). Washington, D.C.: Superintendent of Documents, U.S. Government Printing Office.

Federal Aviation Administration (FAA AC 61-23C)(1997). *Pilot's Handbook of Aeronautical Knowledge* (7/10/97). Washington, D.C.: Superintendent of Documents, U.S. Government Printing Office.

Federal Aviation Administration (FAA AC 61-65C)(1991). Certification*: Pilots and flight Instructors* (2/11/91). Washington, D.C.: Superintendent of Documents, U.S. Government Printing Office.

Federal Aviation Administration (FAA AC 61-126)(1997). Qualification *and Approval of Personal Computer-Based Aviation Training Devices* (5/12/97). Washington, D.C.: Superintendent of Documents, U.S. Government Printing Office.

Federal Aviation Administration (FAA-S-881-1A)(1987). *Private Pilot Practical Test Standards*. Washington, D.C.: Superintendent of Documents, U.S. Government Printing Office

Herold, D.M., Parsons, C.K., & Fedor, D.B. (1997, May). *Individual feedback propensities and their effects on motivation, training success, and performance*. Final Report to the Army Research Institute. Atlanta Georgia: Georgia Institute of Technology, School of Management.

Koonce, J.M. (1984). A brief history of aviation psychology. *Human Factors, 26*(5), 499-506.

Koonce, J.M. (1984, December). Training and certification of the general aviation pilot, In *Automation Workload Technology: Friend or Foe?- Proceedings of the Third Aerospace Behavioral Engineering Technology Conference* (pp. 83-87). Warrendale, Pennsylvania: Society of Automotive Engineers, Inc., P-151.

Koonce, J.M. (1997). Automation issues in general aviation, In M. Mouloua & J.M. Koonce (Eds.) *Human-automation interaction: Research and practice*, (pp. 157-166). Mahwah, New Jersey: Lawrence Erlbaum.

Koonce, J.M. (1998). The history of human factors in aviation, In J. Wise & D.J. Garland (Eds.) *Human factors in aviation* (In Press). Mahwah, New Jersey: Lawrence Erlbaum.

Koonce, J.M. & Bramble, W.J.,Jr. (1998). Personal computer-based flight training devices. *International Journal of Aviation Psychology, 8*(3), (In Press).

Koonce, J.M., Moore, S.L., & Benton, C.J. (1995). Initial validation of a Basic Flight Instruction Tutoring System (BFITS). *Proceedings of the Eighth International Symposium on Aviation Psychology* (pp. 1037-1040). Columbus, Ohio: The Ohio State University.

Koonce, J.M. & Weller, M.H. (1991, October). *Multimedia Aviation Radio Communications Simulator (MARCS): An IBM innovations program project.* Urbana-Champaign, Illinois: University of Illinois Computer Fair.

Sadlowe, A.R. (1991). *PC-Based Instrument Flight Simulation - A First Collection of Papers,* (TS-Vol.4). New York, New York: American Society of Mechanical Engineers.

Human Automation Interaction - A Military User's Perspective

Robert R. Tyler, Ph.D.
Dowling College

INTRODUCTION

For the past thirty years I have been a user of complex weapon systems provided to the Armed Forces for service members to train with and ultimately use on the modern battlefield. Part way through my military career I became a procuring official for some of these systems and their associated training devices. Additionally, I have been actively involved in the command and control of U.S. Marine Corps forces on both real and simulated battlefields. Based on these experiences and illuminated by my scholarly pursuits in Human Factors Psychology, I intend to present my personal observations of human-automation-interaction in military settings. Accordingly, this presentation is without empirical studies or statistical analysis, and it is not controlled for my individual bias or variability.

UBIQUITOUS AUTOMATION

Automation is woven into the fabric of modern existence. Our automatic alarm clock invites us to begin our day. Our automatic coffee maker hastens our awakening. Our house is warmed and cooled automatically. We drive to work in a vehicle that shifts itself, has automatic traction control and automatic anti-skid brakes. Our communications follow us effortlessly throughout the day via self answering fax machines, cellular phones, voice mail, call forwarding, e-mail, etc. Examples of daily automation are endless. In the modern world humans perform the majority of their daily activities aided by, and often times oblivious of, automation. We depend on it.

Since automation is essential to daily life, when we transport humans from their peacetime domicile and workplace to the combat environment we transport the automation as well. In fact, the American military is noted for considering advanced technology (read: automation) to be a force multiplier, a force extender, and a force enabler. The daily existence on the battlefield is every bit as automated as a state-of-the-art industrial plant. Automation effects everything on the battlefield including combat vehicles, communication, weapons systems, intelligence gathering/processing, and command and control.

Good and bad automation affects the battlefield just as it does Main Street, U.S.A. What frustrates you with your office telephone will frustrate the soldier as he communicates to his combat network. Similarly, those wonderful enhancements that automation has provided to mankind are equally effective on the battlefield. Consider the dependence on and effectiveness of a plethora of sensors, electronic landing aids, automatic range finders, and autopilots to name but a few. Clearly, those human factor issues encountered in everyday life carry over into the military environment. If performance is enhanced at home, it can be expected to enhance performance on the battlefield. The converse is just as true. If performance is impeded at home, it can be expected to exasperate the performance in combat.

THE MALEVOLENT BATTLEFIELD IS THE DIFFERENCE

From the user's perspective it is the very nature of warfare and its impact on automation that most concerns the military planner. Simply put, the purpose of warfare is to locate, close with, and *destroy* the opponent. By design, the battlefield is a hostile environment. It is dirty, noisy, hot, cold, wet, dry, and uncertain. But most of all it is lethal to men and machines. Therefore, it is essential that combat automation perform as expected. It must work - every time.

For automation to be reliable it needs designed protection from three things: the physical environment, enemy counter measures, and enemy intrusion of our defenses. To explain, most automation depends on computer processing and most computers experience performance degradation when subjected to temperature extremes, dirt, and high humidity. Therefore, automated systems must be hardened to protect themselves from these hostile variations of the physical environment. Operationally, opponents expend considerable effort to disrupt and interfere with the operation of each other's combat systems. As combatants try to neutralize each other, the interruption of automated systems becomes a primary target. Obviously, to be successful on the battlefield, the automation that contributes to our combat effectiveness must be protected from jamming, interception, hacking, or other intrusions. Conversely, to survive on the battlefield, our systems need to be able to disrupt the enemy's weapon systems while being impervious to their countermeasures. Again, battlefield automation must work as intended every time despite the weather, the location, or the ingenuity of the enemy.

BATTLEFIELD INNOVATION

It has been said that necessity is the mother of invention. History records technological advancements precipitated by armed conflict. However, to many the "innovation" addressed here would most likely be described as abuse or misuse. Weapons and systems are employed in ways never envisioned, nor intended, by their designers. I have observed combat rations (food) being cooked in helicopter engine bays, drinks being cooled with CO_2 fire extinguishers, and runways being "swept" by helicopter rotor wash, but the most poignant example of automation circumvention/misuse I am aware of involves the venerable CH-46 helicopter.

Designed as a civilian logging helicopter, the Boeing-Vertol CH-46 tandem rotor helicopter was pressed into service early in the Vietnam conflict. It had a semi-automated speed trim system that tilted the rotorheads (according to selected positions) to maintain a more level cabin attitude throughout the flight envelope. As combat pilots we discovered that if one selected "hover-aft" at faster airspeeds, the rotorheads programmed aft and acted like two 52-foot wide speedbrakes. Armed with this knowledge, we developed informal tactics for use when spiraling into "hot" zones. We maintained a high airspeed until on a short button-hook final and then select "hover-aft" such that the aircraft would perform a beautiful quick-stop into the zone. What we did not appreciate was that the two bolts that held the aft transmission, gearbox, and rotorhead in the aircraft were not stressed for this maneuver. In short order, we began to tear the back-end off of our aircraft.

The engineers came to our rescue. They strengthened the back of the aircraft; they added a few more bolts; and they redesigned the speed trim system. They connected it to the airspeed indicator system such that the rotorheads would not program to "hover-aft" above 70 knots - an aerodynamically safe speed. Not to be foiled, we improvised. Now when approaching really "hot" zones we still kept our airspeed up, selected "hover-aft", and on short final one of the pilots would extend his hand outside the cockpit to obstruct the airflow over the airspeed sensor. This fooled the enhanced speed trim system into believing the airspeed was below 70 knots and allowed the rotorheads to program aft. We got our speedbrake back! Clearly, our improvising circumvented the automation in ways few designers could have anticipated.

Another reality of the battlefield is that the trained user may become a casualty. The next person who "falls in" on the weapon system may have little, if any, formal training on the systems operation. That soldier will improvise to make it work. It is our task as designers to ensure that the system is designed with appropriate affordances that lead the uninitiated through the process to effective employment. Quick and easy "way-finding" is a tactical imperative.

Finally, recalling that this is a malevolent environment, will the system work in a degraded condition? What happens when the opponent succeeds in damaging our weaponry? Will the operator be able to invent a combat effective use for this "combat modified" piece of equipment? Beyond use of the damaged system, we must ensure that, at a minimum, the damaged vehicle, aircraft, etc., can be safely recovered. Not only does this enhance operator survivability, it enables commanders to repair and reconstitute their combat strength.

INFORMATION MANAGEMENT

A major benefit of automation is the handling of vast amounts of information. Transportation companies like Federal Express advertise that individual packages are tracked and monitored from pick up to delivery. It is this automation that has allowed industry to adopt just-in-time supply philosophies. For some items, the entire process happens without human involvement. An inventory threshold is crossed; a resupply order is automatically placed; the item is shipped; appropriate billing happens; and the item arrives at the intended destination on time. In a similar fashion, as a senior watch officer at Marine Corps Headquarters in Arlington, Virginia, I watched the logistical flow of personnel and equipment into the Marine Corps' sector of Desert Shield/Storm. From our desks in that room, using standard desktop PCS, we communicated in real time to forces in Saudi Arabia. Additionally, we knew the location and cargo of every ship or aircraft that was carrying supplies into theater. Our personnel lists were every bit as extensive. When we needed a particular language skill for a special mission, within hours we had identified appropriately trained individuals in theater who could serve on this mission. From these casual observations it is clear that modern military operations involve the management of vast quantities of information. This information includes everything from intelligence reports to the logistical status of one's own forces.

Military commanders throughout the ages have sought the answer to three questions.

Where am I?

Where are my troops and my supporting arms?

Where is the enemy?

Automation of the modern battlefield promises to answer these questions. Hand held Global Positioning System (GPS) receivers can pin point (within feet), and update several times a minute, a soldier's location anywhere in the world. Transmitting that individual data to the command center, where it is consolidated and collated, can inform the commander as to the dispersion of his forces. Likewise, through the wonders of automation and sensor technology, we can locate and present enemy positions with equal accuracy. The obvious concern is the need to protect and isolate these automated systems from incapacitation or unauthorized manipulation (misinformation) while providing the operators and commanders a clear picture of the battle space. Inundating the decision-makers with raw data does not guarantee their *understanding* of the battlefield. It is important to select and present that data as knowledge bits and decision aids. The goal is to enable the combatants to develop and maintain their combat situational awareness.

CONCLUSION

A role of earlier automation was to extend human capabilities through the application of mechanical leverage. Quickly automation progressed from a mechanical endeavor to a more inclusive one by embracing the enhancement of all human activities. In so doing it minimized our human limitations and extended our capabilities. Continuing to improve on this automation, we can now "see" and shoot beyond visual range. We "know" the answers to the three questions above - or do we really? The purpose of decision aids and information management systems (automation) should be to reduce uncertainty. As mentioned earlier, if we do not present the gathered data in *meaningful* ways, we have not served to reduce the commander's uncertainty. If we have not protected these systems from our adversary's intrusions, we certainly have not reduced the commander's uncertainty. The automation concerns of operator familiarity, specific system enemy countermeasures, and degraded performance pale in comparison to the military's concern over their ability to create, protect and maintain an accurate picture of the battle space. Carl VonClausewitz pointed out in his 1832 book *On War* that the "difficultly of *accurate recognition* constitutes one of the most serious sources of friction in war, by making things appear entirely different from what one had expected." Battlefield automation must reduce that friction by providing knowledge and flexibility to the combatants. It must enhance the collective and individual real-time understanding of the battle space. That is the military user's concern - and our challenge.

REFERENCES

Von Clausewitz, C. (1976). *On war*. (M. Howard & P. Paret, Trans.). Princeton, NJ: Princeton University Press. (Original work published 1832)

Life, Liberty, and the Design of Happiness

P.A. Hancock
Liberty Mutual Research Center

PREAMBLE

As I turned off of Interstate 264 onto Waterside Drive, I had a distinct feeling, as Yogi Berra would have it, of 'deja vu all over again.' and this feeling grew even stronger as I turned for the Omni Hotel. As I stepped through its doors, the experience was complete - I knew I had been here before. As some who read this will know themselves from personal experience, the Omni, Norfolk was the site of the 1983 Annual Meeting of the Human Factors Society. It is now, as it was then, a wonderful location for a meeting. In 1983, I was a first year Assistant Professor with less than three months employment under my belt. In wondering what that individual would have made of his future self and, in turn, wondering what I now would have made of the fresh-faced version, I recalled an experience at the 1983 meeting that has been a constant lesson since. I use it here as an introduction to my present work.

INTRODUCTORY STORIES

A First Norfolk Story

Someone has to be the last presenter at a Conference and, in 1983, my turn had come. In a rather refined twist of cruelty, my session finished at 12:30, while the rest of the meeting had terminated by 12:00. Not then being a major hub, it was understandable that flights from Norfolk, especially to the West Coast were somewhat limited and in a rather piquant way, it was suggested to me that the 'last plane for civilization was leaving at high noon.')[1] It was clear that presenters in my session were painfully aware of this as they each in sequence gave their talks, made their profuse but understandable apologies and headed for the door and their respective journeys. Soon after eleven, the audience was dwindling precipitately and it did not need significant mathematical expertise to work out that what were left were largely presenters and chairman.

As each individual presented and then left, the room grew a little more silent and as my turn approached I realized that this would not be a talk to a substantial number of the academy. The chairperson, who I am now pleased to number among my friends, got up to introduce me and then was forced to excuse himself since he too also had a deadline which could not wait (although he may have different version, please ask him, you can look it up). As I stood to announce my stunning scientific findings to the world, I was left with one audient. However, true to my self, I resolved to give full measure and for the next twenty minutes, with unstinting effort, I endeavored to lay before my solitary listener the nuances of 'space-time and motion study' (Newell & Hancock, 1983). It is, I must add, one of the few conference proceedings papers for which I have ever had a reprint request. At the end of the twenty minutes when I had finished, the lady walked up to me and said in clear and ringing tones - 'Well, young man, I've only had the chance to hear one or two talks at this Conference but yours was clearly the best.' My dragging spirits rose a little and I inquired from which institution of higher learning she came. Her reply echoes in my mind even today:

'Oh! I'm not from any University, I'm here to clean the room, can you leave now please!'

---

[1] I apologize to residents of the area, since it is one that I personally like a lot. It is a quote of someone else's words. However, if you have ever been around at the end of a major meeting when all the evidence disappears in a virtual instant, you can empathize with the sentiment.

There are many putative lessons which I might draw from this experience but the one that has stayed most closely with me is the necessity for humility and humor. It does not take a philologist to recognize their common linguistic root. We are each us much more important to ourselves than our merits support and yet I am persuaded today that perhaps one or two more individuals might like to read this present effort and so, as much in hope as in expectation, I offer a brief written account of the cryptic observations I made at the present Norfolk Conference. However, I cannot begin before giving a brief account of a second and equally instructive incident which took place in the lobby of the Omni Hotel just prior to the Conference commencement.

A Second Norfolk Story

Before the present Automation Conference, I had been at an earlier meeting in Dayton, Ohio and due to an administrative problem was forced to switch to a very early flight from Dayton to Washington, DC, whence I proceeded to Norfolk in a rental car. I arrived mid-morning and had been comforting my fatigued self with the thought of a nap on a soft bed only to be told that the McDonald's All-American basketball game meant that no rooms would be available before mid-afternoon. Not normally the most equitable of individuals, I found myself on the verge of significant mayhem and turned from the obsequious but intransigent desk clerk, determined to do damage to the first living organism encountered. Fortunately, the last remnants of English reserve meant that I was still partly civil as I sat next to an older gentleman in a track suit. He and I began to talk and I knew, even as I sat, that I was conversing with John Wooden, the legendary coach of the past UCLA Bruin Championship winning basketball teams. He turned out to be a very simple, quietly spoken and engaging individual who was kind enough to sign a treasured autograph and to converse with a strange Englishman on the problems of building teams to function under significant stress. I think the lesson here is that even when we do not get what we want, it may not necessarily be bad and may actually lead to unexpected and unanticipated benefits. If you are with me to this point, you will realize that I am telling stories and I want to pursue that theme since my presentation was, as all presentations in reality are, a threaded story along which the beads of fact and buttons of conjecture are ranged so that the weaver can eventually provide whole cloth.

Figures in a Timescape

At the beginning of my presentation, I asked the rhetorical question whether you can kill yourself by holding your breath? In reality it would be exceptional if you could, but it serves to stimulate most individuals into considering the physiological system and the way it serves to protect us, even from ourselves. It turns out that one might be able to pass out from breath-holding, but having done so respiration resumes and consciousness is slowly returned to the abuser. I hasten to add that I would not advocate trying this for yourself since there are always other potential complications. The point being made was that we already have very successful automated systems, not merely those in the engineered realm, but those already resident in nature which have functioned for many millennia to support life and evolution. Human beings should be, and presumably are, very glad that they don't have to engage in volitional control of the many support systems of the body and indeed, life as we know it would not be possible without the 'automated' accomplishment of respiration, heart function and the like. Thus the first point was that automation in different forms had been around and working for many millennia, we should not see automation as something new or frightening.

The second point built upon this sense of history through an oblique reference to the characteristics of an existing 'automated' system. With a little modernization of language, I described the capacities of the Golem which was an entity created by the Rabbi of Prague to protect his flock. Since I have been writing about this elsewhere, I shall not burden the reader with repetitive details. However, I did list some of the

Golem's characteristics and these are noted in Table 1 below.

Table 1.
Characteristics of a Flexible Function Automated System

System Capabilities: Descriptive and Non-Exhaustive Listing

- Full Natural Language (Speaker Independent) Interface
- Full Surrogate Biological Motion (with Supra-Human Strength)
- With Correct Programming - Zero Failure Rate
- Domain Knowledge - Extensive.
- Out-Of Domain Knowledge - Zero
- Self-Intention - Zero
- Out-Of-Domain Operations - Guaranteed Failure
- Absence of Expected Command - Unexpected Failure
- No Affective Abilities, No Voice Output Abilities.

To put these abilities in brief context, the Golem was raised (or created) by Rabbi Yehuda Loew, the Maharal of Prague to help defeat the machinations of those determined to see the demise of the Jewish community there. Joseph Golem, sometimes referred to as Yosell the mute, can only act in pursuit of these community concerns and makes laughable errors when sent on secular errands. Joseph is endowed with limited knowledge and no self-intention and when the Maharal on Friday forgets to tell Joseph to guard the community during the Sabbath, Joseph runs amok, which is one element in his eventual 'de-commissioning.' He is mute and has no emotion since it is said that such emotion would cause him significant problems. I purposely framed these capabilities in modern terms, since many bear a strong relationship to the character of Mr. Data on 'Star-Trek: The Next Generation' referred to by Mark Scerbo in the very first Automation Conference (Scerbo, 1996). The difference being that the Golem was recorded as being created in the 1580's. This story allows us to see that the conceptions of automation, even down to surrogate human beings, is not a new concept. Indeed, many ancient cultures employed automation-based devices so that priests could present apparent acts of the Gods in the Temples of the ancient world in order to impress and oppress the common people. However, the Golem story itself has additional threads. Many of those directly involved in the development of the modern day computer grew up on Golem stories as folk tales. Indeed, the father of cybernetics, Norbert Weiner, compared himself to the Golem in some of his writings, it was a central motif in his life. There is a further thread to science in that the English Necromancer John Dee visited Prague in the 1580's and it is in his 'Mathematical Preface' that we find reference to the lowest form of magic which is the manipulation of the environment, essentially by human engineering. I tried to weave these strands a little tighter through reference to the Compte de St. Germain, an intriguing figure in history who was thought to have lived many times the normal life span. When asked whether his master was six hundred years old, his servant is supposed to have replied 'I cannot say, since I have only been with him three hundred years.' My point being that John Dee, the Maharal and crucial figures to my later argument such as Rousseau and Hobbes, are actually very close to us in time, a mere seven generations. It is only our limited personal life-span that prevents us from seeing them as the colleagues they rightfully are.

In part though Dee's insightful exposition, independent of its connections with Hermeticist philosophy, I sought to show that science supports purpose. To illustrate this, I referred to Francis Bacon's 'Novum Organum.' Part of a greater but unfinished work, the 'Great Instauration.' Bacon following Dee, advocates for the systematic exploration of the world. In reaction to the authoritarian nature of Scholasticism,

Bacon's approach relies on observation not on authority. It was Bacon who articulated the great clearing of the detrius of acceptance rather than experimentation and thus preempts Descartes in his role of Doubter General. Bacon's precepts are echoed in the English Royal Society, one of the first recognized scientific societies, who motto 'Nulla in Verba' can be roughly translated as 'Take No One's Word For It.' What does not pass empirical muster is discardable.

THE PURPOSES OF LIFE

The confluence of each of these themes reaches its fruition in my attack on the question of purpose. Is life merely a never-ending sequence of *parturition, preservation, procreation*, and *putrefaction*? In asserting that there *can* be more than this, I turned to a linkage between design at all levels, at the level of the human brain, at the level of human technology, at the level of human morality, and finally at the level of human society. These links are outlined in skeletal form in Table 2 below.

Table 2.
Nested Self-Similarities in the Substructure of the Purposes of Life

Brain	Technology	Morality	Society
Neocortex	SIMS	Augustine	Happiness
Paleocortex	Automation	Rousseau	Liberty
Brainstem	Safety	Hobbes	Life

As with all forms of nested self-similarity, one can start in any column and transition to others. In the present case, I start at the societal level since it will be that perhaps most familiar to most individuals. The three-part distinction, and indeed essentially, the title of this piece of work comes from words contained in the U.S. Constitution. However, it must be recognized that they did not originate there. The fundamental right to life is the subject of part of the writings of Thomas Hobbes in Leviathan. Hobbes, an under-rated and original intellect, asked fundamental questions about morality well before Descartes crucial questions of knowledge. Hobbes opined that the basis of moral structure was an individual's own right to existence. Thus, the fundamental right, even to the extent of taking the life of another was the defense of self. We still adhere to this today. In technical systems, the equivalent is safety. One of the first things we demand of our creations is that they not destroy. The epithet: *primum non nocere* (first do no harm) applies to the design of most manufacturanda. Indeed, the whole process of certification concerns this issue (see Hancock, 1993). Bringing us back to an earlier point, the brainstem is sight of many of the processes which support life. So, the defeat of the intention to die by holding one's breath is resident in the brainstem, which with associated structures is also the center for affect. In France, if one commits murder under the driving force of the limbic system, there is a possibility of the 'crime of passion.' Under such circumstances, one might even be acquitted. However, if the murder is planned by the frontal cortex 'in cold blood' one is doomed to punishment. This is one of the very many examples in which we treat the origin of intention differently, according to our knowledge of brain structure and function. The insanity defense, of course, being another example.

The next stage in each sequence is, in contemporary developments, the most pertinent. Liberty is, of course, the watchword of American society, but like so much of Jefferson there is some Montesquieu involved and I personally see Rousseau as the philosopher of freedom. His stunning phrase that 'man was born free, and he is everywhere in chains' still resonates today (Rousseau, 1762). At a fundamental level, automation is about freedom. In particular, it is about doing away with the drudgery of repetition and the soul-grinding stupidity of invariant action simply to achieve a living wage. As always, the specter of unemployment and its related cousin de-skilling, hovers ominously in the background of any such consideration. However, hopefully, within a short period of time, we will look on repetitive physical and even repetitive cognitive work as such a repugnant anachronism that it will be conceived then as we conceive of slavery today. Indeed, this is a form of technical slavery to other insensate machines and it is not a reasonable human state. At the level of brain function, many such processes are concerned with support of abilities such as locomotion, which like other over-learned or 'automated' (see Schneider & Shiffrin, 1977) motor abilities, allow us to accomplish a number of higher level goals while still interacting with the predictable elements of the world. Thus most of us are able to walk and chew gum at the same time. Consequently, although we can perform rote motions relatively independent of cognitive involvement, the question is: should we?

THE DESIGN OF HAPPINESS

It is the final component which needs much more detailed elaboration, denied here by space restrictions but more realistically my own poor capabilities. Indeed, it is the goal of happiness toward which much of my contemporary thought is directed. Happiness and its near relation contentment, are not sole properties of the neocortex. Indeed, all the vertical relations I have shown in Table 2 are better seen as nestings, in which achievement at higher levels are predicated upon fulfillment of lower levels. One cannot have liberty without life and similarly one cannot achieve true happiness without liberty as a precondition. In respect of brain structure, the three levels are actually *physically* built on top of each other and like the societal level, are interdependent for full functioning. Until time flows backward, Augustine cannot depend upon Rousseau. However, I like to think of the problems that they addressed rather than the individuals themselves. In this sense, the mind is atemporal, having one of its great but unrecognized advantages the ability to sweep across vast tracts of time, unfettered by the necessity to consider all things in a causal framework. Indeed, it may be this ability to plan and 'see' ahead which is a key differentiate of humans themselves. In technology, our vertical integration is less clear. Indeed, it is this very question that we have been struggling with as a community interested in automation. At the highest level, I have put SIMS, or self-intentioned machine systems, an acronym and concept I developed in more detail in the previous Automation Conference (Hancock, 1997). However, whether the full emancipation of machines, if indeed this is possible, is equated with human happiness is a doubtful proposition.

John Stuart Mill is quoted as having once said: 'Ask yourself if you are happy and immediately you cease to be so.' Rousseau described the happiest time of his life as a brief sojourn on an island in which the cares of existence could be forgotten. Clearly, happiness is a process not a thing and a crucial part of that process is dissociation from want. In our society, we have confused the negative with the positive. That is, we have confused the absence of material need with the presence of material necessity. This confusion has been aided and abetted and perhaps even fostered by material capitalism. We are bombarded into a fog of confusion between what we need (that being air, water, food, shelter, and the like), with what we desire (that being a new appliance, or the latest sparkling bauble to be touted as that with which we cannot live without). Lest you should doubt me on this point, I recommend that you stand between a group of youngsters and a sale of the latest 'must have' toy! In this gross misservice, psychology stands condemned. The tenets of paired-associate learning and early behaviorist theory were pedaled by Watson in his second career as an advertising executive. Rarely can so much damage have been done to a society so surreptitiously. Today, we are regaled in our communication media with omnipresent reminders that we can only experience the wonders of existence through material gain. A recent advert for a well-known product implies that one can only feel the emotion of true love by buying the latest model automobile! It is an obscene perversion of human possibility to tie our future hopes to material gain. Indeed, we know that such gain does not bring happiness but the lotteries of our country continue to pour millions into the Governmental coffers.

If happiness is truly pluralistic, then specification of happiness will be an arduous endeavor indeed. However, if we are able to specify happiness as a process and what represents greater or lesser degrees and that technology can help us to achieve those degrees, independent of the detriment of any other individual, then hope does exist. For example, virtual world experiences need affect no other individual and can provide opportunities for 'convivial' interaction (see Illich, 1973). As we each take small steps along the path of life, we each need to stand up and designate the purposes of life. We have no requirement to be eternally correct and although it is tempting, we should not specify such purposes as absence alone, such as the absence of food or oppression. Rather, we need to articulate collective, global, societal goals if the future is not to be the haphazard quilt of happenstance that marks our essential past. Those who mediate between human and technology must have a crucial voice in this debate.

ACKNOWLEDGMENT

I'd like to thank Dr. Ernest Volinn for his comments on an earlier version of this paper. The views expressed are my own and do not necessarily reflect the opinions or policy of any agency or group with which I have affiliation.

REFERENCES

Hancock, P.A. (1993). Certifying life. In: J.A. Wise, V.D. Hopkin, & D. Garland. (Eds.), *Human factors certification of advanced aviation technologies*. Berlin: Springer-Verlag.

Hancock, P.A. (1997). Men without machines. In M. Mouloua & J.M. Koonce (Eds.), *Human-automation interaction: Research and practice* (pp. 61-65). Mahwah, NJ: Lawrence Erlbaum.

Illich, I. (1973). *Tools for conviviality*. New York: Harper & Row.

Newell, K.M., & Hancock, P.A. (1983). Space-time and motion study. *Proceedings of the Human Factors Society, 27,* 1044-1048.

Rousseau, J.J. (1752). The social contract. (Penguin Edition, London, 1972).

Scerbo, M. W. (1996). Theoretical perspectives on adaptive automation. In R. Parasuraman & M. Mouloua (Eds.), *Automation and human performance: theory and applications* (pp. 37-63). Mahwah, NJ: Lawrence Erlbaum.

Schneider, W., & Shiffrin, R.M. (1977). Controlled and automatic human information processing. *Psychological Review, 84,* 1-66.

DRIVING SYSTEMS

AND

DRIVER PERFORMANCE

Driver Support and Automated Driving Systems: Acceptance and Effects on Behavior

Dick de Waard and Karel A. Brookhuis
Centre for Environmental and Traffic Psychology
University of Groningen, The Netherlands

INTRODUCTION

From manual control to full automation five different grades of automation can be discerned (Endsley & Kiris, 1995). Starting from (1) no automation, via (2) decision support, (3) consensual Artificial Intelligence (AI), (4) monitored AI to (5) full automation, authority of the human being decreases, while the role of the system increases. In (1) a (cognitive) task is accomplished without any assistance, while in (2) automation provides advice. In (3) the system proposes actions and the operator's consent is required while in (4) the operator can only veto actions initiated by the system. In full automation the operator can and does not interact with the automated system at all (see also Endsley, 1996). Similar to aviation, automation in surface travel ranges from manual control to completely automated driving. Most systems that are low in level of automation do not restrict the driver's behavior. A system such as RDS-TMC (Radio Data System – Traffic Message Channel) provides the driver with information on congestion. Such a system is an indirect route-guidance system, leaving the decision to take another route to the driver. Automation is at the level of "decision support", according to Endsley's classification. At the other extreme there is automated driving, such as in the Automated Highway System (AHS, e.g., Congress, 1994). In the AHS cars exchange information with the road infrastructure and with each other, enabling automatic lateral (steering) and longitudinal (speed and headway) control. Only when cars are in the AHS or on dedicated lanes, they will be under automatic control. Advantages of AHS include increased road capacity, increased operation in bad weather and reduction or even elimination of driver error. The latter claim is based on the assumption that drivers will not be allowed to take over control while driving on the AHS, which is still a point of discussion (Tsao et al., 1993). As one of the aims of AHS is to increase road capacity, cars will probably drive at short following distances, at time headways that are beyond human reaction time. AHS is likely to leave very little, if any, behavioral freedom in actual driving (i.e., in headway and speed control), and is the highest level of automation.

Michon (1985) has described driving as a task with processes at three levels. At the top level, the *strategic level*, strategic decisions are made, such as the choice of means of transport, setting of a route goal, and route-choice while driving. At the intermediate level, *the maneuvering level*, reactions to local situations including reactions to the behavior of other traffic participants take place. At the lowest level, *the control level*, the basic vehicle-control processes occur, such as lateral-position control. Automation in driving can take place at all three levels. The earlier mentioned RDS-TMC provides the driver with strategic information, while the AHS has full operational control. At the tactical level information or tutoring systems can inform drivers about their behavior. These tutoring systems can range from silent black-box policing systems to environmental feedback systems. In Figure 1 some of these systems are categorized according to the level of automation and the level of process that are automated.

Goals of Automation

According to Wickens (1992) there are 3 goals of automation, each serving a different purpose. There is 1) automation to perform tasks humans cannot perform at all, or 2) automation to perform tasks

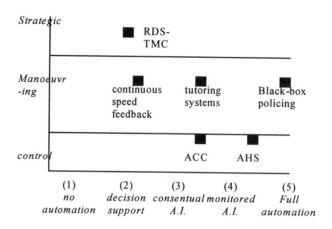

Figure 1. Position of different car control and in-car systems in a two-dimensional space of level of automation and process that is automated. The level of automation (x-axis) is based on Endsley & Kiris (1995), the y-axis is the level of automated process (Michon, 1985). ACC = Adaptive Cruise Control.

humans cannot perform very well or only at the cost of high workload, or 3) automation that assists humans by performing tasks in which humans show limitations. Stanton and Marsden (1996) focus on the second and third goals mentioned by Wickens, and give as arguments in favor of driver automation improved well being, and enhanced road safety. Driver well-being is enhanced by reducing driver stress and workload, while safety is directly increased by offering different automated solutions for different driver errors (Stanton & Marsden, 1996). Navigation systems and collision avoidance systems are expected to reduce errors in terms of taking the wrong exit, hitting cars in front and so on.

In the following sections the effects on driver behavior and acceptance of a tutoring system and the Automated Highway System (AHS) will be discussed. In the AHS the control processes of car driving are automated, as are most of the maneuvering level decisions (e.g., react to other traffic slowing down). Driving in the AHS at 0.25-s time headway to a car-in-front at high speed is a task that humans cannot perform (although some think they can), in particular if the lead car suddenly slows down. AHS would fall into Wickens' first category and would work in favor of driver workload and safety. Tutoring systems aim to help the driver obey the law and to reduce the number of violations in order to enhance safety. In principle this is a task the drivers can perform, although there are many conditions in which it is difficult. A local speed limit, for instance, is not always immediately clear from the road and its environment (see, e.g., De Waard et al., 1995). A tutoring system could be classified as an assisting system, although it is questionable whether a genuine in-car policing system is a support system. The tutoring systems that will be described, however, aim to reduce driver error by giving feedback. The Highway Code regulations are taken as starting point for that. The tutoring systems give feedback on maneuvering-level behavior, leaving the driver in control.

Social Acceptance

The social context for introduction of new technology is of major importance (Rothengatter et al., 1991). A system may function perfectly in the technical sense, if it is not accepted by the public, it will not be used. In particular interventionist systems, systems that take over control or reduce freedom to a large extent, are disliked by the public (Van der Laan et al., 1997). Elderly people's opinion about systems is also important as elderly have been found to be reluctant to use and accept new technology (Hancock &

Parasuraman, 1992). It has been found, however, that after experience with a system elderly drivers' opinion can change from restraint to welcoming (e.g., Van der Hulst, et al., 1994, see also below). Nevertheless, using new technology could be a source of stress for the elderly (Matthews & Desmond, 1995) while in particular elderly should and could benefit from automation, just as they do in case of automatic transmission (e.g., Brouwer, 1996).

Tutoring/Policing systems

In Europe studies into the effectiveness of in-vehicle law-enforcement and feedback systems have been performed from 1989 onwards. Research concentrated on automatic policing (AUTOPOLIS, see Rothengatter, 1992, 1993 for an overview) as well as on the detection of impaired driving behavior (DREAM, Thomas et al., 1989, De Waard & Brookhuis, 1991). Automatic policing systems have been proposed as police surveillance techniques are only temporarily effective in increasing law-abiding behavior (e.g., Rothengatter, 1997). In-vehicle monitoring of driver behavior and providing feedback about law compliance ('tutoring') is more successful in changing driver behavior as indicated by a study of De Waard et al. (in press), and is accordingly expected to be more beneficial for traffic safety (Rothengatter, 1991). In the DETER program (Brookhuis, 1993, 1995) the DREAM and AUTOPOLIS projects merged into one, and efforts concentrated on designing a system that aimed to deter drivers from making violations. Although on-site systems (such as automatic speed cameras and Variable Message Signs) were envisaged, the ultimate goal was an in-car system that could communicate with the road environment. Inside the car local speed limits and other traffic regulations were "known" by the system. Whenever the drivers did not obey the local restrictions, they would receive warning messages about their behavior. These systems have been subjected to experimental tests with samples of drivers in a driving simulator, and in an instrumented vehicle.

A first test (De Waard et al., in press) was performed in the driving simulator of the Centre for Environmental and Traffic Psychology. The simulator is a high fidelity fixed-base simulator (see Van Wolffelaar & Van Winsum, 1995, for more details). In the experiment relatively young (30-45 years) and elderly (60-75 years of age), all experienced drivers participated. Drivers' behavior with respect to speed compliance, behavior at a stop sign, red light running and illegal entrance of one-way streets was monitored. Subjects completed four test rides, during the second and the third of the rides feedback was provided in case a violation was detected. It was found that drivers made significantly fewer violations when they received tutoring messages. Elderly drivers continued to make fewer violations after the tutoring system was switched off during the final ride, whereas young drivers returned to their "baseline violation"-level. These results make clear how important inclusion of specific groups of drivers in experiments can be. Elderly drivers used the system as a driver support system, while young drivers looked on the system as a policing system. With increasing level of automation, elderly are increasingly prone to complete reliance or even dependency on the system to provide information. In case of a system failure, e.g., due to failure in communication with the infrastructure, drivers could receive incorrect information and rely on that. The difference in appreciation of the system of the young and elderly drivers also became apparent in ratings of acceptance that were assessed before and after experience with the system. After reading a description of the tutoring system both young and elderly drivers expected a useful in-car device. After the test rides this opinion did not change for the young, while it was enhanced for the elderly. Before the test, the elderly did not expect a satisfying system while the young driver's opinion was neutral in this respect. After experience the elderly were very positive about this aspect of the system, while the young had become more negative than beforehand.

The same tutoring system was additionally tested in an instrumented vehicle, on-the-road (De Waard & Brookhuis, 1997). Again, no actual enforcement was applied, for not obeying the law there were no other penalties than auditory warnings. In the on-the-road test behavior of two groups of young experienced drivers were compared: both groups first completed a trial without feedback, after that in a second trial one group

Figure 2. Continuous speed limit feedback.

received tutoring messages whenever they violated a traffic rule while the other group was told that offenses would be registered in a black box without feedback. Both groups made less violations during the second trial, but the feedback group made significantly less violations than the black-box group. Again, the tutoring system was considered useful although the messages were disliked. Therefore, in a final experiment the feedback system was altered into a support system (Brookhuis & De Waard, 1996). On the dashboard of the car, in the neighborhood of the speedometer, a display was installed on which the current speed limit was depicted (see Figure 2). The color in which the speed limit was displayed depended upon the driver's speed behavior. If speed was below the speed limit, color would be green. If drivers drove between 0 and 10% above the posted limit, color would be amber, while driving faster than the limit + 10% the color in which the speed limit was depicted would be red, while in addition an auditory warning was issued (e.g., "You are driving too fast, the local limit is 50"). Again it was found that feedback on behavior was effective, although the effects on speeding behavior were not as large as in the previous on-the-road experiment. Reason for this was that some of the subjects "played" with the display feedback to test its behavior and/or stay in the amber (violation) area. Acceptance of the display was, however, more favorable than acceptance of auditory feedback. Over 80% of the subjects who had experienced the display considered it acceptable, the rest were neutral towards it. Auditory feedback was considered acceptable by 40%, 50% considered it unacceptable (see Brookhuis & De Waard, 1996, for details).

Automated Driving

In recent years the industry has put a lot of effort into the development of different driver assistance systems. Amongst these are adaptive cruise controls, vision enhancement systems, collision avoidance systems, route navigation systems, etceteras. In principle the Automated Highway System (AHS) makes use of all these existing systems and merges them in one vehicle. On the AHS the vehicle is under automated control, and it maintains its own lateral and longitudinal position. As soon as the car is on the AHS the driver has little more to do than to enter the destination. Nevertheless, entering and exiting the AHS may not be that simple. Bloomfield et al. (1995), Levitan and North (1995) and De Vos and Hoekstra (1997) showed that giving over control to the system, regaining control, and mixing with traditional traffic all have their own problems. Also, carry-over effects of high speed automated driving can be expected to endanger traffic safety after leaving the AHS to non-AHS lanes and roads (Bloomfield et al., 1995, De Vos et al., 1996).

Research with respect to driver acceptance of platoon driving has shown that one of the claims of AHS, increased road capacity, has to deal with driver resistance to short headway driving. De Vos et al., (1996) found that drivers rated an average time headway of 1.1 seconds as most comfortable, while De

Waard et al. (1999, in press) found that the majority of drivers preferred 1 second time headway rather than 0.25 seconds. The advantage of reduced travel times due to the possibility of high speed driving on the AHS is also subject of discussion, as calculations by Levitan and North (1995) have shown that a *lower* driving speed leads to a *higher* vehicle throughput. This paradoxical finding is due to entering and exiting traffic, a low speed differential between the AHS and non-automated lanes is most beneficial for throughput. An important finding for AHS acceptance is also that most drivers dislike the fact that they have to give over control to some system (e.g., De Waard et al., 1999, in press).

Apart from driver preference and optimal throughput, operator behavior in case of system failure is very important in automated driving. As designers cannot anticipate all possible scenarios of failure, the flexible human operator will probably have to remain in control (Meshkati, 1996). Therefore, the possibility that drivers can reclaim control in automated driving is to be expected. Levitan and North (1995) simulated dynamic vehicle behavior in the AHS after control actuator failure. They found that in most cases the time that was left for corrective actions was below minimum reaction time of an alert driver. Stanton et al. (1997) used a driving simulator and tested driver readiness to reclaim control under an Adaptive Cruise Control (ACC) failure scenario. Eight out of twelve subjects responded effectively, i.e. steered and/or applied the brake, when the ACC failed. However, driving with ACC still requires the driver to remain "in the loop", lateral control is not automated and steering is required. In the AHS both speed and lane position control are automated, and if all is functioning well there is nothing to do for the driver but to (occasionally?) monitor the system's status. One can expect that driver readiness to take over control from such a passive position could be worse than while actively driving with an ACC. In addition to this, we know from aviation that automation affects pilot behavior and malfunctions are sometimes not recognized or are misinterpreted (see, e.g., Parasuraman & Mouloua, 1996). In an experiment in a driving simulator readiness of drivers to interfere was tested in case of AHS failure (De Waard et al., 1999, in press). In the experiment in a final scenario the front sensor failed to detect a vehicle that suddenly merged extremely close in front of the AHS vehicle. AHS failure was not very salient, i.e. there were no auditory warnings of system failure, failure could only be deduced from non response by the automated system and a dashboard control light that switched off. This system could therefore be called a "Strong and Silent" system (Woods, 1996). Drivers could only take-over control by pressing the brake. All the time their physiology (heart rate) was measured, while after the test self-reports on mental effort, activation, safety and risk were collected. It was found that 50% did not respond to the emergency situation at all, while an additional 14% of the drivers responded very late. Driver's heart rate clearly reflected the point at which the emergency occurred. However, instead of a startle reaction, heart rate merely reflected an orienting response, surprise. This corresponded with subjective reports collected after the event, many said that they were surprised that the system did not react. General opinion about the situation was that it had been risky.

Desmond et al. (in press) also tested driver response to unexpected events in a driving simulator. In an experiment drivers completed two rides of each 40 minutes (automatic and manual). During the automated rides the automatic driving system failed, leading to a lateral displacement towards the road edge, while during the manual rides there were wind gusts with a similar effect. Desmond et al. (in press) also found that performance recovery after system failure was worse in automatic driving than during manual driving. This happened even though subjects had been alerted to the fact that these events could take place. Laboratory experiments have also shown that failure is faster detected in systems under manual control (e.g., Parasuraman et al., 1996). All these studies show the dangers of full automation and that if drivers are to reclaim control, they should be –in one way or another– kept involved.

DISCUSSION AND CONCLUSIONS

Automation of the different tasks in driving has consequences for the drivers' reaction to them. Systems that provide information and that do not restrict behavioral freedom are in general accepted. Take-over of tasks, however, is often disliked.

With increasing level of automation operators should be able to rely on it, as their role becomes less prominent, or is even eliminated. Silent systems that cannot be overruled can easily endanger safety. In the field of car driving, a major risk of new in-car technology is over-reliance on these systems (complacency). Complacency is related to reliability of a system, trust in a system builds up over time after it has proven to display predictable behavior (Muir, 1987), and it is trust in a system that enhances over-reliance. Negative experience with system reliability, e.g., a large amount of false alarms such as found in aviation in original collision avoidance systems (Kantowitz & Campbell, 1996), also endanger safety as it will lead to ignoring of warnings. In the above discussed tutoring and automated driving studies evidence for complacency was found. Elderly drivers were inclined to trust the tutoring system to provide them with information on speed limits. In the study on the Automated Highway System (AHS) subjects expected the system to react to an emergency, and more than half of them did not reclaim control or did so in a very late phase. In the monotonous environment of the AHS things can get even worse, as stress and fatigue are factors that may enhance complacency (Matthews & Desmond, 1995).

Adaptive automation (e.g., Scerbo, 1996), implying that mode or level of automation can be altered in real-time, is mentioned as one of the solutions to overcome out-of-the loop behavior. Research on adaptive task allocation shows that temporarily returning control to the human operator has favorable effects on detection of automation failures during subsequent automation control (Parasuraman et al., 1996). From aviation it is known that pilots worry about the loss of flying skills and consider manual flying of a part of every trip important to maintain their skills (McClumpha et al., 1991). In the AHS this could be implemented restricting access to the AHS for a limited period of time, and to require manual driving on a non-AHS lane for some minutes before allowing access to the AHS again. However, apart from practical consequences in terms increased weaving in and out of the AHS lane leading to reduced throughput (Levitan & North, 1995), many drivers may resist this idea (why drive manually again while all is going well?).

The issue of driver acceptance is more and more recognized. Behavioral freedom is of major importance for driver acceptance. In the end it is not technology that determines how we will drive, but what we are willing to. How far we should go in automation is a social choice (Sheridan, 1996). Since the AHS demonstration in San Diego in 1997 we know that technically the AHS is possible. As take-over of control is generally disliked the question merely is "do we really want it"? Nowadays, technically almost anything is possible. The different studies discussed in this paper show that it is not what is technically conceivable that should determine what is implemented. Most important is the user's reaction to these systems and their behavioral adaptation to them.

ACKNOWLEDGEMENT

Stephen Fairclough and Frank Steyvers are thanked for giving useful comments on an earlier draft of this paper.

REFERENCES

Bloomfield, J.R., Buck, J.R., Carroll, S.A., Booth, M.S., Romano, R.A., McGehee, D.V., & North, R.A. (1995). *Human Factors aspects of the transfer of control from the Automated Highway System to the driver*. Report FHWA-RD-94-114. McClean, Virginia, U.S.A.: US DOT, Federal Highway Administration, Turner-Fairbank Highway Research Centre.

Brookhuis, K.A. (1993). Detection, tutoring and enforcement of traffic rule violations – The DETER project. In *Vehicle Navigation and Informations Systems, IEEE-IEE* Conference proceedings (pp. 698-702). Ottawa, Canada: Trico Printing.

Brookhuis, K.A. (1995). *DETER, Detection, Enforcement & Tutoring for Error Reduction. Final report*.

Report V2009/DETER/Deliverable 20. Haren, The Netherlands: University of Groningen, Traffic Research Centre.

Brookhuis, K.A. & De Waard, D. (1996). *Limiting speed through telematics –Towards an Intelligent Speed Adaptor (ISA)*. Report VK 96-04. Haren, The Netherlands: University of Groningen, Traffic Research Centre.

Brouwer, W.H. (1996). Older drivers and attentional demands: consequences for human factors research. In K.A. Brookhuis, C.M. Weikert, J. Moraal, & D. de Waard (Eds.) *Aging and Human Factors, Proceedings of the Europe Chapter of the Human Factors and Ergonomics Society 1993 meeting* (pp 93-106). Haren, The Netherlands: University of Groningen, Traffic Research Centre.

Congress, N. (1994). The Automated Highway System: an idea whose time has come. *Public Roads: a journal of highway research, 58*, 1-7.

Desmond, P.A., Hancock, P.A., & Monette, J.L. (in press). Fatigue and automation-induced impairments in simulated driving performance. *Transportation Research Records*.

De Vos, A.P., Theeuwes, J. & Hoekstra, W. (1996). *Behavioral aspects of Automatic Vehicle Guidance (AVG): the relationship between headway and driver comfort.* Report TM-96-C022. Soesterberg, The Netherlands: TNO Human Factors Research Institute.

De Vos, A.P. & Hoekstra, W. (1997). *Behavioral aspects of Automatic Vehicle Guidance (AVG); Leaving the automated lane.* Report TM-97-C010. Soesterberg, The Netherlands: TNO Human Factors Research Institute.

De Waard, D. & Brookhuis, K.A. (1991). Assessing Driver Status: a demonstration experiment on the road. *Accident Analysis & Prevention, 23*, 297-307.

De Waard, D., Jessurun, M., Steyvers, F.J.J.M., Raggatt, P.T.F., & Brookhuis, K.A. (1995) Effect of road layout and road environment on driving performance, drivers' physiology and road appreciation. *Ergonomics, 38*, 1395-1407.

De Waard, D. & Brookhuis, K.A. (1997). Behavioral adaptation of drivers to warning and tutoring messages: results from an on-the-road and simulator test. *Heavy Vehicle Systems, International Journal of Vehicle Design, 4*, 222-234.

De Waard, D., Van der Hulst, M., & Brookhuis, K.A. (in press). Elderly and young drivers' reaction to an in-car enforcement and tutoring system. *Applied Ergonomics*.

De Waard, D., Van der Hulst, M., Hoedemaeker, M., & Brookhuis, K.A. (1999, in press). Driver behavior in an emergency situation in the Automated Highway System. *Transportation Human Factors*.

Endsley, M.R. (1996). Automation and Situation Awareness. In R. Parasuraman & M. Mouloua (Eds.) *Automation and Human Performance: Theory and Applications* (pp. 163-181). New Jersey: Lawrence Erlbaum Associates.

Endsley, M.R. & Kiris, E.O. (1995). The out-of-the-loop performance problem and level of control in automation. *Human Factors, 37*, 381-394.

Hancock, P.A. & Parasuraman, R. (1992). Human Factors and safety in the design of intelligent vehicle-highway systems (IVHS). *Journal of Safety Research, 23*, 181-198.

Kantowitz, B.H. & Campbell, J.L. (1996). Pilot workload and flightdeck automation. In R. Parasuraman & M. Mouloua (Eds.) *Automation and Human Performance: Theory and Applications* (pp. 117-136). New Jersey: Lawrence Erlbaum Associates.

Levitan, L. & North, R.A. (1995). *Human Factors design of Automated Highway Systems: preliminary results.* Compendium of Technical Papers from the 65th Annual Meeting of the Institute of Transportation Engineers, Denver, CO, U.S.A.

Matthews, G. & Desmond, P.A. (1995). Stress as a factor in the design of in-car driving enhancement systems. *Le Travail Humain, 58*, 109-129.

McClumpha, A.J., James, M., Green, R.G., & Belyavin, A.J. (1991). Pilots' attitudes to cockpit automation. *Proceedings of the Human Factors Society 35th annual meeting* (pp. 107-111). Human Factors and

Ergonomics Society: Santa Monica, Ca, U.S.A.

Meshkati, N. (1996). Organizational and safety factors in automated oil and gas pipeline systems. In R. Parasuraman & M. Mouloua (Eds.) *Automation and Human Performance: Theory and Applications* (pp. 427-446). New Jersey: Lawrence Erlbaum Associates.

Michon, J.A. (1985). A critical view of driver behavior models: what do we know, what should we do? In L. Evans & R.C. Schwing (Eds.), *Human behavior & traffic safety* (pp. 485-524). New York: Plenum Press.

Muir, B.M. (1987). Trust between humans and machines, and the design of decision aids. International Journal of Man-Machine Studies, 27, 527-539.

Parasuraman, R., Mouloua, M., & Molloy, R. (1996). Effects of adaptive task allocation on monitoring of automated systems. *Human Factors, 38*, 665-679.

Parasuraman, R., Molloy, R., Mouloua, M. & Hilburn, B. (1996). Monitoring of Automated Systems. In R. Parasuraman & M. Mouloua (Eds.), *Automation and Human Performance: Theory and Applications* (pp. 91-115). New Jersey: Lawrence Erlbaum Associates.

Parasuraman, R. & Mouloua, M. (Eds.). (1996). *Automation and Human Performance: Theory and Applications*. New Jersey: Lawrence Erlbaum Associates.

Rothengatter, J.A. (1991). Automatic policing and information systems for increasing traffic law compliance. *Journal of Applied Behavior Analysis, 24*, 85-87.

Rothengatter, J.A. (1992). *Automatic Policing Information Systems*, report V1033/AUTOPOLIS/ FIN92. Haren, The Netherlands: University of Groningen, Traffic Research Centre.

Rothengatter, J.A. (1993). Violation detection and driver information. In A.M. Parkes & S. Franzén (Eds.), *Driving future vehicles* (pp. 229-234). London: Taylor & Francis.

Rothengatter, J.A. (1997). The effects of media messages, police enforcement and road design measures on speed choice. *IATSS Research, 21*, 80-87.

Rothengatter, J.A., De Waard, D., Slotegraaf, G., Carbonell Vaya, E., & Muskaug, R. (1991). *Social acceptance of automatic policing and information systems*. Report 1033/AUTOPOLIS/Deliverable 7. Haren, The Netherlands: University of Groningen, Traffic Research Centre.

Scerbo, M.W. (1996). Theoretical perspectives on adaptive automation. In R. Parasuraman & M. Mouloua (Eds.), *Automation and Human Performance: Theory and Applications* (pp. 37-63). New Jersey: Lawrence Erlbaum Associates.

Sheridan, T.B. (1996). Speculations on future relations between humans and automation. In R. Parasuraman & M. Mouloua (Eds.), *Automation and Human Performance: Theory and Applications* (pp. 449-460). New Jersey: Lawrence Erlbaum Associates.

Stanton, N.A. & Marsden, P. (1996). From fly-by-wire to drive-by-wire: safety implications of automation in vehicles. *Safety Science, 24*, 35-49.

Stanton, N.A., Young, M., & McCaulder, B. (1997). Drive-by-wire: The case of driver workload and reclaiming control with adaptive cruise control. *Safety science, 27*, 149-160.

Thomas, D.B., Herberg, K.-W., Brookhuis, K.A., Muzet, A.G., Poilvert, C., Tarrière, C., Norin, F., Wyon, D.P., Scievers, G, & Mutschler, H. (1989). *Demonstration experiments concerning driver status monitoring*. Report V1004/DREAM. Köln, Germany: Technische Überwachungs-Verein Rheinland e.V.

Tsao, H.-S.J., Hall, R.W., & Shladover, S.E. (1993). Design options for operating Automated Highway Systems. In *Proceedings of the IEEE-IEE Vehicle Navigation & Information Systems '93 Conference*. Piscataway, NY, USA: IEEE.

Van der Hulst, M., De Waard, D., & Brookhuis, K. (1994). New Technology and behavior of older drivers, mental effort and subjective evaluation. In K. Johansson and C. Lundberg (Eds.), *Aging and Driving, Effects of aging, diseases, and drugs on sensory functions, perception, and cognition, in relation to driving ability* (pp. 53-57). Stockholm: Karolinska Institutet.

Van der Laan, J.D., Heino, A., & De Waard, D. (1997). A simple procedure for the assessment of acceptance of Advanced Transport Telematics. *Transportation Research -C, C5*, 1-10.

Van Wolffelaar, P.C. & Van Winsum, W. (1995). Traffic Modelling and driving simulation – an integrated approach. In *Proceedings of the Driving Simulator Conference 1995 (DSC '95)*. Toulouse, France: Teknea.

Wickens, C.D. (1992). *Engineering Psychology and Human Performance.* New York: HarperCollins.

Woods, D.D. (1996). Decomposing Automation: apparent simplicity, real complexity. In R. Parasuraman & M. Mouloua (Eds.), *Automation and Human Performance: Theory and Applications* (pp. 3-17). New Jersey: Lawrence Erlbaum Associates.

The Impact of Automation on Driver Performance and Subjective State

P.A. Desmond, P.A. Hancock and J.L. Monette
University of Minnesota

INTRODUCTION

The effect of automation on the driver's performance and subjective state is a critical issue in the light of future transportation developments such as Intelligent Vehicle-Highway ,Systems (IVHS). In such systems, in-vehicle navigation and collision avoidance systems are integrated with the control software of the vehicle to automate many aspects of the driving task. Automated systems may have negative consequences for the driver's performance and subjective state, Studies in contexts other than driving have demonstrated some of the adverse effects of automation on performance. For example, Endsley and Kiris (1995) showed that individuals who are forced to manually operate a system following a breakdown in automation show impaired performance when compared with individuals who perform the task in a manual mode of operation. Two explanations have been proposed to account for automation-induced performance impairments. Endsley and Kiris (1995) argue that such impairments can be explained in terms of a loss of situation awareness (SA). Situation awareness is "the perception of elements in the environment within a volume of time and space, the comprehension of their meaning, and the projection of their status in the near future" (Endsley, 1988, p.97). Endsley and Kiris (1995) propose that individuals who have lower SA are slower to detect difficulties and are slower to focus on relevant features of the system which will allow them to identify the problem and regain manual performance. Operator complacency has also been proposed as a major factor that relates to degraded vigilance in monitoring performance under automation. Studies of aircraft cockpits suggest that performance may be impaired by complacency resulting from the pilot's confidence in automated systems (Singh, Molloy & Parasuraman, 1993). As well as impairing performance, we may expect automated tasks to elicit stressful subjective reactions in individuals. As Matthews et al. (this volume) point out, automated tasks are similar to vigilance tasks since they usually demand sustained monitoring but infrequent response from the individual. Vigilance tasks have been found to lead to symptoms such as fatigue, boredom and a reduction in motivation (e.g. Scerbo, 1998). This chapter outlines two studies that have focused on the impact of an automated system on drivers' performance and subjective state. Implications of the research for future transportation developments are also discussed.

STUDY 1

In the first study, 40 drivers performed both a manual and an automated drive on the Minnesota Wrap-around environment simulator (WES). Both drives lasted for 40 minutes. In the manual drive, drivers had fall control over the vehicle throughout the drive. In the automated drive, the vehicle's velocity and trajectory was under system control. Three perturbing events occurred at early, intermediate and late phases in both drives. In the manual drive, the perturbation took the form of a "wind gust" that caused the vehicle to drift towards the edge of the right-hand driving lane. In the automated drive, the perturbing event was a failure in automation that caused the vehicle to veer towards the edge of the right-hand lane at the same magnitude as the wind gusts. Following a failure in automation, drivers were required to manually control the vehicle for three minutes until the automated driving system became operational again. While the automated system was operational, drivers were instructed to keep their hands off the steering wheel. The driver's lateral control of the vehicle, as indexed by *heading error,* was the principal performance measure that was recorded following the perturbing events in both drives. Heading error is the mean deviation between the direction of the road and the direction of the vehicle measured in degrees. A variety of subjective scales were administered to drivers before and after both drives. Three fatigue scales (Desmond, Matthews & Hancock, 1997, Matthews & Desmond, in press) were used to measure physical fatigue, perceptual fatigue and boredom/apathy. Mood was assessed with the UWIST Mood Adjective Checklist (UMACL; Matthews, Jones & Chamberlain, 1990) which assesses energy, tension and hedonic tone (pleasantness of mood). A modification of Sarason, Sarason, Keefe, Hayes and Shearin's (1986) Cognitive Interference Questionnaire was used to assess intruding thoughts. Its scale comprises items relating to (1) taskrelevant interference and (2) task-irrelevant personal concerns. Scales developed and validated at Dundee University were used to assess motivation, perceived control, concentration and active effortful coping (Matthews, Joyner, Gilliland, Campbell, Huggins & Falconer, in press).

Figure 1 shows standardized change scores for state measures across manual and automated drives. Both drives appeared to induce negative mood changes such as increases in tension, anger, depression, and also increases in boredom, perceptual fatigue and cognitive interference. In addition, decreases in energy, concentration, motivation, perceived control and active coping were also found following manual and automated driving conditions. The manual drive produced a significantly higher level of physical fatigue than the automated drive.

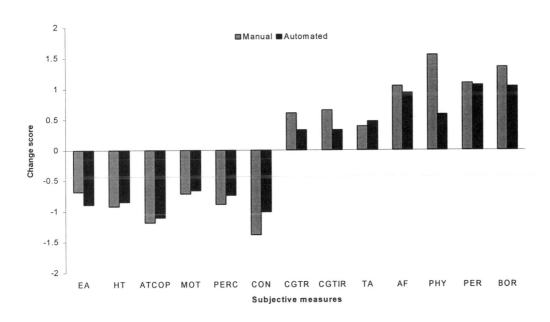

Figure 1. Standardized change scores for state measures across manual and automated drives. (EA=Energetic arousal, HT=Hedonic tone, ATCOP=Active coping, MOT=Motivation, PERC=Perceived control, CON=Concentration, CGTR=Task-relevant cognitive interference, CGTIR=Task-irrelevant cognitive interference, TA=Tense arousal, AF=Anger/frustration, PHY=Physical fatigue, PER=Perceptual fatigue, BOR=Boredom).

Figure 2 shows mean heading error across early, intermediate and late phases of manual driving during both automated and manual drives. These phases represent manual control of the vehicle immediately following automation failure, and perturbing wind gusts in the manual drive. Each phase of manual driving lasted for three minutes. For early, intermediate and late phases of manual driving, mean heading error was analyzed across nine 20-second periods in both drives. The nine 20-second blocks of manual driving performance were recorded at early, intermediate and late phases of the drive immediately following automation failure, and wind gusts in the manual drive. Heading error is greater immediately following automation failure than following wind gusts in the manual drive during early, intermediate and late phases of driving performance. Thus, performance recovery appears to be better in the manual driving mode than in the automated mode during the first 20-second period ($t(39)=5.07, p<.001$).

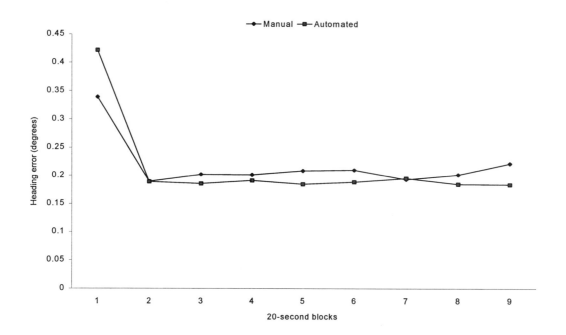

Figure 2. Mean heading error across early, intermediate and late phases of manual driving in automated and manual driving conditions.

STUDY 2

In the second study, 11 drivers completed automated and manual drives as in Study 1. The aim of Study 2 was to test whether the impairment in performance found in the automated drive could be reduced or eliminated if drivers were physically engaged in the driving task throughout the drive. Thus, in the second study, drivers in the automated drive were instructed to keep their hands on the steering wheel while the automated system was operational. Study 1 revealed that automation-induced impairment in lateral control occurred within the first 20 seconds following system failure. Thus, in the second study, drivers' performance was analyzed for 30 seconds following each perturbing event. Figure 3 shows heading error across six 5-second periods for manual and automated driving conditions. The results replicate the findings of Study 1: heading error is greater immediately following automation failure than following wind gusts in the manual drive.

CONCLUSION

The findings from the subjective state measures in the first study suggest that monitoring an automated driving system may be just as effective in inducing subjective fatigue and stress states as prolonged driving under monotonous driving conditions. At a performance level, the research highlights some of the potential

60

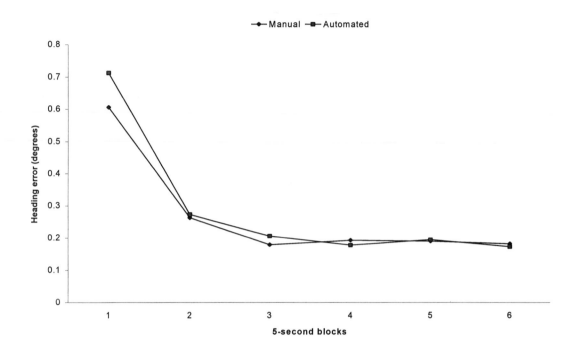

Figure 3. Mean heading error across early, intermediate and late phases of manual driving in automated and manual driving conditions.

problems associated with automated systems in driving. Both studies showed that following a failure in automation, drivers are slower to recover from system failure than drivers who are manually engaged in driving. The findings have important implications for automated highway systems (AHS). The studies suggest that AHS should attempt to keep the driver within the driving loop in order to guard against performance impairments. The research reinforces Hancock et al.'s (1996) view that human centered design approaches are superior to full automation of a task.

REFERENCES

Desmond, P.A., Hancock, P.A. & Matthews, G. (1997). Development and validation of the driving fatigue scale. In C. Mercier-Guyon (Ed.), *Proceedings of the 14th International Conference on Alcohol, Drugs and Traffic Safety*.

Endsley, M.R. (1988). Design and evaluation for situation awareness enhancement. *Proceedings of the Human Factors Society 32nd Annual Meeting*.

Endsley, M.R. & Kiris, E.O. (1995). The out-of-the-loop performance problem and level of control in automation. *Human Factors, 37*(2), 381-394.

Hancock, P.A., R. Parasuraman., & E.A. Byme. (1996). Driver-centered issues in advanced automation for motor vehicles. In R. Parasuraman & M. Mouloua (Eds.), *Automation and human performance: theory and applications* (pp 337-364). Mahwah, NJ: Lawrence Erlbaum Associates.

61

Matthews, G., Jones, D.M., & Chamberlain, A.G. (1990). Refining the measurement of mood: The UWIST Mood Adjective Checklist. *British Journal of Psychology, 81*, 17-42.

Matthews, G., Joyner, L., Gilliland, K., Campbell, S.E., Huggins, J., & Falconer, S. (in press). Validation of a comprehensive stress state questionnaire: Towards a state 'Big Three'? In I. Mervielde., I.J. Deary, F. De Fruyt, & F. Ostendorf (Eds.), *Personality psychology in Europe* (Vol. 7). Tilburg: Tilburg University Press.

Matthews, G. & P.A. Desmond. (in press). Personality and multiple dimensions of task-induced fatigue: a study of simulated driving. *Personality and Individual Differences*.

Sarason, I.G., Sarason, B.R., Keefe, D.E., Hayes, B.E., & Shearin, E.N. (1986). Cognitive interference: Situational determinants and traitlike characteristics. *Journal of Personality and Social Psychology, 51*, 215-226.

Scerbo, M.W. (1998). What's so boring about vigilance? In R.R. Hoffinan, M.F. Sherrick, & J.S. Warm (Eds.), *Viewing psychology as a whole: The integrative science of William N. Dember* (pp. 145-166). Washington, DC: American Psychological Association.

Singh, I.L., Molloy, R. & Parasuraman, R. (1993). Individual differences in monitoring failures of automation. *Journal of General Psychology, 120*, 357-374.

Using Signal Detection Theory and Bayesian Analysis to Design Parameters for Automated Warning Systems

Raja Parasuraman
Catholic University of America

P. A. Hancock
Liberty Mutual Insurance Co.

INTRODUCTION

Alarms and warning systems are proliferating with the rapid pace of automation at work and in the home (Parasuraman & Mouloua, 1996). Alarms are meant to signal threats or emergencies that demand an immediate response. But often they turn out to be false. Excessive false alarms abound in civil (Patterson, 1982) and military aviation (Tyler, Shilling, & Gilson, 1995), nuclear power plants (Seminara, Gonzalez, & Parson, 1977), and other high-technology industries (Pate'-Cornell, 1986). As a result, people tend to ignore warning systems much of the time. Often they may turn off an alarm before verifying that it was indeed false. Can warning systems be designed so as to be useful? We describe a quantitative, analytical solution to this problem that provides step-by-step procedures for designing effective warning systems.

We describe the problem and the proposed solution in the context of collision-warning systems for motor vehicles (Parasuraman, Hancock, & Olofinboba, 1997), although our analysis applies to any warning system. Rear-end collisions constitute approximately 25% of all traffic accidents that are reported to the police (Knipling et al., 1993). Inattention is one of the leading causes (Wang, Knipling, & Goodman, 1996). Collision-warning systems have been developed in response to this problem. Proponents claim that such systems can alert inattentive motorists in time to take evasive action.

THE TECHNOLOGICAL RESPONSE

Numerous collision warning system systems are currently being developed (ITS America, 1996). Several have already entered the market. An example is the Forewarn™ system by Delco Electronics. This system uses very high-frequency microwave object-detection sensors to measure the position and relative speed of objects ahead of (77 GHz) and behind (24 GHz) the vehicle. Caution and emergency alarms are issued at driver-preset distances. Alarms are visual (icons on a head-up display), auditory (audio system), or tactile (pulsed vibration of the brakes).

What has fueled the explosive growth of collision-warning systems? Clearly, the annual toll on the road of approximately 40,000 fatalities and more than 6 million collisions (National Safety Council, 1996) represents a compelling problem in need of a solution. Unfortunately, *perfect* detection of impending collisions cannot be achieved.

Although many apparently accurate systems have been fielded, a crucial fact that is neglected is that collisions, particularly ones that cause loss of life or disabling injury, are very rare events. The frequency of impending collisions, or those that are avoided by appropriate action on the part of the driver, is generally higher than that of actual collisions. Nevertheless, the *base rate* or prior probability of either actual or impending collision events is likely to be very low. Collision-warning systems are designed to detect a subset of these events, that is, those scenarios that would result in a collision if the driver did not take appropriate action. Thus, for most drivers, the base rate of traffic events leading to impending collision is likely to be quite low. This fact has major implications for the design and implementation of warning systems, as we demonstrate analytically.

Although sensitive alarm systems with high detection rates and low false alarm rates have been developed, the *posterior probability* of an impending collision given an alarm can be quite low because of the low base rate of impending collision events.

SETTING THE DECISION THRESHOLD

The algorithm for any warning system can be set with a detection criterion or threshold—the minimum level of evidence that is needed to indicate an impending collision. In choosing a criterion value, the designer must balance the need for early detection with the avoidance of false alarms. Signal detection theory (SDT) provides a basis for determining the appropriate decision threshold or criterion that balances these two needs (Swets, 1992). Let S represents the environmental event or signal associated with an impending collision, N a non-collision or noise event, and R a positive alarm response of the warning system. $p(R|S)$ and $p(R|N)$ are then, respectively, the probabilities of correct detections (hits) and false detections (false alarms) associated with the alarm system. The accuracy (sensitivity) of the system (d') and the detection criterion (β) can be derived from these probabilities using standard SDT computing formula:

$$d' = z[p(R|S)] - z[p(R|N)] \qquad [1]$$

$$\beta = y[p(R|S)] / y[p(R|N)] \qquad [2]$$

where $z[\ ]$ and $y[\ ]$ are the normal deviate and ordinate, respectively, of the normal distribution.

The performance of an alarm system can be evaluated by examining the relationship between the detection rate $p(R|S)$ and false alarm rate $p(R|N)$. The depicted curve is known as the Receiver Operating Characteristic (ROC). The ROC represents the locus of all correct and false detection rates associated with a particular alarm system having a given sensitivity as the decision threshold (β) is varied. In theory, the decision threshold can be set at any minimum value of $\beta = \beta_f$ for any specified maximum false alarm rate $p(R|N) = f$. For a system with accuracy $d' = s$, and for $p(R|N) = f$, and substituting for $p(R|S)$ in equations [1] and [2] gives

$$\beta_f = y[z^{-1}[s + z[f])]] / y[f] \qquad [3]$$

where z^{-1} represents the inverse of the normal deviate of the Gaussian distribution. To better appreciate these relationships, consider an example of a very sensitive system with a d' of 6. Suppose it is determined that drivers would use the system appropriately if the false alarm rate is no more than .1% ($p(R|N) = .001$). Then the decision threshold for the system should be set such that $\beta_f = 1.71$ or higher. If β_f is set at 1.71, the detection rate is a very high 99.82%. Thus, only a very small .18% of impending collision events will be missed, and only .1% of non-collision events will be responded to and hence present a false alarm.

DESIGNING FOR HIGH POSTERIOR PROBABILITY

The previous analysis and the numerical example illustrate the initial design consideration for the development of usable alarm systems. However, setting a minimum decision threshold is insufficient by itself, because the *posterior probability* of a collision event given an alarm response must also be considered. One fact may conspire to limit the effectiveness of alarms: the low a priori probability or base rate of potential collision scenarios. This may result in the posterior probability of a true alarm being quite low. The posterior probability $p(S|R)$ can be determined from the detection and false alarm rates $p(R|S)$ and $p(R|N)$ and the a priori probability of the signal or the base rate p using Bayes' Theorem.

$$p(S|R) = p(R|S) / (p(R|S) + p(R|N)(1 - p)/p) \qquad [4]$$

64

For collision events with a low prior probability, the posterior probability of a true alarm can be quite low, and sometimes embarrassingly close to 0. This can be the case even for very sensitive warning systems with very high detection rates and low false alarm rates, as in the hypothetical alarm system with a d' of 6 discussed earlier.

Equation [4] can be used to examine the relationship between the posterior probability of a true alarm and the base rate. This can be done by selecting a particular value for d' and then varying β as a parameter. Figure 1 shows posterior probability values as a function of a priori probability for a sensitive detection system of given accuracy $d' = 5$. The family of curves represents the different posterior probabilities that can be achieved with different decision thresholds. As Figure 1 shows, while $p(S|R)$ increases with the a priori probability p, it approaches 1.0 only for relatively high values of p. Conversely, for low values of p, the posterior probability is quite low. For low values of p, ß needs to be high for high posterior probability.

The implications of these relationships can be better appreciated by computing the odds for specific cases. Assume a collision-warning system with a d' of 5. Suppose the design procedure described earlier is followed: the designer specifies a maximum permissible false alarm rate f and computes the minimum decision threshold β_f. If f is .5%, then β_f must be set at a minimum of 1.47. This gives a near-perfect detection rate for the system of 99.23%, with a false alarm rate of only .5%. These are seemingly impressive performance statistics. But in reality, a driver could find that the posterior odds of a true alarm with such a system can be astonishingly low, so that the system may not usable at all. Even with such a sensitive detection system, and with the decision threshold set with $\beta_f = 1.47$, when the a priori probability is low, say .001, only 1 in 6 alarms that the system emits represents a true alarm. If the a priori probability is .0001, then only 1 in 50 alarms will be true. If the designer is tempted to use a more liberal decision threshold in order to achieve an almost-perfect detection rate of .999, then when p is .0001, the posterior odds of a true alarm is a staggeringly low 1 in 286!

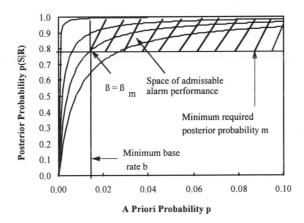

Figure 1. Posterior probability functions that can be used to determine warning system parameters.

Consistently true alarm response occurs only when the a priori probability of a signal is relatively high. There is no guarantee of this in many real systems. More generally, these results suggest that designers of collision-warning systems must take into account both the decision-threshold at which these systems are set and the a priori probabilities of the collision types.

To further illustrate the importance of posterior probabilities, consider another example, taken from the work of Farber and Paley (1993), which allows for estimation of the base rate for particular collision scenarios. Certain portions of Interstate Highway I-40 in Albuquerque, New Mexico have "loop detectors" installed in the road that can determine the speed of vehicles passing over them. These were used to collect data on the speed, closing rate, and headway (distance) between pairs of motor vehicles traveling over this stretch of the highway. Sufficient data were collected for a simulation analysis of 35, 689 pairs of vehicles. The simulation, a Monte Carlo analysis, was conducted in order to evaluate the potential effectiveness of different collision-warning systems. Without collision warning, Farber and Paley estimated a crash rate of 173 for every million lead vehicle stops. This gives a prior probability or base rate p of .000173. Note that this figure refers to impending collisions, or traffic events that would lead to a collision if the driver took no evasive action. However, because drivers took appropriate action, the data are filtered by human response, so that the base rate of events when the human takes no action is not known. As a first approximation, and in the absence of other data, we assume that the base rate of all impending collisions can be represented by this value. Given this base rate, a detection system with $d' = 5$, would yield posterior odds of a true alarm of only 1 in 3. Even for an extraordinarily sensitive system, with a hit rate of .9999 and a false alarm rate of .0001 ($d' = 7.4$), the posterior odds of a true alarm would be only about 2 in 3. These numbers attest to the powerful influence of the base rate on the true alarm rate.

How high should the posterior probability be? Certainly as high as possible; and it seems clear that $p(S|R)$ should be greater than .5 for people to attend to alarms. But how much greater? It is known that low posterior probabilities discourage user action (Parasuraman & Riley, 1997) and also delay speed of response (Casey, 1993; Getty, Swets, Pickett, & Gounthier, 1995). Assume that the minimum acceptable value of the posterior probability $p(S|R) = m$ lies somewhere between .5 and 1.0. The design question now is, given a system with accuracy s and a measured base rate of the collision event of $p = b$, what should the appropriate decision threshold (β_m) be to achieve a posterior probability of at least m? Equations [1], [2], and [4] can be used to answer this question. Since z[ ] and y[ ] represent non-linear functions, a simple, tractable analytical solution is not possible. But a graphical analysis easily points to the correct solution.

Fig. 1 shows posterior probability functions for a system with $d' = 5$. The stippled area represents the space of desired performance: posterior probability greater than the minimum m for an a priori probability that is at least b. The decision threshold β_m that guarantees a posterior probability of p_m can then be determined by iteration and forward solution of equations [1], [2], and [4]. For example, to achieve $m = .8$ when $b = .001$, β_m must be at least 181.1.

Given a specified base rate, our analysis shows how to set $\text{ß} = \beta_m$ in order to optimize performance by achieving a high desired posterior probability m. Of course, the base rate will vary with collision type and with the definition of a impending collision event as a signal. As was mentioned earlier, many collision-warning systems are designed so that the driver can adjust or pre-set the warning distance to be relatively long or short.

CONCLUSIONS

The rapid growth of automation has led to a proliferation of warning and alarm systems. Unfortunately, many warning systems are prone to false alarms, and it is not clear that existing collision-warning systems can be used effectively by drivers to prevent crashes. The quantitative, analytical approach we have presented, based on SDT and Bayesian statistics, provides step-by-step procedures for setting alarm parameters to avoid these problems.

The first step is to specify a maximum false alarm rate for the system. The minimum decision threshold β_f that an alarm system with given sensitivity d' must be set at can then be computed. However, setting a minimum alarm decision threshold to achieve a maximum permissible false alarm rate is necessary but not sufficient for effective alarm performance. In addition, alarm parameters must be designed so that the posterior probability of a true alarm is relatively high for collision events associated with particular base rates of occurrence. Functions relating the required decision threshold β_m to achieve a posterior probability of at least m can be derived for alarm systems of given sensitivity d'.

The analysis we have presented provides a set of standards against which the performance of collision-warning systems can be tested. Although sensitive alarm systems with high detection rates and low false alarm

66

rates have been developed and marketed, our analysis shows that these systems can be ineffective if the posterior probability of an impending collision given an alarm is low. Thus, alarm systems must not only be designed for high sensitivity (d') and set with a decision threshold β_f to achieve a low false alarm rate (f), alarm parameters must also be set with a decision threshold β_m that maximizes the posterior probability of a true alarm.

REFERENCES

Casey, S. (1993). *Set phasers on stun*. Santa Barbara, CA: Aegean.

Farber, E., & Paley, M. (1993, April). Using freeway traffic data to estimate the effectiveness of rear end collision countermeasures. *Paper presented at the Third Annual IVHS America Meeting*, Washington DC.

Getty, D. J., Swets, J. A., Pickett, R. M., & Gounthier, D. (1995). System operator response to warnings of danger: A laboratory investigation of the effects of the predictive value of a warning on human response time. *Journal of Experimental Psychology: Applied, 1,* 19-33.

ITS America (1996). *Collision warning systems.* (Fact Sheet No. 3). (ITS America, Washington DC).

Knipling, R. R., Mironer, M., Hendricks, D. L., Tijerna, L., Everson, J., Allen, J. C., and Wilson, C. (1993). *Assessment of IVHS countermeasures for collision avoidance: Rear-end crashes.* (NTIS No. DOT-HS-807-995). (National Highway and Traffic Safety Administration, Washington DC).

National Safety Council (1996). *Accident facts.* Ithaca, IL: Author.

Parasuraman, R., Hancock, P.A., & Olofinboba, O. (1997). Alarm effectiveness in driver- centered collision-warning systems. *Ergonomics, 39,* 390-399.

Parasuraman, R. & Mouloua, M. (Eds.). (1996). *Automation and human performance: Theory and applications.* Mahwah, NJ: Erlbaum.

Parasuraman, R. & Riley, V. (1997). Humans and automation: Use, misuse, disuse, abuse. *Human Factors, 39,* 230-253.

Pate'-Cornell, M. E. (1986). Warning systems in risk management. *Risk Analysis, 6,* 223-234.

Patterson, R. D. (1982). *Guidelines for auditory warning systems on civil aircraft.* (CAA Paper 82017). (UK Civil Aviation Authority, London).

Seminara, J. L., Gonzalez, W. R., & Parson, S. O. (1977). *Human factors review of nuclear power plant control room design.* (Rep. No. NP-309). (Electric Power Research Institute, Palo Alto, CA).

Swets, J. A. (1992). The science of choosing the right decision threshold in high-stakes diagnostics. *American Psychologist, 47,* 522-532.

Tyler, R. R., Shilling, R. D., & Gilson, R. D. (1995). *False alarms in Naval aircraft: A review of Naval Safety Center mishap data.* (Special Rep. No. 95-003). (Naval Air Warfare Center, Training Systems Division, Orlando, FL).

Wang, J. S., Knipling, R. R., & Goodman, M. J. (1996). The role of driver inattention in crashes: New statistics from the 1995 Crashworthiness Data System. In *Proceedings of the 40th Annual Meeting of the Association for the Advancement of Automotive Medicine.* Vancouver, British Columbia, October 7-9.

Monitoring Driver Impairment Due to Sleep Deprivation or Alcohol

Stephen H. Fairclough and Robert Graham
The HUSAT Research Institute
Loughborough University, UK

INTRODUCTION

The related areas of real-time monitoring and diagnosis are central to the development of Intelligent Transport Systems (ITS). The provision of timely and relevant traffic information is dependent on monitoring the drivers' activities relative to an information infrastructure. For example, an Autonomous Intelligent Cruise Control (AICC) system must be capable of adjusting host vehicle speed by monitoring and predicting variations in the speed of other vehicles. Similarly, the provision of collision avoidance warnings is reliant on the accurate monitoring and diagnosis of an potential rear-end collision situation. The capability of technological systems to adapt appropriately and in real time is restricted by the proficiency of monitoring and diagnosis sub-systems.

The evolution of telematic systems which monitor and diagnose the presence of driver impairment (i.e. degraded driving performance due to the influence of fatigue, alcohol, drugs or illness) represents the most ambitious research and development programme in this field. The monitoring system under development on the current CEC-funded project SAVE (Brookhuis, De Waard, & Bekiaris, 1997) employs a multi-sensor approach, where data from an array of vehicle sensors are combined with psychophysiological information on eye closure and the presence of long duration eye blinks (collected via remote means using a machine vision algorithm). These systems must be capable of a high degree of accuracy if they are to function as accident countermeasures (Fairclough, 1997). In practical terms, system accuracy must be achieved against a background of high variability, both on an intra- and inter-driver basis and within a fluctuating driving environment. Impairment monitoring system must be capable of producing highly accurate performance on the basis of a low signal-to-noise ratio.

The aim of the SAVE project is to produce a prototype telematic system capable of diagnosing and discriminating between impairment due to fatigue, alcohol and sudden illness. The system functions on a cybernetic basis. If the performance of the driver deviates from a normative template, the system provides a warning message. The driver receives visual and verbal feedback at two levels of differentiation, i.e. a standard warning and a severe warning. If the driver is assessed to be dangerously impaired, the SAVE system activates a longitudinal and lateral control device (known as the Automatic Control Device or ACD). The ACD takes vehicular control from the impaired driver, provides an alert to other drivers in the vicinity and attempts to bring the vehicle to a safe and stationary position, The monitoring system is the decision-making component of the SAVE system architecture, providing warning feedback or automatic control on the basis of a real-time diagnosis.

It is essential that the decision whether or to evoke automatic intervention or warning feedback is based on reliable diagnostic criteria. The dilemma for system designers and researchers alike is to establish the criteria that define impaired driving in both quantitative and qualitative terms. In mental workload research, the estimation of unacceptable levels of workload is known as determination of the workload redline. It is proposed that there is little conceptual difference between the assessment of a workload redline and an impairment redline. Both are concerned with the setting of a lower limit for acceptable performance and conditions of operation.

The derivation of criteria for warnings and interventions into a classification system is a complicated affair. Under extreme circumstances where the vehicle weaves out of lane, steering input ceases and the driver slumps in the seat - it is obvious that the system should intervene. In fact, the SAVE system includes sensors to detect driver head position and grip force on the steering wheel to cover this eventuality. However, it is anticipated that real-time diagnosis will not be so unequivocal, and any classification system for driver impairment will be characterized by a large number of grey areas. The difficulty of diagnosis is not the detection of a characteristic driver error per se, so much as estimating the severity of that error within a classification system. Within this conceptualization, the lousiness of diagnosis concerns an estimation of magnitude(i.e. the extent of a deviation from normative driving) as a first step, followed by matching the magnitude of the error to the response of the system, i.e. large and/or sustained errors are penalised by system intervention whereas smaller and/or temporary errors are corrected by warning feedback.

The obvious candidate for an impairment criteria is the point at which primary driving performance begins to decline. This may be problematic from the perspective of system sensitivity. There is some evidence that primary performance levels may sustained at a normative level, but only at an energetical cost to the operator (De Waard, 1996; Hockey, 1997). The energetical cost associated with performance under duress may be characterized as a stress response, from increased levels of sympathetic nervous activity to higher levels of subjective mental workload (Hockey, 1997). However, criteria based around driving measures are associated with a higher degree of validity than more sensitive alternatives such as psychophysiology, i.e. performance impairment is indexed directly rather than using a proxy measure. However, the use of primary task measures still fails to resolve the issue of redline definition, i.e. how to quantify an unacceptable level of impairment in terms of driver performance?

The solution used in this study was to compare the effects of a fatigue manipulation (sleep deprivation) with impairment due to alcohol . The use of alcohol for comparative purposes is attractive because: (a) alcoholic intake may be quantified on a linear scale of Blood Alcohol Content (BAC), (b) there is a linear relationship between BAC and accident involvement (Walls & Brownlie, 1985) and (c) alcoholic intake is associated with a limit above which it is unlawful to drive a vehicle. The use of an alcohol manipulation is not an argument that impairment due to alcohol and fatigue are synonymous. But from the perspective of primary task performance, both fatigue and alcohol have similar effects on driving performance: impairing reaction time, car following and lateral control (i.e. Attwood, Williams, & Madill, 1980; De Waard & Brookhuis, 1991; Laurell, McLean, & Kloeden, 1990; Ryder, Malin, & Kinsley, 1981). Therefore, the legal limit for alcohol may be used as a comparative redline for driver impairment due to fatigue, at least with respect to key task components such as lateral control. There are other reasons which justify a comparison between impairment due to alcohol or fatigue on driving performance. Advanced monitoring systems such as the SAVE system wish to differentiate the intoxicated driver from the fatigued driver with respect to system feedback and ACD activation. Therefore, it is equally important to emphasise those points of dissociation between both categories of impairment.

The aim of the study was to compare the effects of a fatigue manipulation (2 levels of sleep deprivation) on a prolonged (i.e. 120 minutes) simulated drive with impairment produced by an illegal level of alcohol (above 0.08% BAC in the UK at the time of writing).

METHOD

Design

The study was designed to compare the effects of a between-subjects factor (treatment group) across two within-subjects factors (time-on-task and driving scenario). The four treatment groups are described as follows: *Control group:* These subjects received a full night of sleep before the trial and did not receive alcohol. *Partial sleep deprivation (PartSD) group:* These subjects were instructed to sleep for four hours between midnight and 0400 hours on the night before the trial. This group did not receive alcohol. *Full sleep deprivation (FullSD) group:* These subjects were instructed to remain awake throughout the night before the trial and did not receive alcohol. *Alcohol-impaired (Alcohol) group:* These subjects had a full night of sleep on the night before the trial and received an alcoholic drink (a mixture of vodka and lemonade) prior to the experimental session. The amount of alcohol administered was calculated based on the subject's body weight, according to the Widmark Equation (as described by Walls & Brownlie, 1985). Subjects received sufficient alcohol to approximate a peak Blood Alcohol Content (BAC) level in the range of 0.08 - 0.1 per cent (mg/g). The time-ontask associated with the journey was measured at three levels, each corresponding to a cumulative forty minute period within the two hour journey.

Subjects

Sixty-four subjects participated in the experiment. All subjects were male and had normal or corrected 20/20 vision. Each subject was designated into one of four experimental group, each containing sixteen subjects. The allocation of subjects into groups was performed to balance demographic variables of age, driving experience, annual mileage, driving frequency, average alcohol intake and average hours sleep per night across the four groups. On average, the subject group as a whole drove everyday reported an annual mileage of greater than 10000 miles and less than 15000 miles and slept for an average of 8 hours per night. All were paid for their participation.

Apparatus

The experiment was performed using a fixed-base driving simulator. Subjects were seated inside a Ford Scorpio and viewed a large projector screen (approximately 3 x 4m). A Pentium PC was used to simulate the vehicle model and to generate the driving scene. The computer-generated scene was projected onto the screen via a Sony Multiscan projector. The vehicle interior and the experimenter's location were linked via an intercom system. Electrocardiograph (ECG) data was collected via a analogue to digital converter connected to a Macintosh Powerbook™ computer.

Experimental Task

The simulated driving scene comprised of a straight, two lane, left-side drive road flanked by road signs (which advised a 70 mph speed limit), vegetation and marker posts. Off-road areas were coloured a uniform green on either side of the road boundary. The upper portion of the scene was coloured blue to emulate daytime driving. Vehicle dynamics included a small "wind gust" factor which was activated randomly throughout the session. Every alternate five second period, one of three outcomes occurred randomly: (a) no wind gust, (b) a wind gust approximating two steering degrees to the left sustained for five seconds and (c) a wind gust approximating two degrees to the right sustained for five seconds. Unfortunately the simulator software was not capable of providing interactive sound. Subjects were exposed to a recording of in-vehicle sound collected from a real vehicle travelling at an approximate speed range of 60-70 mph on a motorway. The subjects completed three forty-minute journeys in the course of an experimental session. No data was collected curing the first ten minutes of each forty minute block.

Experimental Measures

Primary Driving Task. This group of variables comprised measures of lateral control (i.e. lane tracking) and measures of longitudinal control (i.e. speed, time headway). Lateral control was measured et several levels of criticality: (1) frequency of accidents as defined when all four "wheels" left the left side lane edge, or when the subject either travelled across the right lane boundary and collided with a passing vehicle, or when the subject left the road on the right side lane edge and (2) frequency of lane crossings when two vehicle wheel made contact with the left lane edge or the right lane boundary, (3) frequency of incidents when the minimum Time-To-Line Crossing (TLC) (Godthelp, Milgram, & Blaauw, 1984) was less than 2 seconds. In order to index subjects' steering input to lateral control (the number of zero-crossings of the steering wheel as defined by McLean & Hoffman, 1975) was measured. Longitudinal control was measured by mean speed, mean time headway to a lead vehicle and speed variability. All measures of primary driving performance were originally sampled at 10Hz and reduced to 2Hz for the purpose of analysis.

Psychophysiology. ECG data was collected via three disposable electrodes attached to the subject's chest. it-peaks of the ECG trace were detected and corrected for artifacts. The resulting data was analysed with respect to mean heart rate expressed as mean Inter-Beat Interval (IBI) and heart rate variability (HRV). The HRV variable was subjected to spectral analysis, to decompose the HRV signal into three bandwidths (Murder, 1979): a low frequency band (0.02-0.06Hz) identified with the regulation of body temperature, a mid frequency band (0.07-0.14Hz) related to short-term blood pressure regulation and a high frequency band (0.15 - 0.50Hz) influenced by respiratory regulation.

Subjective Measures. The study employed a battery of subjective measurement questionnaires. Subjects completed multiple administrations of: (a) the UWIST Mood Adjective Scale (Matthews, Jones, & Chamberlain, 1990), (b) a raw version of the NASA Task Load Index (RTLX) (Byers, Bittner, & Hill, 1989; Hart & Staveland, 1988), (c) the Karolinska Sleepiness scale (Kecklund & Akerstedt, 1993), (d) a Cognitive Interference scale (Sarason, Sarason, Keefe, Hayes, & Shearin, 1986) and (e) an eight-point sobriety scale devised for the study.

Experimental Procedure

The subjects came to the Institute on two occasions, the first being a Practice session followed by a Test session on a different day. On most occasions, the Practice session took place in the morning prior to the day of the Test Session. During the Practice session, subjects were weighed, completed a test of visual acuity and performed a twenty-two minute "training" journey followed by a twenty minute baseline journey to index subjects' "normal" behaviour in the simulator. On completion of the baseline journey, subjects completed a second set of subjective questionnaires. The two sleep deprivation subject groups were instructed to stay awake on the night before the Test Session. PartSD subjects were instructed to sleep for four hours between midnight and 0400. To ensure subjects remained awake, they were required to leave a time-stamped message on a telephone answering machine once every sixty minutes between 0400 and an hour before their Test Session was due to commence. Those subjects in the FullSD group followed an identical procedure except that time-stamped messages were left on the answering machine every sixty minutes between 23:00 until an hour before their Test Session.

All subjects were provided with transport to and from the Institute on the day of their Test Session. This arrangement was made to ensure the integrity of the alcohol placebo. On arrival for the Test Session, all subjects were provided with a drink. The Control group and the two sleep-deprived groups received a placebo containing lemonade with a tablespoon of vodka floated on the surface to provide an alcoholic smell. The Alcohol group received a mixture of vodka and lemonade sufficient to induce a peak Blood Alcohol Content (BAC) of approximately 0.08 - 0.1%. In addition, the Alcohol group received specific instructions to eat a light meal at least an hour before the Test Session. The concentration of alcohol required for each subject was formulated by body weight based on the Widmark equation (i.e. %BAC = consumed alcohol/(body weight x Widmark factor) where Widmark factor = ratio of water content in body). During the test session, the subjects were told that they would receive a modest financial punishment for every accident which occurred during the session (i.e. either if the car left the road or collided with another vehicle). This instruction was a deception intended to motivate the subjects to maintain performance (no subject actually suffered any financial penalties regardless of performance). Then the subjects completed the first forty minute block of the simulated journey. On completion of the journey, subjects completed a second set of subjective questionnaires and were breathlysed once more. Both the subjective questionnaires and breathalyser tests were administered following completion of the second and third forty minute blocks. The recess between each journey block was restricted to five minutes duration. The combined duration of the three journey blocks was approximately two hours. On completion of the experimental session, subjects were debriefed, paid for participation and transported to their homes.

RESULTS

Alcohol Manipulation

Repeated testing with an alcometer revealed a descending%BAC curve throughout the 120 minute journey. The initial alcometer reading sampled ten minutes after ingestion showed a peak average at 0.081 (*s.d.* = 0.03 7). Subsequent samples taken 50 minutes following ingestion ($M = 0.076$, *s.d.* = 0.019) declined to a minimum average of 0.055 (*s.d.* = 0.013).

Lateral Control

The lateral control of the vehicle was scored at three levels of severity. The frequency of accidents is illustrated in Figure 1. It was apparent that the FullSD and Alcohol subjects ran the vehicle off the road more frequently than the Control group. No accidents were recorded for the PartSD group. However, only 10 subjects out of the total sample accounted for all accidents and therefore, no detailed statistical testing was performed.

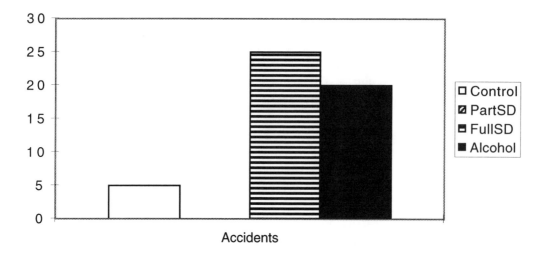

Figure 1. Frequency of off-road Accidents for each experimental group.

The frequency of lane crossings were scored within a five minute window. MANOVA testing revealed that the FullSD group made more frequent lane crossings than either the Control or the PartSD groups [$F(3,59)$ = 4.58, $p < 0.01$]. The Alcohol subjects made more frequent lane crossings than either Control or PartSD groups, but were not significantly different. These findings are illustrated in Figure 2. The frequency of lane crossing showed a significant increase over time [λ_w (2,111) = 0.73, $p < 0.01$]. However, this increase was modest in real terms, i.e. rising by 1 lane crossing per 5 minutes across the whole 120 minute journey. An interaction effect revealed the detrimental influence of full sleep deprivation to be exacerbated curing the final forty minutes of the journey. During this period, the lateral control of the FullSD group was significantly poorer than the Alcohol group [λ_w (6,118) = 0.79, $p < 0.05$].

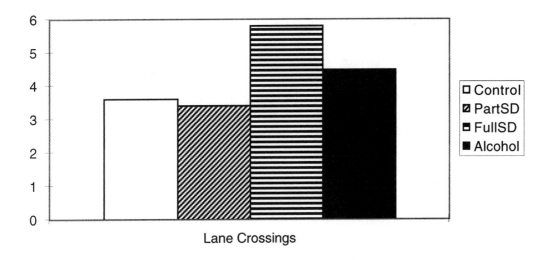

Figure 2. Frequency of Lane Crossings (per 5 minutes) for each experimental group.

The frequency of near-lane crossings was scored as the frequency of minimum Time-To-Line Crossing (TLC) incidents where the driver had 2 seconds to make a corrective steering action in order to avoid a lane crossing. MANOVA testing revealed an effect of marginal significance [$F(3,59)=2.60, p=0.06$], that indicated the PartSD group made a higher number of near-lane crossings than the other experimental groups. This finding is is illustrated in Figure 3.

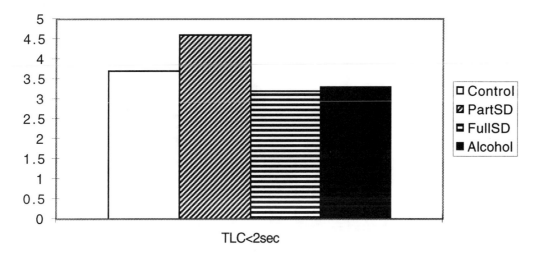

Figure 3. Frequency of Near-Lane Crossings (per 5 minutes for each experimental group).

The analyses of steering wheel reversal rate revealed a significant main effect for experimental condition [$F(3,59)=26.6, p<0.01$]. Post-hoc testing revealed that mean steering wheel reversal rate was higher for the Control and Alcohol groups ($M = 15.6$ and 14.3 respectively) compared to both the PartSD and the FullSD groups ($M = 11.2$ and 10.9 respectively).

Longitudinal Control

There were no differences between the experimental groups with respect to mean speed. Although mean speed did significantly increase over time-on-task [$\lambda_w (2,120) = 0.88, p < 0.05$] by approximately 1.5

mph. The analysis of speed variability (indexed as speed standard deviation) showed a significant main effect due to the experimental group [$F(3,59) = 4.56, p < 0.01$]. Post-hoc testing revealed that speed variability was lower for the Alcohol group ($M = 4.2$ mph) compared to either the PartSD or the FullSD groups ($M = 5.7$ mph for both).

An analysis of mean time headway to a lead vehicle revealed a marginal effect due to the experimental condition [$F(3,54) = 2.22, p = 0.08$]. Post-hoc tests revealed a significant difference between the Alcohol ($M = 3.10$s) compared to the FullSD group ($M = 3.98$s).

Psychophysiology

Psychophysiological data was lost from 7 subjects across the 4 subject groups due to measurement artifacts. It is known that alcohol has a confounding effect on mean IBI (Brookhuis & De Waard, 1993), usually decreasing mean IBI. In addition, GonzalezGonzalaz, Llorens, Novoa, & Valeriano (1992) found effects of low alcohol ingestion on the mid-range component of the HRV signal. Descriptive data from the present study revealed that the net effect of alcohol was to decrease mean IBI by approximately 65 ms. Therefore, ECG data from the remaining 15 subjects in the Alcohol group was not subjected to statistical testing. The raw IBI data was subjected to analysis using CARSPAN software (Murder & Schweizer, 1993) to analyse the three bandwidths of the HRV signal. This data was subjected to a natural log transform and a baseline conversion prior to parametric MANOVA analysis. As in the previous analysis, the same 7 subjects were rejected as outliers. There were significant main effects due to experimental condition [$F(2,32)=2.52, p=0.05$] end time-on-task [$\lambda_w (2,64) = 0.64, p < 0.01$]. Both main effects are illustrated in Figure 4. It was apparent that the mid-frequency component was significantly suppressed for the PartSD group compared to the Control group ($p<0.05$). In addition, mean power in the mid-range frequency increased between the first and third period of the journey ($p<0.05$).

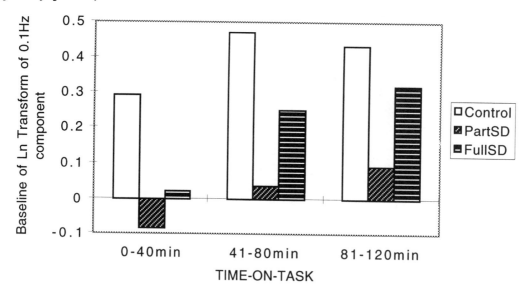

Figure 4.. Ln transform of mean power in mid-range frequency band of HRV analysis (expressed as a baseline) across experimental groups and time-on-task.

Subjective Measures

The subjective data was analysed via a series of MANCOVAs with data collected from the end of the practice session used as a covariate. Note that subjects completed the questionnaires on four occasions through

74

the Test Session including a pretest administration. The results of these analyses are shown in Table 1.

Table 1
Analyses of subjective data.

		CONDITION		TIME ON TASK	
		$F_{(3,60)}$	P	λ_w (2,120)	P
MOOD	Energetic arousal	5.88	<0.01	0.54	<0.01
	Tense arousal	1.43	NS	0.94	NS
	Hedonic Tone	1.34	NS	0.65	<0.01
WORKLOAD	Mental demand	1.09	NS	0.92	NS
	Physical demand	4.41	<0.01	0.5	<0.01
	Time demand	4.06	<0.05	0.95	NS
	Performance	2.73	0.05	0.80	<0.01
	Effort	3.16	<0.05	0.87	<0.05
	Frustration	4.51	<0.01	0.61	<0.01
	Mean workload	2.63	0.05	0.68	<0.01
COGNITIVE INTERFERENCE	Task Relevant	4.23	<0.01	0.60	<0.01
	Task Irrelevant	1.29	NS	0.45	<0.01
SLEEPINESS	Karolinska scale	11.37	<0.01	0.39	<0.01
SOBRIETY	Self-Rating	12.11	<0.01	0.81	<0.05

Measurement of subjective mood revealed a significant decline of energetic arousal for the two sleep-deprived groups compared to control subjects. It was also found that energetic arousal and hedonic tone declined with time-on-task. Analyses of the subscales of RTLX revealed that sleep-deprivation reduced self-rated estimates of good performance and increased perceived levels of effort invested in the task. For the FullSD group, subjective workload was also increased by rises in both the physical and temporal demands of the driving task. The effect of Alcohol on subjective workload was to significantly reduce the level of frustration experienced by subjects. This was not replicated with respect to the hedonic tone scale, however alcoholic intoxication may produce increased relaxation and well-being (as indicated by the Sobriety scale). All timeon-task effects indicated rising subjective workload between the first and last test administration. It was found that the frequency of Task Relevant thoughts was higher for the FullSD group compared to either the Control or Alcohol groups. The PartSD group exhibited a similar pattern, but were only significantly higher than the Control group. The companion factor of Task Irrelevant thoughts did not show any significant differences between groups. An analysis of time-on-task revealed how the frequency of Task Relevant thoughts declined over time whilst Task Irrelevant thoughts increased. Both subjective ratings of sleepiness and sobriety were included to reference the experimental manipulations. Subjective sleepiness increased over time and was significantly higher for both sleep-deprived groups compared to the Control group. Selfrated sobriety was lowest for the Alcohol group.

DISCUSSION AND CONCLUSIONS.

The analysis of lane crossings showed clearly that the FullSD group exhibited poorer lateral control than either the Control or PartSD group. The effect of alcohol was to degradel ane-keeping, but not to the same extent as a full night without sleep. It was significant that a restriction of four hours sleep on the night prior to testing had no detrimental effect on subjects' lane-keeping performance. In addition, there were no accidents for the PartSD group. These data indicate that either the partial sleep deprivation manipulation had no effect on driving behaviour or subjects were able to compensate to sustain an adequate level of task performance. The cumulative effect of time-on-task was modest in real terms (frequency of lane crossings increased by one per five minutes) but did achieve statistical significance.

The PartSD group exhibited a level of lane-keeping performance that was equivalent to the Control group (i.e. in terms of accidents and lane crossings). The singularity of the lateral control of the PartSD group is highlighted when the level of steering input is considered. The lateral position of the vehicle may be considered as the *output* resulting from the level of steering *input.* With this interpretation in mind, a lower level of steering input translated into poorer lateral control for the FullSD group. However, the PartSD group sustained a level of lateral control equivalent to the Control group whilst operating on the low level of steering input that characterized the FullSD group. A reduction of steering activity combined with an absence of any performance decrement shows an increased level of steering efficiency and may be indicative of an anticipatory lane-keeping strategy which required a low level of precise steering input. Circumstantial support for this interpretation is provided by the analysis of near-lane crossings (Figure 2) where incidents when TLC fell below 2 seconds were significantly higher for the PartSD group. In other words, PartSD subjects showed a tendency to correct the lateral trajectory of the vehicle on a proactive basis, on most occasions anticipating vehicle drift across the lane boundary by at least 2 seconds. By contrast, the Alcohol group exhibited a comparative level of steering input to the Control group which was combined with a poorer level of lateral control. These findings illustrate the importance of using multiple variables of performance to diagnose the specific category of impairment.

Measures of longitudinal control and speed control revealed a number of differences between the Alcohol group and the sleep-deprived subjects. Both sleepdeprived groups exhibited an increase of speed variability compared to the Alcohol group (inspection of the mean values for speed variability indicated that the Control group was similar to the Alcohol group in this regard). The absence of any significant interaction effect revealed that increased speed variability was a general characteristic of the sleepdeprived subjects. This may have resulted from a tendency to relax the foot on the accelerator and then to correct as fatigue increased or be indicative of a general time lag in the speed adjustment of sleep-deprived subjects. It was apparent that the FullSD group slightly increased average time headway to a lead vehicle. This may have been a risk compensation strategy prompted by increased sleepiness and reduced confidence in their own performance. The Alcohol subjects reduced mean time headway in comparison to the other three groups (but were only significantly different from the FullSD group). It is hypothesized that the contrasting effects of alcohol and full sleep deprivation on following behaviour are connected to different levels of self-assessed impairment, i.e. alcohol subjects did not perceive their performance to be impaired whilst sleep-deprived subjects did.

The data from the primary driving task failed to disclose any impairment of lateral control as a result of partial sleep deprivation. However, analyses of subjective data (Table I) and psychophysiology (Figure 4) revealed that the partial sleep deprivation manipulation had a pronounced effect on the subjects. These data are in line with the predictions outlined on cognitive-energetical compensation and control (Hockey, 1993; Hockey, 1997, Mulder, 1986). According to this view, primary task performance is resilient to the influences of increased task demands and stressors such as sleep deprivation. Under difficult circumstances, the operator invests additional mental effort to protect performance. But the act of effort investment is associated with a number of costs including increased subjective workload and negative affect. In line with these predictions, subjective data revealed an equivalent level of impairment for both PartSD and FullSD groups. The net effect of sleep deprivation for both groups was to increase subjective sleepiness and to degrade the perceived quality of performance. The FullSD group had the additional burden of increased subjective mental workload with respect to higher physical and temporal task demands. The former effect may have been due to increased muscular lethargy whereas the latter is indicative of a declining efficiency of psycho-motor co-ordination (therefore creating a subjective increase of task pacing). Both groups of sleep-deprived subjects attempted to compensate for the influence of sleep deprivation by increasing mental effort. This mechanism was apparent in the psychophysiological data vie the suppression of the O.1 Hz component (Figure 4) and the subjective attempt to increase effort and to reduce cognitive interference by increasing the frequency of task-relevant

76

thoughts (Table 1). The compensatory strategy of the PartSD subjects effectively prevented any significant degradation of lateral control performance, but this was not the case for the FullSD group. Aside from a decline in sobriety, the Alcohol group did not perceive any detrimental effects of impairment (Table 1). The contrast between the PartSD group (who perceived high levels of subjective impairment but performed adequately) and the Alcohol subjects (who perceived no subjective impairment but performed comparatively poorly) was striking as it confirmed an obvious connection between the awareness that one is impaired and the initiation of compensatory strategies to minimise the impact of impairment on performance.

The sensitivity of primary task measures was largely dependent on the safetycritical connotations of specific variables under investigation. For example, safetycritical measures such as frequency of lane crossings only revealed a significant degradation for the FullSD group. However, less critical differences in levels of steering input and speed control showed clear differences between the Control group and both groups of sleep-deprived subjects. It was apparent that those vehicle variables which had a secondary rather than a primary relationship to safety were the most sensitive to the influence of sleep deprivation. This hypothesis suggests there is a degree of risk compensation inherent in any compensatory response to fatigue. If the PartSD and FullSD groups are considered as points on a continuum of sleep deprivation, it was apparent that less safety-critical changes in steering strategy (including an increase of efficiency) preceded a decline of safety-critical incidents such as lane crossings and accidents. For the Alcohol group where no compensatory response was forthcoming, this incremental pattern of impairment from secondary to primary safety critical measures was not apparent, i.e. the only difference between Alcohol and Control subjects was with respect to lane crossings.

A distinction between an intoxicated driver and a tired driver may rest on a multivariate profile of the driving task. In the first instance, a very tired driver and an intoxicated driver may exhibit degraded lateral control. However in the case of the tired driver, a decline of lateral control is accompanied by a reduction of steering activity and more variable speed control. Where other vehicles are present, the tired driver may show a slight increase of time headway whereas the opposite pattern would be expected in the case of the drunk driver. Therefore, it was possible to diagnose between both categories of impairment purely on the basis of primary performance.

This findings suggest that at elematics system that is solely reliant on data from the primary driving task must be capable of monitoring and diagnosing both safety critical and non-safety critical changes in vehicle control. It is postulated that the latter category may comprise the earliest and most sensitive indicators of incremental driver fatigue. In the case of the Alcohol group, impairment did not appear to follow the same pattern and only safety-critical changes were apparent. Therefore, it may not be possible to detect non-critical impairment in the case of alcoholic intoxication, teased on measures of the performance alone. The study demonstrated a number of differences and similarities between driving impairment induced by sleep deprivation and alcohol in terms of both quantitative and qualitative data. The use of alcohol as a redline was contentious in certain respects, particularly due to the influence of time-on-task and the placebo procedure. With respect to time-on-task, subjects in the Alcohol group experienced a descending alcohol absorption curve and therefore the redline was not fixed. In practice, the alcohol manipulation manifested itself as a fuzzy redline bandwidth falling from an upper to a lower limit over the two hour test session. The protocol for sleep deprivation provided those subjects with important cues with which to gauge subjective impairment and to formulate compensatory strategies. In comparison, the placebo manipulation provided subjects in the Alcohol group with minimal cues. Whilst all subjects in the Alcohol group were aware of the influence of alcohol (Table 1), they had no idea of the quantity of alcohol imbibed. This dearth of feedback may have been responsible for an absence of any compensatory response. On the other hand the well-known inflation of confidence associated with alcohol may have placed these subjects under the illusion that performance was not impaired. The implications for fatigue monitoring systems from the study are that: (a) non-critical measures of primary vehicle control may be sufficiently sensitive to diagnose the early symptoms of fatigue, and (b) that primary task measures were sufficiently diagnostic to discriminate the case of the fatigued driver from the intoxicated driver.

ACKNOWLEDGEMENTS.

This work was funded by CEC within the Transport Telematics programme and under the SAVEproject (TR 1047).

REFERENCES

Attwood, D. A., Williams, R. D., & Madill, H. D. (1980). Effects of moderate blood alcohol concentrations on closed-course driving performance. *Journal of Studies on Alcohol, 41(7)*, 623-634.

Brookhuis, K. A., & De Waard, D. (1993). The use of psychophysiology to assess driver status. *Ergonomics, 36(9),* 1099-1110.

Brookhuis, K. A., De Waard, D., & Bekiaris, E. (1997). Development of a system for the detection of driver impairment. In C. Mercier-Guyon (Ed.), *Proceedings of the 14th International Conference on Alcohol, Drugs and Traffic Safety* (Vol. 2, pp. 581-586). Annecy, France: CERMT.

Byers, J. C., Bittner, A. C., & Hill, S. G. (1989). Traditional and raw task load index (TLX) correlations: are paired comparisons necessary? In A. Mital (Ed.), *Advances in industrial ergonomics and safety* (Vol. 1, pp. 481-488). London: Taylor and Francis.

De Waard, D. (1996). *The measurement of driver mental workload*. Rijksuniversiteit Groningen, Groningen, The Netherlands.

De Waard, D., & Brookhuis, K. A. (1991). Assessing driver status: a demonstration experiment on the road. *Accident Analysis and Prevention, 23(4)*, 297-301.

Fairclough, S. H. (1997). Monitoring driver fatigue via driver performance. In Y. I. Noy (Ed.), *Ergonomics and safety of intelligent driver interfaces* (pp. 363-380). Mahwah, NJ: Erlbaum.

Godthelp, J., Milgram, P., & Blaauw, G. J. (1984). The development of a time-related measure to describe driving strategy. *Human Factors, 26*, 257-268.

Gonzalez-Gonzalaz, J., Llorens, A. M., Novoa, A. M., & Valeriano, J. J. C. (1992*).* Effect of acute alcohol ingestion on short-term heart rate fluctuations. *Journal of Studies on Alcohol, 53(1)*, 86-90.

Hart, S. G., & Staveland, L. E. (1988). Development of the NASA-TLX (Task Load Index): results of empirical and theoretical research. In P. A. Hancock & N. Meshkati (Eds.), *Human mental workload* (pp. 139-183). Amsterdam: North Holland.

Hockey, G. R. J. (1993). Cognitive-energetical control mechanisms in the management of work demands and psychological health. In A. Baddeley & L. Weiskrantz (Eds.), *Attention: Selection, awareness and control.* (pp. 328-345). Oxford: Clarendon Press.

Hockey, G. R. J. (1997). Compensatory control in the regulation of human performance under stress and high workload: a cognitive-energetical framework. *Biological Psychology, 45*, 73-93.

Kecklund, G., & Akerstedt, T. (1993). Sleepiness in long distance truck driving: an ambulatory EEG study of night driving. *Ergonomics, 36(9)*, 1007-1018.

Laurell, H., McLean, A. J., & Kloeden, C. N. (1990). *The effect of blood alcohol concentration on light and heavy drinkers in a realistic night driving situation* (Report 1/90): NHMRC Road Accident Research Unit, University of Adelaide.

Matthews, G., Jones, D. M., & Chamberlain, A. G. (1990). Refining the measurement of mood: The UWIST Mood Adjective Checklist. *British Journal of Psychology, 81*, 17-42.

McLean, J. R., & Hoffman, E. R. (1975). Steering reversals as a measure of driving performance and steering task difficulty. *Human Factors, 17,* 248-256.

Mulder, G. (1979). Mental load, mental effort and attention. In N. Moray (Ed.), *Mental workload: Its theory and measurement* (pp. 327-343). New York: Plenum Press.

Mulder, G. (1986). The concept and measurement of mental effort. In G. R. J. Hockey, Gaillard, A. W. K. & Coles, M. G. H. (Eds.), *Energetical issues in research on human information processing* (pp. 175-198).

Dordrecht, The Netherlands: Martinus Nijhoff.

Mulder, L. J. M., & Schweizer,D. A. (1993). *Carspan: Cardiovascular experiments analysis environment:* iec ProGamma, Groningen, The Netherlands.

Ryder, J. M., Malin, S. A., & Kinsley, C. H. (1981). *The Effects of Fatigue and Alcohol on Highway Safety* (DOT-HS-805-854). Washington DC: National Highway Traffic Safety Administration.

Sarason, l. G., Sarason, B. R., Keefe, D. E., Hayes, B. E., & Shearin, E. N. (1986). Cognitive interference: situational determinants and traitlike characteristics. *Journal of Personality and Social Psychology, 57,* 691-706.

Walls, H. J., & Brownlie,A. R. (1985). *Drink, drugs & driving.* London: Sweet& Maxwell.

AIR TRAFFIC CONTROL

Conclusions from the Application of a Methodology to Evaluate Future Air Traffic Management System Designs

Philip J. Smith, David Woods, Charles Billings,
Rebecca Denning and Sidney Dekker
Cognitive Systems Engineering Laboratory, The Ohio State University

Elaine McCoy
Department of Aviation, Ohio University

Nadine Sarter
Institute of Aviation, University of Illinois at Urbana-Champagne

INTRODUCTION

We have been exploring the use of a general methodology to predict the impact of future Air Traffic Management (ATM) concepts and technologies. In applying this methodology, our emphasis has been on the importance of modeling coordination and cooperation among the multiple agents within this system, and on understanding how the interactions among these agents will be influenced as new roles, responsibilities, procedures and technologies are introduced. To accomplish this, we have been collecting data on performance under the current air traffic management system, trying to identify critical problem areas and looking for exemplars suggestive of general approaches for solving such problems. Using the results of these field studies, we have developed a set of concrete scenarios centered around future system designs, and have studied performance in these scenarios with a set of 40 controllers, dispatchers, pilots and traffic managers. This paper provides a brief overview of the methodology employed, and then provides a summary of the major recommendations that have resulted from its application.

METHODS FOR THE DEVELOPMENT OF FUTURE SYSTEM REQUIREMENTS

An important methodological question is: How can we identify system requirements for new ATM concepts and technologies, recognizing the potential for these changes to create new roles and procedures for individual participants, new forms of coordination across personnel and organizations, and new types of information to communicate, assess and integrate? This is difficult in part because the system of interest does not yet exist. A further challenge arises because many details of this future system design may be under-specified. To design and test these new roles before committing large pools of resources, we need a model method to quickly prototype and study how the people and technologies will coordinate in realistic operational scenarios for different ATM concepts.

Different developers, stakeholders and decision-makers are each likely to develop their own insights and views on how the whole system will work in the future. Consequently, another challenge is how to rigorously explore all of the implications of these different viewpoints. For without doing so, it would be easy to oversimplify the impact of a new system on the roles, decisions, coordination needs, and information requirements of the people involved in the ATM system.

These characteristics--a still under-specified system design, accompanied by the potential to oversimplify the impact of design decisions on people's roles and activities--create a difficult methodological challenge. How can we assess the impact of new ATM concepts and technologies on the individual and collective performances of controllers, dispatchers, flight crews and traffic managers?

In our work we have been using methods that try to balance cost, timeliness, degree of control, face

81

validity and environmental richness to provide useful input to system developers early in the design process. The methods that we have been exploring to support requirement identification are:

- empirical-collecting data about how people may carry out new roles and utilize possible new systems, thus helping to identify potential problems and define requirements for system development.
- scenario driven-rigorously exploring the implications of new ATM concepts and technologies by having different kinds of expert practitioners explore them in the context of concrete situations.
- iterative and converging-using multiple approaches that build 'on each other and converge on results that can support design decisions.

The end result for us has been an emphasis on different forms of CONCEPTUAL WALKTHROUGHS. In these conceptual walkthroughs, we posit a future ATM system, or parts of it, through development of a concrete scenario that instantiates a situation within that future system. We then collect data by asking a set of aviation practitioners with different areas of expertise to evaluate or play out potential roles and interactions under that scenario, identifying how it could work and where it might be vulnerable.

Research Tools for Studying Future Systems. Since the system of interest does not yet exist, we have chosen to use conceptual walkthroughs of possible future ATM worlds to study the impact on human roles, decisions, coordination requirements and information needs. In preparing for these conceptual walkthroughs, we develop a scenario that instantiates various generic issues or challenges for air traffic management and posit a hypothetical ATM world, or parts of it, in terms of proposed technologies or procedures. Different experts representing different roles and perspectives within the ATM system then think through or play out their potential roles, interactions and information needs within this scenario. We do not provide detailed simulations of particular interfaces or systems; rather we ask the participants to describe in detail how such systems would have to function to provide support so that they could accomplish their tasks successfully.

Critical to this method is the use of concrete scenarios to anchor the participants in the details of the coordination, communication, decision making and information exchange requirements necessary to handle the situation successfully. In addition, we have found that having multiple participants representing different perspectives about the ATM system maximizes the information generated by such scenario-driven conceptual walkthroughs. Anchoring people in concrete situations quickly reveals ambiguities about what their roles are and about how they would carry out those roles. Having people with different perspectives explore how to deal with these roles and interactions provides insights about how different people and organizations can coordinate to achieve all of the parties' goals for a safe, efficient and economical ATM system.

We have constructed a number of scenarios and used them in several conceptual walkthroughs to demonstrate the value of this approach. The scenarios developed to date address several kinds of issues for the development of future ATM systems such as transitions in method or locus of control (e.g., the transition from free flight rules to controlled airspace) and factors limiting system capacity (e.g., crossing traffic) among others. However, this set represents only an initial exploration of the kinds of factors that should be considered.

Using These Tools to Evaluate System Designs. This initial scenario set can be used to support many different activities. We have used them as the basis for several different kinds of investigations to explore the potential impact of new ATM technologies on individual performance and on the coordination among multiple parties in the ATM system.

Scenario-driven knowledge elicitation using subject-matter experts. One method that we have applied is to use our scenarios to structure interviews with subject-matter experts, either singly or in a group. These subjects have been asked to comment on the likelihood of certain events occurring given alternative future system designs, to predict the effects of such events on their operations, and to discuss ways in which they would avert the occurrence of such events or compensate for them if they did occur. In applying this method, an effort is made to utilize subjects from different specialties, in order to elicit their different perspectives. Several scenarios have been explored (and refined) using this approach.

Conceptual walkthroughs using future "incident reports." Scenarios have also been used as the basis for "incident reports" in which a future incident is predicted, and is presented in the form of a formal report investigating that incident. These incident reports, with supporting documentation, are presented to participants (we have used groups of air traffic controllers, dispatchers and pilots) to consider as if they had actually occurred in some future system. The technique is used to structure a conceptual walkthrough by the participants, eliciting the ways in which the incidents might have been avoided, and how the system might be insulated against such occurrences or their effects. We have used this method to study how cooperative problem-solving in a hypothetical system can be facilitated, and how roles, responsibilities, procedures, policies and technologies must be designed to enhance performance and to make the system as error-tolerant as possible.

"Role-playing" conceptual walkthroughs. Another scenario was constructed to permit the observation of cooperative problem-solving more directly. Using this method, subjects are given the background and context of a scenario and the rules under which the system is operating. They are presented with the onset of an event and are then asked to "play out" the scenario as it occurs. The role playing is supported by a gaming board that represents the aircraft in particular ATC sectors. The participants can manipulate the gaming board to play out how the situation could evolve given different contingencies, actions, and interventions. Multiple participants debate among themselves different strategies for handling the situation. The methods by which they jointly resolve the problem are the data of interest. We believe this approach has considerable potential as a second method to conduct a conceptual walkthrough, eliciting additional insights concerning how various human and machine elements would interact in some future system.

CONCLUSIONS

We have applied all three of the methods described above, working with a total of 40 controllers, dispatchers, pilots and traffic managers. A sample of the resultant conclusions is presented below.

Changing Roles and Information Requirements. Changing roles by re-distributing authority (locus of control) has strong implications for the kinds of information and information displays needed to support these new roles. New ATM concepts change the roles of many of the people involved in the system. Under some current proposals under consideration, dispatchers will have more flexibility in route planning; flight crews will play a greater role in ensuring separation; and controllers will act more as monitors, making new kinds of decisions about when to intervene. If the changes created by these shifts in the locus of control, and in decisions concerning whether and when to intervene, are not accompanied by a corresponding shift in access to information, problems can arise.

Airlines and Flight Planning. One of the major problems with the current system is that the ATM system often has no access to information about the impact of its decisions on airline business concerns. As a result of this separation of authority and information, many of the decisions made by traffic managers and controllers are based solely on considerations regarding traffic flows and separation. Even when two solutions to a traffic problem are equally acceptable in terms of safety and traffic flow management, FAA staff generally do not have the information necessary to select a solution that is preferable to an airline in terms of its business concerns.

As a response to such problems, the FAA has been shifting the locus of control to the airlines where

83

possible. One example is with the expanded NRP, where (subject to certain constraints) the airlines are now allowed to file the routes that they prefer. The assumption is that, since the airlines have the information about their business needs, they are in a better position to make such decisions (with the ATM system then monitoring these flights to detect and deal with any potential safety or system capacity constraints). On the other hand, although the airlines have information about their own business priorities, they have only limited information about air traffic bottlenecks. As a result, they must make decisions based on inadequate information.

As several airline air traffic coordinators and dispatchers have commented in our studies:

- "Under the expanded NRP, it's like shooting ducks in the dark."
- "The problem with the expanded NRP is that there's no feedback to the AOCs. Nobody's getting smarter. Someone has to be responsible for identifying and communicating constraints and bottlenecks."
- "It used to be that weather was the biggest source of uncertainty for flight planning. Now it's the air traffic system."

Thus, whoever is given the authority to make strategic decisions about routing flights needs access to all of the pertinent information. The implication is that, when exploring future designs for the aviation system, one of the most important questions to be considered is how to effectively distribute and display the information needed to support decisions.

Flight Crews and Separation. This same general issue arises in a. tactical setting because of proposals to give flight crews more authority to change routes and altitudes while en-route. How do changing roles for flight crews and controllers affect who should have access to what types of information for tactical air traffic control decisions?

If, in a free flight environment, pilots are sometimes given responsibility for maintaining separation, then they and any available support software will have to play part of the role that controllers currently play (keeping in mind that, under current proposals, controllers will at a minimum still be monitoring the situation). They will have to detect potential confliction points in a timely fashion, generate solutions, and coordinate their actions with other aircraft.

Various studies of controllers indicate that there are a number of complex factors that they consider, including weather, the intentions of other aircraft, available contingencies, and positional uncertainty. In addition to considering the implications of these factors for their own flight, this new environment will require pilots to think about these factors as they impact surrounding aircraft. Interesting new patterns of communication will also be required, as there will no longer be a single authority approving route and altitude changes. Furthermore, pilots will have less knowledge than controllers currently do about typical traffic patterns in particular sectors (since pilots won't see the same sector day after day), potentially making the cognitive demands on flight crews even greater.

Thus, our research has raised provocative questions about the roles of flight crews, controllers and support software in such an en-route free flight environment. It is clear that careful consideration must be given about who should have access to what information, about how wide a "field of view" each participant should have, and about how this information should be displayed.

Coordination During Transitions in Level or Locus of Control. Another characteristic of many of the proposed future concepts for ATM is flexibility. Under these proposals, flight paths and plans should be adjusted dynamically to best meet the rapidly changing demands and circumstances of the air traffic environment. ATC should intervene only when circumstances demand; otherwise individual operators should be able to plan flights and manage flight paths as they see fit based on their perspective and goals.

This flexibility creates demands for coordination in several ways. For example, flight crews and ground controllers will have joint responsibility for positive separation. Locus of control will shift as circumstances demand. For example, traffic density in terminal areas will require transitions from free flight rules to greater control by ATC and its supporting computers. How will these transitions be made smoothly? One of the major challenges for a more flexible and less centralized traffic management environment is the need to be able to handle transitions in locus or level of control in order to cope with highly dynamic and unpredictable factors such as weather, emergencies and system failures. How will people recognize when circumstances create the need to shift locus of control? How will people communicate the change in management strategy and transition to the new form of control? What are the alternate levels of control? Are there intermediate levels between full ATC control and full en-route free flight? the alternative levels of control? Are there intermediate levels between full ATC control and full en-route free flight?

Several of the scenarios we developed include situations where the locus or level of control needed to change in order to accommodate an event or environmental condition, thus allowing us to explore issues in successful coordination and cooperative problem-solving. Such transitions require highly effective coordination and communication between different people in the system and different computer based support systems. We also included a variety of elements that complicated communication and coordination, using these elements as probes to trigger discussion among the participants in the meeting about how to avoid or cope with such problems by means of new procedures, technologies or protocols.

Paradigms for Distributed Control. There is a variety of paradigms for distributing control. It is useful to consider three distinct paradigms as anchor points for exploring possible strategies in ATM. An actual system is likely to be based on intermediate strategies or a mixture of these. The first is "control by directive", where an agent in the ATM system (a controller, for example) simply issues an instruction which is to be followed (unless there is some overriding concern that prevents this). The second is "control by permission", in which the ATM system initially specifies a solution, but will consider and sometimes give permission to requests for alternatives submitted by system users. A third is "control. by exception", in which system users are allowed to select and act on their own solutions or plans, which are then monitored by the ATM system for potential problems as these plans are enacted.

Within these paradigms there are a number of important variations that need to be carefully considered. For example, one extreme version of "control by exception" is to restrict interventions to localized, reactive responses by the ATM system. Under this variation, the ATM system would leave flights alone until and unless serious safety or system capacity problems were imminent. The assumption is that the users would generally select plans that avoided such situations so that interventions by the ATM system would be infrequent, functioning only as a backup safety net. Under another variation, the ATM system might play a more proactive role, dynamically setting certain constraints and communicating these constraints for the users to consider when they are comparing alternative plans. This could include examples like constraining the number of flights allowed to arrive at a particular airport when weather has impacted its operations, or curtailing en-route free flight in sectors with complex or high density traffic patterns. A third alternative would be for the ATM system to communicate a constraint by listing an explicit set of options for selection by the user. A fourth would be true collaborative control, where the users and the ATM system jointly assess the situation and consider alternatives.

Numerous human factors issues arise in considering these alternatives, including:

- How do we ensure adequate involvement of various participants so that they maintain situation awareness, including continuous awareness of who is in control?
- How can technology support the individual performances of controllers, dispatchers, pilots and traffic managers (assessing situations and monitoring for problems; generating and selecting from alternative solutions), as well as cooperative performance?
- How will workload be predicted and managed given greater variability in the behavior of the system?
- In situations where there are goal conflicts, how do we ensure that adequate "refereeing" occurs (especially if the conflict has safety implications)?

Thus, this methodology serves to raise important questions to guide design decisions.

Adaptive Information Management in Future Air Traffic Control

Jacqueline A. Duley and Raja Parasuraman
The Catholic University of America

AVIATION AUTOMATION

Over the last several decades the aircraft flightdeck has evolved in its use of automation. The earliest implementations of automation involved providing assistance to the pilots in the form of maintaining aircraft control. For example, the autopilot was developed because of the inherent instability of the aircraft during flight. With the introduction of electronic displays, automation provided assistance to pilots in navigating their aircraft with the addition of the plan view display, which incorporated much of their flight planning information into one integrated display. Fadden (1990) termed these two types of aviation automation, control and information automation, respectively.

There have been considerable benefits to aviation from the introduction of these and other forms of flight deck automation (Billings, 1997). Nevertheless, several incidents and accidents have demonstrated the problems that can arise when poorly-designed automation is implemented (Parasuraman & Riley, 1997). Some forms of automation which Wiener (1989) has termed "clumsy" have resulted in the pilot often being off-loaded during periods of low workload and being given additional tasks to perform in periods of high workload. As a result, the anticipated workload reducing benefits of automation have not been realized. Other documented problems of some forms of cockpit automation include mode errors, opaque interfaces, and automation complacency (for reviews, see Parasuraman & Riley, 1997; Sarter, 1996).

In contrast to the flight deck, automation of air traffic control has not yet reached a high level of complexity and authority (Hopkin, 1995; Wickens, et al., 1997). However, large projected increases in air traffic are driving several efforts to introduce advanced automation tools for controllers in the National Airspace System (NAS). Currently, air traffic controllers working en-route sectors use plan view displays to guide aircraft traversing the sectors en-route to the receiving Terminal Radar Approach Control (TRACON). In an en-route sector, the controller monitors the ongoing flights, responds to various pilot requests, and adjusts to weather conditions or other anomalies by instructing pilots to alter their aircraft airspeeds, flight levels, and often headings in order to maintain safe and efficient flow of traffic throughout the current as well as adjacent sector(s) (Wickens, et al., 1997). Under this air traffic control philosophy, the controller has complete authority and responsibility for maintaining the minimum allowable separation among aircraft in the assigned sector.

FUTURE AIR TRAFFIC MANAGEMENT

The volume of air traffic is projected to increase by up to 100% over the next decade. As a result, increasing capacity while maintaining or enhancing safety has become a priority for air traffic management. Responses to this increased demand such as constructing new airports, improving aircraft flow management, or expanding existing facilities, can only provide a partial solution because of their limited applicability or high cost. Two other responses have been proposed (Wickens, et al., 1998). The first response to such demands is to increase the use of automation in the air traffic control environment in order to aid controllers under high traffic load conditions. A recent report by the National Research Council Panel on Human Factors in ATC Automation recently noted the following as areas of potential controller vulnerability: monitoring for and detection of unexpected low-frequency events; expectancy-driven perceptual processing; extrapolation of complex four-dimensional trajectories; and use of working memory to either perform complex cognitive problem solving or to temporarily retain information (Wickens, et.al, 1997). This proposal still allows for the controller to have the authority for traffic separation and management, unlike the second proposal.

The second proposal, Free Flight (FF), represents an operating capability under instrument flight rules (IFR) in which pilots will be allowed to choose, in real time, optimum routes, speeds, and altitudes (RTCA, 1995). Air traffic restrictions will be imposed only as needed, not continuously, as they are now. FF can be thought of as having three components, Free Scheduling, Free Routing, and Free Maneuvering, with the latter aspects representing more advanced stages of FF. At all stages, however, it is envisaged that traditional ground-based control will be replaced by a cooperative arrangement whereby both air- and ground-based systems will share responsibility for the safe and expeditious transit of aircraft. Proponents claim that FF will provide the entire aviation user community the flexibility of visual flight rules while maintaining the safety net and protection of IFR. The goal is to provide each user maximum flexibility and increased access to the entire airspace system.

In either case, new technology will be implemented that increases the information available to the controller as well as potentially incorporate automation in the form of decision aids, task restructuring, and/or enhanced radar and navigational representations. The potential result of these implementations is that the en-route controller may become more of a supervisor than an active participant just as the pilot's direct control of the aircraft has decreased with the increase in automation complexity in the modem aircraft. To reduce the potential for such peripheralization of the controller, a human-centered ATM system with workstation tools to *assist* the human controllers in carrying out their goals for safe and efficient management of the airspace is required. A human-centered ATM system will allow the controller to maintain an up-to-date picture of the traffic situation, an understanding of the automation status and resolution processing, as well as permit the controller to adjust his/her workload when necessary.

User Requirements

With the implementation of new technology, the capabilities of the NAS will increase. Meanwhile the management of the resulting additional information will become more complex and air traffic controller tasks may become more demanding. In order for controllers to meet the demands of the NAS, they must have the appropriate tools and knowledge to execute their authority. Such tools can be designed only after the information requirements of controllers are well understood and analyzed. To this end, we recently conducted a survey of en-route (military and civilian) air traffic controllers in the US and UK regarding their information requirements (Duley, et al., 1997).

This study examined the information requirements of controllers for the following tasks: separating aircraft; assessing and resolving aircraft-airspace conflicts; determining projected routes; providing clearances to aircraft/pilots; assessing and resolving aircraft (non)conformance to controller instructions and air traffic regulations; accepting and initiating hand-offs to adjacent sector controllers. Controllers were asked to respond to each of 69 provided information parameters with respect to how often the information should be presented and its importance to the completion of the tasks. Frequency of information presentation was categorized into four types. Two of the levels are the binary form of either the information parameters *are* or *are not* needed: it is information that is always needed by the controller so it should be presented continuously or it is information that is never needed by the controller thus, it should be eliminated. The remaining two levels of presentation frequency suggest a necessity for a controller centered system where the human and the automation are coupled such that the advanced ATM system would neither be fully automated nor fully manual: information parameters should be available at the controller's discretion either through a verbal request of the information from the pilots or a request of the system, and the information should be presented to the controller at computer/system discretion.

The results revealed that the en-route controllers preferred to have the information presented either at all times or to have it available at their discretion. However for a few parameters within individual controller tasks, the controllers preferred that the information be provided at the discretion of the system; for example they preferred that the "projected sector load" be provided at system discretion when they are separating aircraft and when they are assessing/resolving conflicts. However when performing the task of assessing/resolving conflicts, the controllers suggested that some of the information parameters that they "always need" when separating aircraft are *not needed* at all in this case.

The implications of such user preferences are that en-route controllers require a large amount of information to do their job. Further complicating this issue is that not all of the information is needed at all times for all controller tasks. A certain task may require one type of information while another task may not require it. In addition, these tasks are not performed sequentially but concurrently. To conform to these user requirements, therefore, the interface between the processed flight and radar data and the controller should be able to provide the controller all of the necessary *relevant* information for a task at the appropriate time, in a format that can be easily perceived. Such an interface has been recommended by Smith, et al. (1997) for the airline dispatcher environment where there is also a "large data space". Smith, et al. found in their evaluations of proposed flight planning tools that dispatchers often failed to view valuable information at the appropriate times. For example, 3 of the 20 dispatchers failed to view a display separate from the route planning display regarding the turbulence in an area directly affecting the proposed route.

These considerations suggest the use of information management tools for the controller in the future NAS. Information management is likely to be necessary to allow the controller to operate in a highly saturated airspace in a safe and efficient manner. Good information management has the potential to bridge the "information transfer gap" among (1) the information *available* to the controller, (2) the information *needs* of the controller, and (3) the information presentation *capability* of the system (Miller, 1998). However in order to fully assist the controller, the information manager must be able to present the information in a manner that is adaptive to the needs of the controller. As stated by a controller with Eurocontrol regarding the role of the task display, "tell me what I don't know, but don't clutter the screen with data" (Nordwall, 1998).

Adaptive Information Management

Information management systems can be static, or pre-determined by the designer, or adaptive, or flexible to varying contextual factors and user requirements. There may be several advantages to an adaptive system. In adaptive systems, information displays vary with the current information needs of the human operator. To accomplish this, displays must change the appearance of their information presentation according to context. This process is moderated by a knowledge-based system such as an information manager (Rouse, et al., 1988; Miller & Funk, 1997) in cooperation with the operator. The information manager incorporates information automation as well as adaptive aiding in the form of decision aids for the operator. These decision aids may then act as control automation, if they are given high-level authority.

In order to achieve a level of *cooperation* between the air traffic controller and the system, the algorithm chosen to manage information presentation must be adaptive as well as adaptable (Billings, 1997). In other words the system must be able to adapt to the information requirements of the controller but also the system must be adaptable by the controller according to his/her preferences. To be *adaptive* the system must be context-sensitive for example, the system may change the information presentation when a controller recognizes a potential conflict between a pair of aircraft and the controller must utilize information that pertains to flight planning as well as to projected aircraft positions; information which the controller may not have needed prior to detecting the conflict. It is such a case that the information manager may implement an adaptive aiding tool such as a conflict detection and resolution aid.

The system becomes *adaptable* in this example when the controller either delegates to the automated system to take action in resolving the conflict or the controller may decide to view the automated system's recommended resolution and then consent to that option or select an alternative resolution. Permitting the controller to choose the preferred level of automation (management by delegation or by consent; Billings, 1997) and therefore the preferred personal level of involvement in the air traffic control task creates a cooperative human-machine system where the controller remains in the center of the system and is able to maintain a comprehensive level of awareness of the air traffic. The system may also be adaptable in the sense that the controller may communicate to the automation his/her information presentation, action, and decision-aid preferences when initiating control of the sector at the beginning of a shift just as in the Lockheed Pilot's Associate (Banks & Lizza, 1991).

A recent study by Hilburn, et al. (1997) demonstrated the potential for successful implementation of such an adaptive information automation system. They tested air traffic controllers in a simulation of Amsterdam airspace using the Center TRACON Automation System (CTAS), an automated tool aimed at assisting air traffic controllers with managing and controlling arriving traffic into terminal airspace. CTAS is composed of three tools: the Traffic Management Advisor, the Descent Advisor (DA), and the Final Approach Spacing Tool. Hilburn et.al. used only the DA in this study to provide the controllers with estimated time of arrival for aircraft and a traffic plan to sequence and schedule the traffic into the terminal airspace. The automation detected arrival time conflicts and projected separation conflicts, and offered advisories aimed at solving these conflicts. Their results showed that when the automated system was operated on an adaptive schedule such that it was activated only under a high task load period, the workload of the controller as measured subjectively and physiologically decreased.

To further substantiate the need for an adaptive methodology, a study by Galster, et.al. (1998) demonstrated that if the ATM system evolves to a FF architecture and does not include adaptive information management as well as adaptive aiding tools, the controllers' workload levels will be impacted. The results of this study of simulated high traffic load revealed controller operational errors in detecting probable conflicts between pairs of aircraft as well as an increase in perceived workload.

SUMMARY

The future of ATM includes the addition of automation for that of the air traffic controller. Assistance in the form of information automation as well as automated decision aid tools will be necessary for alleviating the extreme levels of workload predicted to occur for a controller working in a saturated traffic environment. In implementing this automation, designers must first consider the information needs of the controller and a system whereby relevant information is given to the controller at the appropriate time. An information manager that is both adaptive and adaptable is a possible solution.

It is also important to consider that novice and expert controllers are individual users of the ATM system and therefore it is necessary to design an interface for both types of users (Hancock & Chignell, 1988). In considering both types of air traffic controllers, the interface will be able to ease the transition and training of controllers from the current system of air traffic control to the future system of air traffic management.

ACKNOWLEDGMENTS

This research is supported by National Aeronautics and Space Administration Grant No. NAG-2-1096 from NASA Ames Research Center (Kevin Corker is the program manager), and Grant No. NGT-1-52137 from NASA Langley Research Center (Paul Schutte is the technical monitor). The views presented here are those of the authors and are not necessarily representative of NASA.

REFERENCES

Banks, S. & Lizza, C. (1991). Pilot's associate: A cooperative knowledge-based system application. *IEEE*

Expert, 6 (3), 18-29.

Billings, C.E. (1997). *Aviation automation: The search for a human-centered approach.* Mahwah, NJ: LEA.

Duley, J.A., Galster, S.M., Masalonis, A.J., Hilburn, B., & Parasuraman, R. (1997, October). *Enroute controller information requirements from current ATM to free flight.* Paper presented at 10th CEAS Conference on Free Flight, Amsterdam, The Netherlands.

Galster, S.M., Duley, J.A., Masalonis, A.J., & Parasuraman, R. (1998, March). *Effects of aircraft selfseparation on conflict detection and workload in Free Flight.* Paper presented at the Third Automation Technology & Human Performance Conference, Norfolk, VA.

Fadden, D.M. (1990). Aircraft automation challenges. In *Abstracts of AJAA -NASA -FAA -HFS Symposium, Challenges in Aviation Human Factors: The National Plan.* Washington, DC: American Institute of Aeronautics and Astronautics.

Hancock, P.A. & Chignell, M.H. (1988). Mental workload dynamics in adaptive interface design. *IEEE Transactions on Systems, Man, and Cybernetics,* 18(4), 647-658.

Hilburn, B., Jorna, P.G., Byrne, E.A., & Parasuraman, R. (1997). The effect of adaptive air traffic control decision aiding on controller mental workload. In M. Mouloua & J.M. Koonce (Eds.), *Human automation interaction: Research and practice (pp.* 84-91). Mahwah, NJ: LEA.

Hopkin, V.D. (1995). *Human factors in air traffic control.* London: Taylor & Francis.

Miller, C.A. (1998, January). *Bridging the information transfer gap: Measuring the goodness of information fit.* Paper presented at Intelligent User Interfaces, San Francisco, CA.

Miller, C.A. & Funk, H.B. (1997). Knowledge requirements for information management: A rotorcraft pilot's associate example. In M. Mouloua & J.M. Koonce (Eds.), *Human automation interaction: Research and practice (pp.* 186-192). Mahwah, NJ: LEA.

Nordwall, B.D. (1998, February 2). Europe makes progress on ATC harmonization. *Aviation Week & Space Technology,* 148, 60-63.

Parasuraman, R. & Riley, V. (1997). Humans and automation: Use, misuse, abuse. *Human Factors,* 39 (2), 230-253.

Rouse, W.B., Geddes, N.D., & Curry, R.E. (1988). An architecture for intelligent interfaces: Outline of an approach for supporting operators of complex systems. *Human-Computer Interaction,* 3, 87-122.

RTCA. (1995). *Report of the RTCA board of director's select committee on free flight.* Washington DC: Author.

Sarter, N.B. (1996). Cockpit automation: From quantity and quality, from individual pilot to multiple agents. In R. Parasuraman & M. Mouloua (Eds.), *Automation and human performance: Theory and applications (pp.* 267-280). Mahwah, NJ: LEA.

Smith, P.J., McCoy, C.E., & Layton, C. (1997). Brittleness in the design of cooperative problem solving systems: The effects on user performance. *IEEE Transactions on Systems, Man, and Cybernetics,* 27(3), 360-371.

Wickens, C.D., Mavor, A.S., & McGee, J.P. (1997). *Flight to the future: Human factors in air traffic control.* Washington, DC: National Academy Press.

Wickens, C.D., Mavor, A.S., Parasuraman, R., & McGee, J.P. (1998). *The future of air traffic control: Human operators and automation.* Washington, DC: National Academy Press.

Wiener, E.L. (1989). *Human factors of advanced technology ('glass cockpit) transport aircraft* (Tech. Rep. 117528) Moffett Field, CA: NASA Ames Research Center.

Free Flight in Air Traffic Control: Human Factors Implications

Donald O. Weitzman
TRW Information Services Division

INTRODUCTION

Free flight (FF) is a flight operating capability under instrument flight rules (IFR) in which aircraft will have the freedom to select their route and speed in real time. FF will allow pilots to select optimum routes and altitudes and control their own schedules through the application of communications, navigation, and surveillance technologies and the establishment of air traffic management procedures that maximize flexibility while guaranteeing safe separation.

The major difference between today's direct route (or off-airways) clearance and FF will be the pilot's ability to operate the flight without specific route, speed, or altitude clearances. FF would apply en-route unless air traffic becomes so congested that FAA restrictions must be imposed. For example, when aircraft approach congested airspace, where airport capacity is limited, when unauthorized entry into restricted airspace is imminent, when potential maneuvers may interfere with other aircraft, and when safety appears to be compromised, air traffic control (ATC) will again be imposed.

Currently, air traffic control is based on a highly centralized, ground-based, human intensive system. This system is graphically illustrated in Figure 1. To carry out the control task the controller has a variety of information presented to him/her, the most important parts being the flight plans for the aircraft under his/her control and the present traffic situation as shown by radar. On the basis of the present position and altitude as seen by radar, the controller will mentally predict where the aircraft will be in the future and whether two or more aircraft will violate the predetermined minimum safe separation standard (5 nautical miles).

FF calls for a more highly automated, technology-intensive system. In FF, aircraft will operate in an environment that minimizes ground-based control while making full use of cockpit technologies, as shown in Figure 2. The Global Positioning System (GPS) provides precise position, velocity and time estimates for all GPS equipped aircraft while Automated Dependence Surveillance (ADS) automatically broadcasts position, velocity and time data from one aircraft to another and to ground-based controllers using digital data communication (data link). Furthermore, the cockpit display of traffic information (CDTI) makes viable the division of responsibility between ground and air for maintaining safe separation among aircraft. Accordingly, only position and short-term intent information will be provided to the ATC controller who will perform monitoring functions and serve as an emergency backup in the event of technical system failure. This short-term intent information will also be provided to nearby affected aircraft. The ATC controller will only intervene to resolve conflicts that he/she is able to predict; that is, short-term restrictions will be imposed only when two or more aircraft are in contention for the same airspace. In normal situations, aircraft maneuvering will be unrestricted. Therefore, a greater emphasis will be placed on recovery from incidents than on their prediction. FF will be reactive rather than proactive. This design philosophy would seem to eliminate the human component on the ground from active, ongoing participatory roles in the system. Similar design philosophies have been studied and have received strong criticism from the human factors community (Perrow, 1984; Endsley & Kris, 1995; Billings, 1996; Weitzman, 1997).

Supporting Technologies

Questions regarding FF are generally directed toward the roles that technology will play rather than towards human factors concerns. Because of the substantial advances in the underlying technologies in detection and resolution devices (CDTI) and automation systems for flow management and separation assurance, many decisions for FF are being heavily driven by the underlying technologies. The emphasis today is on the technologies rather than how to deploy the power available through these technologies to assist human performance, i.e., the problem of how to design a system that aids the controller (and pilot) in the process of solving problems.

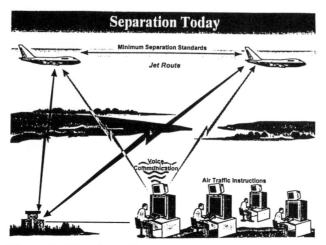

Figure 1. The air traffic control system today.

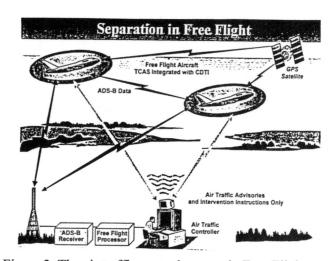

Figure 2. The air traffic control system in Free Flight.

Because automation systems are not very good problem solvers when situations depart from those the designer expected, automation is inherently "brittle." While there are many ways to plan for and prevent unexpected situations from arising, it is still difficult to build mechanisms into automation that can cope with unexpected, novel, special or under-specified conditions or recover from errors or bugs in its plans. For both pragmatic and theoretical reasons it is not likely that the automation will overcome these limitations anytime soon. And it is these factors -- unexpected situations or surprises, special conditions or contexts, bugs in the software -- which are ubiquitous in actual serious incidents in difficult technical problem solving environments.

While some environments are more prone to these factors than others and the quality of the technology affects the frequency with which they arise, the potential for these factors to arise would seem to be high in FF. It is in the nature of complex, tightly coupled, highly dynamic and interactive environments, where the degree of uncertainty is high, and where there are risky consequences to potential outcomes that nasty surprises most often occur. In FF, all of these factors would seem to be high suggesting that the resulting problem solving environment could make the human highly vulnerable to error. Experience with complex human-technical systems indicate that performance failures occur when designers unintentionally create cognitive environments that are difficult to manage due to the demands of the real world itself.

FREE FLIGHT: A JOINT HUMAN AND TECHNICAL SYSTEM

The quality of human performance will remain the key contributor to maintaining safety in FF, particularly when dealing with unexpected events. Yet when performance deficiencies are noted, they are usually attributed to the human element alone. The label "human error" is often used as a residual category and implies responsibility and blame while focusing changes on more training, better motivation, improved procedures, or different people. From a pure technology point of view, the obvious solution is more automation. While this approach is likely to achieve some success, it is fundamentally limited by two simple facts. One is that all systems must ultimately rely on humans. Increased automation simply shifts human responsibilities from one control mode to another. The other fact is that perfectly competent, reliable automation is not immediately possible. The key questions for FF are how far are we from this realization and what the consequences are from departures from the ideal? An interesting (but largely unavailable) statistic bearing on the second of these questions is the number of system failures that were avoided because humans reacted skillfully and rapidly to unexpected events.

One way to move outside the purely technical world of FF is to consider the human-technical system as a problem solving system (Roth, Bennett, & Woods, 1988). From a problem solving perspective, controller and pilot performance and safety in FF will depend upon the knowledge available to them, on how that knowledge is activated and utilized and on the cognitive demands placed on controllers and pilots under actual conditions. In such a dynamic environment, the effects that can be achieved by practice, training, and pre-planned procedures or routines will be limited. The most significant aid to the problem solving and decision making process will probably come from improvements or innovations that reduce these cognitive demands.

Computer-Based Performance Aiding in Free Flight

FF technology will depend more and more on human decision making skills and will be vulnerable to error in that decision making. As an example, consider the application of computer technology for aiding air traffic control decision making such as an automated decision support system that outputs some form of problem resolution. Because the controller will still have responsibility for the outcome, he/she will have the authority to override machine output, in principle. However, if the controller is to override machine solutions, he/she must be able to comprehend the air traffic situation and have an adequate understanding of the automated decision making process. If not, then the human operator is likely to be placed in a very stressful situation where the only sensible strategy is to "opt-out" rather than risk overriding the automation.

The question is how good will controllers be at filtering correct from incorrect machine solutions when the dynamic control of traffic in FF will be moving further and further away from the controller? It must be recognized that maintaining and improving performance by doing applies to cognitive as well as to perceptual-motor skills.

Take the critical issue of the controller's role in the automated system for FF.; human-in-the loop or human-out-of-the-loop architecture. The usual course of increases in automation are to move the human's role further from direct contact with the managed process. For FF this means that accompanying the change in the level of automation there is a change in the role of the controller - he or she is moved further "out-of-the-loop." Experience with the effects of human-out-of-the-loop automation suggests this is a poor architecture for human-technical control or management of complex processes. Our experience suggests the need for new architectures where the level of automation is high but where the human plays a continuously active role in the control and management of the process where the controller is more "in-the-loop."

A WAY FORWARD

A productive approach to FF is to focus on factors that underlie effective joint human-technical performance. This approach leads to an emphasis on person-computer performance, on cognitive demand/resource matching, on distributed management of decision making, on the ability of the human to detect and correct both human and machine errors and to handle unexpected or novel situations. Accordingly, what factors are important to support human performance in FF (e.g., decision/memory aiding, error detection and recovery aids, data integration), what forms they can take (e.g., automation, computer-based displays, procedures, training, staffing), how to measure the potential for error and predict their frequency, and which tasks are most vulnerable to error are essential topics for FF design. After all, in the FF environment it is the performance of the operational personnel that should be the essential guide for how to find and use the potential available in technology.

Research on human performance in risky systems emphasizes that skill is the capability to adapt behavior to changing circumstances in pursuit of a goal. This means that the effectiveness of technology to support human performance in FF (where changing circumstances will be the norm rather than the exception) is related to its ability to increase the human problem solver's adaptability to the problems that will arise in FF. There are several steps that can be taken to convert the power of technology into an effective joint human-technical problem solving system for FF:

a. One is to build displays that improve the ability to share knowledge between the controller, the pilot, and the computer and thus enhance the interaction between them. Similarly, information that makes explicit the automation's understanding of the problem state, its intentions, the hypotheses that have been rejected or accepted and the hypotheses that are still to be pursued can help avoid following unproductive problem solving paths (i.e., the phenomena of being led down the garden path).

b. A second step is to help the controller (and pilot) recognize if the system is behaving as expected (e.g., are the relevant goals being met? has an error occurred?). In complex domains like ATC, error detection and recovery are inherently distributed over multiple people and groups. By distributing management of the ATC system among human and machine subsystems, system capabilities can expand through a balance of human and computer participation. Enhanced communication, interaction, and feedback between the computer and human operator are the essential requirements to effect true distributed management.

c. A third step - particularly in the face of novel or unexpected contingencies - is to provide more capabilities to aid the controller (and pilot) in the process of solving problems by offering not fixed solutions, but informative counsel such as critiquing techniques, performing information processing tasks (such as massaging the data into a form to more directly answer questions the human problem solver must face) and by using graphic techniques to enhance the controller's understanding of the problem (visualization will aid ATC comprehension). But for such computational aiding to work one must first determine the requirements and potential bottlenecks in cognitive task performance and then build aids needed to support those tasks.

d. A fourth step is to permit the controller (and pilot) to guide the decision support system's reasoning. This includes mechanisms for the controller (and pilot) to add or change information the automation is using about the state of the problem. More ambitious use of computational power can provide the controller

94

with capabilities to explicitly manipulate the attention of the automated system. For example, one needs to provide the controller (and pilot) the ability to direct the diagnostic path pursued by the automation. This will permit the system to capitalize on the insights of the human in guiding problem solving rather than simply having them follow machine directives. One of the big problems in providing automated decision aiding is that too often the automation dominates the human. Experience has shown that breakdowns in performance often occur whenever attempts are made to automate the problem solving process and assign the human the passive role of following instructions.

e. Another step is to permit the human and the automation to act as independent decision makers with tasks allocated to both, thus keeping the human in the loop at all times This means that the automation will act as a backup decision maker that is switched on/off whenever the human is predicted to be unable/able to service all of the required tasks in the system (i.e., adaptive task allocation). It means that the automation will aid but not replace many of the controller's (and pilot's) tasks unless it can perform the tasks completely and reliably, as well as all other tasks that depend on it. Specifically, it means that computer aiding will permit the controller (and pilot) to dynamically vary the allocation of tasks between him/herself and the computer; that is, adaptive task allocation will step in when needed and provide assistance in a form appropriate to the situation. Thus, even as the automated assistance becomes smarter and more numerous, the controller (and pilot) will never be denied the opportunity for active performance of any ATC or flight control function.

f. A logical extension of this work is to extend the work of adaptive computer aiding so that it can be integrated with adaptive displays and expert systems. This will necessitate the development of intelligent systems that can adapt its attentional strategies to dynamic information requirements and resource limitations, adapt its planning strategies to the uncertainty about its environment and the constraints placed on its behavior, and adapt its choices of problem solving strategies and hypotheses to currently available information, to dynamic tactical and strategic objectives, and to challenging competing demands and opportunities for action. Realization of such a vision is highly dependent on new designs where computers can learn from experience, where intelligent programs can "reason" nonmonotonically and where they have commonsense knowledge and show commonsense reasoning. Coming to grips with these capabilities should enable designers to overcome the limitations in current expert systems and is likely to produce new successes for ATC (as well as aviation) automation.

REFERENCES

Billings, C.E. (1996). *Aviation automation: The search for a human-centered approach.* Mahwah, NJ: Erlbaum.

Endsley, M.R. & Kris, E.O. (1995). The out-of-the-loop performance problem and level of control in automation. *Human Factors, 3 7.,* 3 81-3 94.

Perrow, C. (1984). *Normal accidents: Living with high risk technologies.* New York: Basic Books.

Roth, E.M., Bennett, K.B., & Woods, D.D. (1988). Human interaction with an "intelligent" machine. In E. Hollnagel, G. Mancini, & D.D. Woods (Eds.), *Cognitive engineering in complex dynamic worlds (pp. 23-69).* New York: Academic Press.

Weitzman, D.O. (1997). Air traffic control automation: Lessons from Chernobyl and Three Mile Island. *Journal of Air Traffic Control,* January-March, 33-35.

Effects of Aircraft Self-Separation on Controller Conflict Detection Performance and Workload in Mature Free Flight

Scott M. Galster, Jacqueline A. Duley, Anthony J. Masalonis, and Raja Parasuraman
The Catholic University of America, Cognitive Science Lab

INTRODUCTION

The rapid growth in worldwide air travel projected for the next decade will dramatically increase demand for air traffic services. As a result, several new strategies for more efficient air traffic management (ATM) have been proposed (Perry, 1997; Wickens, Mavor, Parasuraman, & McGee, 1998), among them Free Flight (FF). The goal of FF is to allow aircraft under instrument flight rules the ability to choose, in real time, optimum routes, speeds, and altitudes, in a manner similar to the flexibility now given only to aircraft operating under visual flight rules. (RTCA, 1995). To achieve this goal, traditional strategic-based separation (flight-path based) will be replaced by tactical separation based on flight position and speed. The responsibility for separation will gradually shift from the current ground-based system to a more cooperative mix between the air and ground. FF can be viewed as a series of stages on a continuum, along which new technologies and procedures will be added as they prove useful and effective in the attainment of the mature FF goal.

While relinquishing active responsibility for separation, the air traffic controller (ATCo) will become more of a monitor as FF moves towards a mature level. A recent report by a National Research Council (NRC) panel on future ATM raised some concerns regarding this development (Wickens et al., 1998). One concern is how effective a controller can be in a purely monitoring role (Parasuraman & Riley, 1997). Human monitoring of automated systems can be poor, especially if the operator has little active control over the automated process (Parasuraman, Molloy, & Singh, 1993). Because of such concerns, the NRC panel recommended that authority for separation remain firmly on the ground, with a greater emphasis being placed on providing automation support tools for ATCos (Wickens et al., 1998). Nevertheless, the panel recommended that extensive human-in-the-loop simulations be conducted to evaluate different FF concepts for their impact on safety and efficiency. Moreover, no matter what level of FF is eventually implemented, ATCos will remain ultimately responsible for maintenance of safety (ensure separation). As a result, there is an urgent need to investigate the effects of different levels of FF on ATCo workload and performance.

Endsley, Mogford and Stein (1997) recently conducted such a study with full performance level (FPL) en-route civilian controllers under four different procedural conditions: current procedures (baseline), direct routing, direct routing allowing pilot deviations with conveyed intent, and direct routing allowing pilot deviations without conveyed intent. ATCos reported significantly higher subjective workload when they were not advised of pilot intentions in advance of a deviation from the flight plan. They also showed a trend towards committing more operational errors in the direct routing without shared intent condition compared to the baseline condition. This trend suggested that controllers found it difficult to maintain separation standards when they did not have a clear picture of the pilot's intentions, as might be the case in mature FF.

Hilburn, Bakker, Pekela, and Parasuraman (1997) also examined the impact of FF on the performance of U.K. military controllers. Conventional control in structured airspace was compared to FF conditions in which aircraft did or did not share intent information with controllers. These conditions were crossed with low and high traffic levels. ATCos reported higher subjective mental workload for high traffic loads; however, there were no overall differences between the control conditions. ATCos also reported higher mental workload in high traffic scenarios under conventional control conditions than they did under uninformed FF. Physiological measures of workload, including blink rates and pupil diameter, revealed a trend towards lower workload in the FF conditions.

The discrepancy between these two studies may be attributed to several factors, including the different populations tested, U.S. civilian ATCos by Endsley et al. (1997), and U.K. military controllers by Hilburn et al. (1997). Differences between these groups have recently been noted (Hilburn & Parasuraman, 1997; Duley et al., 1997). For example, U.K. military pilots typically do not file flight plans. Hence U.K. controllers are accustomed to working without flight progress strips, which may acclimate them to a certain level of uncertainty regarding flight paths, a by-product of increased levels of FF.

Neither of these previous studies examined the end stage along the FF continuum, or mature FF. The introduction of technologies such as the Global Positioning System (GPS) and Automatic Dependent Surveillance-Broadcast (ADS-B), as well as procedural changes such as airborne self-separation, if successful, will result in an airspace that will be considerably more dense and complex than it is today. Will ATCos still be able to monitor the saturated airspace effectively when they are no longer actively involved in separation? The present investigation addressed this aspect of mature FF in a simulation study with civilian en-route controllers. The characteristics of the simulated airspace included a high traffic load, aircraft initiating and executing evasive maneuvers to avoid conflicts without controller guidance, and separation violations. The controller operated primarily as a monitor of the airspace, as would be the case under mature FF.

METHODS

Participants

Ten currently active FPL en-route ATCos served as paid participants, most from the Washington Air Route Traffic Control Center (ZDC). Their ages ranged from 32 to 42 (mean = 35.6 years) and their experience levels ranged from 8 to 15 (mean = 12.1 years).

ATC Simulation

A medium-fidelity PC-based ATC simulator was used (Masalonis et al., 1997). The simulator included a primary visual display (PVD) of traffic including data blocks, waypoints, and jetways. An adjacent monitor contained a data link interface and an area devoted to electronic flight progress strips. A trackball allowed the ATCo to traverse between the two screens. The ATC simulator had the capability of presenting traffic scenarios in any airspace in North America. However, in the present study a "generic" airspace was used to avoid any possible interactions between degree of airspace familiarity and the experimental conditions.

Four 30-minute scenarios were created to combine moderate (11 aircraft) and heavy traffic loads (17 aircraft) in a 50 mile radius sector with the presence or absence of self-separating and conflicting aircraft. Conflicts (defined as a loss of separation of 5 nm laterally and 1000 ft. vertically) were indicated to the controller by the appearance of a red circle around each of the aircraft involved. A self-separation event was defined as an evasive maneuver (either altitude or speed change) made by one aircraft to avoid a potential conflict with another aircraft. There were two conflicts and four self-separating events in each of the conflict-present scenarios.

Controllers were required to accept all incoming aircraft into the sector, monitor each aircraft's progress over the waypoints listed in its flight plan, and handoff each aircraft. In the conflict-present scenarios, they were also required to indicate verbally any potential conflicts that they judged would occur by stating which two aircraft were involved in the potential conflict and approximately over which waypoint the conflict would occur.

Procedure

After a demonstration and practice trial each controller completed three 30-minute scenarios and was offered a break between successive scenarios after they filled out the NASA-TLX subjective mental workload questionnaire. All 10 controllers completed the moderate and heavy traffic load scenarios without conflicts and self-separating events first. Five controllers then completed the moderate traffic load scenario containing conflicts and self-separating events while the other five completed a similar scenario under high traffic load.

Dependent Measures

Accepting and handing off aircraft were considered primary tasks. Tracking and monitoring aircraft by updating the electronic flight strips provided a realistic secondary-task measure of controller workload. For each aircraft and for all required ATCo tasks, response times (RTs) for correctly executed actions and frequencies of missed actions were tabulated. RT for acceptance of an aircraft consisted of the time it took the controller to read the message from the aircraft on the datalink interface once the message was displayed until the time the controller accepted the aircraft. The monitoring task RT consisted of the time after an aircraft passed a waypoint to when the controller clicked the corresponding waypoint button on that aircraft's flight progress strip. A missed action was recorded if the controller never pressed the corresponding waypoint in the flight strip after the aircraft passed the waypoint. A miss was not recorded if the controller completed a handoff, either correctly or incorrectly, prior to the aircraft reaching the remaining waypoint(s) on its flight plan within the sector. Handoff RTs were calculated as the time it took the controller to complete the handoff procedure once the aircraft had entered the handoff zone shown on the PVD.

In the conflict scenarios an additional measure termed the advanced notification time was also computed. In the case of a conflict, the advanced notification time was defined as the time the separation violation occurred minus the time the ATCo gave a verbal indication of the conflict. Accordingly, the greater the duration, the more in advance the controller recognized that a conflict was about to occur, and the better the controller's performance. If the controller did not detect the conflict prior to its occurrence a miss was registered. The advanced notification time for a self-separation event consisted of the reported time prior to the initiation of the evasive maneuver that the controller reported the potential conflict. A miss in this case was recorded when the controller did not recognize the potential conflict prior to the aircraft initiating the evasive maneuver.

RESULTS

Subjective and Secondary-Task Measures of Mental Workload

When conflicts were not present, controllers reported significantly ($p<.001$) higher subjective workload (mean NASA-TLX score) under the high traffic condition (71.4) compared to the moderate traffic condition (49.3). A similar traffic load effect ($p<.05$) was found for the conflict-present scenarios (high traffic=71.3, moderate traffic=49.5). However, neither the main effect of conflict presence/absence nor the interaction between traffic load and conflict presence/absence were significant. Thus, subjective workload was affected by traffic load in the expected direction, but was relatively unaffected by the addition of conflicts.

The secondary-task measure of ATCo workload (monitoring flight progress) gave similar results. When conflicts were not present, controllers omitted checking off significantly ($p<.01$) more waypoints in the high (31.9%) compared to the moderate traffic condition (4.7%). Also, under moderate traffic, controllers missed significantly ($p<.05$) more events with conflicts present (18.2%) than with them absent (4.7%).

These results suggest that a possible strategy employed by controllers was to shed the secondary monitoring task as traffic increased and also when conflicts were present under moderate traffic. However, under high traffic, the proportion of missed events with conflicts present (33.7%) did not differ significantly from that with conflicts absent (31.9%). Thus the majority of task shedding was due to increased traffic rather than conflict presence.

Aircraft Acceptances and Handoffs

There were no significant effects of traffic load or conflict presence on median RTs for accepting aircraft into the sector. For the handoff task, when conflicts were absent, RTs were significantly ($p<.05$) longer under high (16.8 s) than under moderate (8.1 s) traffic load, but the corresponding RTs in the conflict-present scenarios were not significantly different.

Conflicts and Self-separations

In the moderate traffic condition 10% of conflicts were not reported prior to loss of separation. In the high traffic condition the miss rate climbed to 50%, as indicated in Figure 1. The conflicts that *were* recognized prior to their occurrence had mean advanced notification times of 196 s in the moderate traffic condition and 145 s under high traffic. These results provide a clear indication that under a high traffic load, ATCos were slower to recognize and report impending conflicts and committed more operational errors than they did under a moderate traffic load.

ATCos also had great difficulty in recognizing self-separation maneuvers initiated by aircraft, also shown in Figure 1. Controllers missed 70% of the potential conflicts (that were resolved by self separation) in the moderate traffic condition and 85% under high traffic. For self-separation events that *were* recognized, the mean advance notification time provided by ATCos was markedly reduced under high traffic (157 s) compared to the moderate traffic (337 s) condition.

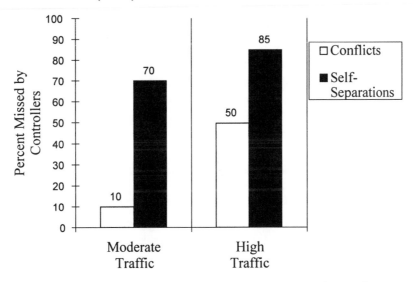

Figure 1. Percentage of missed conflict and self-separation events under moderate and high traffic.

DISCUSSION

Mature FF will require controllers to be monitors of aircraft that will be free to maneuver to minimize fuel burn and to avoid conflicts with other aircraft. Additionally, the future airspace is likely to be much more dense and complex than it is today. The present results show that under such conditions of saturated airspace, ATCos had difficulty both in detecting conflicts and in recognizing self-separating events in a timely manner. Controller workload also increased, as indexed both by subjective and secondary-task measures.

These findings extend those of previous studies examining controller performance under FF (Endsley et al., 1997; Hilburn et al., 1997) by showing the impact of *mature* FF on controller workload and conflict-detection performance. Controllers in the present simulation study were required to monitor under very high traffic loads, did not have intent information, and were not provided any automation support tools. The use of these conditions was deliberate: system safety may be best evaluated under such worst-case scenarios. At the same time, although the results support previously expressed concerns about the ability of controllers to monitor airspace under mature FF, they also provide a benchmark against which performance improvements can be measured through the use of automation tools.

One example of such a tool is the conflict probe which could be used to improve potential conflict detection and allow the controller more time to consider available options. The present results, if replicated and extended to other airspace and traffic scenarios, could be used to set performance parameters for these and other automation support tools. For example, the advance notification time of 157 s we observed for self-separation maneuvers that were detected in the high traffic condition suggests a minimum notification time that a ground-based automated conflict probe should provide under mature FF. Of course, development of automated tools will also need to consider conflict *resolution* (in addition to conflict *detection*) times, as well as the integration of detection and resolution decisions between the air and the ground. Corker, Pisanich and Bunzo (1997) recently used a computational human performance model, Man Machine Integrated Design and Analysis System (MIDAS), to predict flight crew times to initiate evasive maneuvers to maintain separation under different conflict scenarios. For a scenario requiring ATC intervention (because one aircraft was not equipped for conflict detection), they obtained a mean time of 134 s for resolution of a 90 degree encounter geometry, but indicated that longer times might occur for shallower encounter angles. Human performance data such as these, as well as those of the present study, can be used to set design parameters for both air-based and ground-based conflict detection tools.

Higher levels of automation have been proposed as the means by which higher ATM capacity can be achieved without compromising safety. Recently, the NRC suggested that high levels of automation are best suited to tasks relating to information acquisition, integration and presentation to facilitate controller decision making (Wickens et al., 1998). Proposed automated tools should be outlined with clear performance improvement goals and should be modeled and measured to ensure that those goals are attained within the decision making process of the ATCo. The present results provide a baseline assessment of mature FF against which the benefits of such ground-based automation support tools for the ATCo can be evaluated.

ACKNOWLEDGMENTS

This work was supported by Grant No. NAG-2-1096 from NASA Ames Research Center, Moffett Field, California, USA. Kevin Corker was the technical monitor. The views presented here are those of the authors and do not necessarily reflect the views of the sponsor. We thank all of the controllers who participated in this study as well as Mike and Kimberly Connor from the National Aviation Research Institute (NARI) for their assistance in recruiting controllers.

REFERENCES

Corker, K., Pisanich, G., & Bunzo, M. (1997). Empirical and analytic studies of human/automation dynamics in airspace management for free flight. In *Proceedings of the 10th International CEAS Conference on Free Flight*, Amsterdam.

Duley, J.A., Galster, S.M., Masalonis, A.J., Hilburn, B.G., & Parasuraman, R. (1997). En route controller information requirements from current ATM to free flight. In *Proceedings of the 10th International CEAS Conference on Free Flight*, Amsterdam.

Endsley, M.R., Mogford, R., & Stein, E. (1997). *Effect of free flight on controller performance, workload, and situation awareness.* (Technical Report). Atlantic City, NJ: FAA Technical Center.

Hilburn, B.G., Bakker, M. W. P., Pekela, W.D., & Parasuraman, R. (1997). The effect of free flight on air traffic controller mental workload, monitoring and system performance. In *Proceedings of the 10th International CEAS Conference on Free Flight*, Amsterdam.

Hilburn, B.G. & Parasuraman, R. (1997). Free flight: Military controllers as an appropriate population for evaluation of advanced ATM concepts. In *Proceedings of the 10th International CEAS Conference on Free Flight,* Amsterdam.

Masalonis, A.J., Le, M.A., Klinge, J.C., Galster, S.M., Duley, J.A., Hancock, P.A., Hilburn, B.G. & Parasuraman, R. (1997). Air traffic control workstation mock-up for free flight experimentation: Lab development and capabilities. *In Proceedings of the Human Factors and Ergonomics Society 41st Annual Meeting*, Albuquerque, NM.

Parasuraman, R., Molloy, R., & Singh, I.L. (1993). Performance consequences of automation induced "complacency." *International Journal of Aviation Psychology, 3*, 1-23.

Parasuraman, R., & Riley, V. A. (1997). Humans and automation: Use, misuse, disuse, abuse. *Human Factors, 39,* 230-253.

Perry, T.S. (1997). In search of the future of air traffic control. *IEEE Spectrum* (August), 19-35.

RTCA (1995). *Report of the RTCA Board of Director's Select Committee on Free Flight.* Washington, DC: RTCA.

Wickens, C.D., Mavor, A., Parasuraman, R. & McGee, J. (1998). *The future of air traffic control: Human operators and automation.* Washington DC: National Academy Press.

Automation and the Air Traffic Controller: Human Factors Considerations

Bert Ruitenberg
International Federation of Air Traffic Controllers' Associations (IFATCA)

INTRODUCTION

The similarities between Air Traffic Control (ATC) Operations and Aircrew working with electronic flight-deck presentations may seem remote, but there are signs that aircrew operating in such an environment are identifying problems associated with automated controls and data. Aircrews are becoming passive monitors rather than active fliers and there have been numerous reports of pilots trying to 'fix' the computer, rather than fly the airplane. Is this the way ahead for the Air Traffic Controller (ATCO)? Will ATCOs become so reliant on automated systems that active participation is reduced to a monitoring status? And will the ATCO be expected to pick up the pieces, and take over, if the computer cannot provide the appropriate solution to the problem?

There are already areas where ATC is being automated in some way or another. In most cases data exchange is being enhanced by computerized Databases and electronic data displays. The introduction of color radar, for instance, allows for a greater measure of enhancement than monochrome. The computerization of Air Traffic Flow Management (ATFM) is seen as an essential element to efficiently deal with the various flow control rates and increases in traffic demand. However, in the case of the Shanwick Oceanic ATC Centre computer there is a radical move towards a computerized 'control' function, with computer assigned Oceanic clearances being issued. This system has not proved to be fail-safe, nor is it seen as being able to deal with all clearance requests. If it was not for the fact that an ATCO is still needed to provide a clearance that the computer cannot cope with, then it would be quite possible that this system could operate without any ATCO involvement, at all. Air Traffic Control Assistants would input clearance requests and die computer would provide the answer. ATCOs would no longer be necessarily required.

Mode S transponders are already being seen as solution to resolving many ATC problem and difficulties, around the world. Not only to supply the ATCO with seemingly limitless amounts of data, but also to provide a Conflict Alert and Resolution system for aircraft so equipped.

IFATCA policy exists on Airborne Collision Avoidance Systems (ACAS): "IFATCA *recognizes that the development of airborne collision avoidance systems should be encouraged. However, it must be accepted that the primary means of collision avoidance within a controlled airspace environment must continue to be the Air Traffic Control System which should be totally independent of airborne emergency devices such as ACAS, Autonomous airborne devices should not be a consideration in the provision of adequate Air Traffic Services".*

IFATCA also draws attention to the compatibility between a controller's responsibility for providing positive separation and with the controller's ability to discharge them.

The International Federation of Air Line Pilots' Associations (IFALPA) also highlights the inherent requirements of a data-link system such as Mode S. The integrity of information should be assured by system design, together with automatic warning in the event of malfunction. The system should be capable of growth and also be standardized. Data link should not supersede the use of voice where this is the optimum. medium. and voice communication should remain available as a back-up. (IFALPA Annex 10 (COM) Polstat (April 1987)).

Electronic data screens may replace Flight Progress Strips, but when the TV-monitor fails all the data is immediately lost. The need for a fail-safe system, or immediate access to duplicated standby equipment, is paramount. Human Factor research has already identified that automation does not always reduce ATCO workload, in fact it can increase it. Certain aspects of an ATCO's tasks need to remain a manual type operation to maintain job satisfaction. Automation can reduce awareness and alertness. The man-machine interface is extremely important and Psychologists highlight the need for the system to fit the man, rather than the other way around. Human response involves the use and co-ordination of eyes, ears and speech, together with a little common sense. Computers rely on the 'right' button being pressed and the 'right' program installed to ensure the 'right' action is taken. One of the difficulties facing computer programming of ATC Systems is the ability to engineer a program that can deal with every eventuality and situation. If the computer is faced with a problem that it cannot solve, it will fail, but ATCOs are expected to provide solutions to every problem they face.

AUTOMATION

Human Factors research has shown that automated tabular lists of aircraft data impose a major search task when traffic becomes heavy. Alphanumeric labels (Secondary Surveillance Radar, SSR) encounter serious difficulties when label overlap occurs in high-density traffic. Because it cannot all be displayed at once, some information will have to be requested from the computer. This requires ATCOs to undertake extra tasks to use keyboards, or other methods, to retrieve data. In automated systems, in addition to extra data entry and retrieval tasks, the ATCO may have more to remember, because less information is permanently displayed. This may well lead to multi-data displays, or extensive data 'menus' to find the information that is required.

It may still be right to introduce an automated aid, but it should be justified on its own merits, rather than implying benefits purely in workload terms. There is a need to specify exactly how new and updated data is entered into the system If done automatically, there is a need for correct data, as well as a means of highlighting when data changes. If undertaken manually, there must be keyboard and data entry skills to maintain accuracy. Errors can easily occur if personnel are distracted or fail to enter the data correctly. In any such system, whether the data input is automatic or manual, there must be a means of cross checking the data to ensure absolute reliability and confidence.

WORKLOAD

It is often suggested that by automating a system, you can reduce peak workloads, or can increase system capacity This is often far from the truth Additional working practices, such as data entry/retrieval methods can actually increase workload. When ATCOs are working at peak levels of workload, this is usually due to them exercising positive control over the maximum number of aircraft in the constraints of limited airspace Merely by automating certain aspects of an ATC system will not enable the ATCO to necessarily handle more traffic. Any automation should be directed in removing non-essential tasks allowing the ATCO to concentrate on more important tasks. It is not correct to assume that because workload of one kind can be reduced by an automated aid, a commensurate increase in workload of another selected kind must be possible. A reduction in speech does not necessarily lead to more time for decision making. A relative consistent finding in ATC studies, is that an automated aid, such as SSR labeling on a display, may lead to substantial reduction of essential verbal messages, without producing a corresponding improvement in any other measurable aspect of an ATCOs performance.

JOB SATISFACTION

It is generally recognized that educated personnel with higher levels of skill and status (which includes ATCOs) find that their main source of satisfaction in a job lies in its intrinsic interest to them. (Flerzberg, 1957). A satisfying job should require effort, provide a challenge and make use of the person's skills. Automation may well reduce the effort of certain tasks and the stress associated with them, but may lead to loss of job satisfaction by taking away some of the intrinsic interests of the job, and the perceived 'control' over certain functions. It is widely accepted that the challenges of the ATCO's job are one of the main reasons that ATCOs enjoy their profession. Even working under difficult circumstances, sometimes with inadequate equipment and with high traffic levels, ATCOs may well perceive they are working very hard, possibly under great stress, but afterwards they will generally accept that they have achieved a great deal in dealing with a complex and busy situation and will find an element of satisfaction. A reduction of workload to the point that job satisfaction is reduced can lead to boredom and general discontent.

SYSTEM INTEGRITY AND ATCO RESPONSIBILITY

When undertaking their duties, ATCOs are generally fully aware of the responsibilities they carry. They continually check their own and other's work so that safety is not compromised. An ATCO is usually encouraged to be self-critical and to embrace self-analysis of their methods of operation. Because of the inherent safety factors that are associated with the task, ATCOs are trained to expect the unexpected. They are generally cautious with the data they handle. Pilots do not always do what is expected of them and ATCOs are continually forming alternative solutions to problems. The nature of the job requires an ATCO to be able to react quickly and calmly to 'unusual' events. Automation requires great confidence from the operator. Operators must be able to trust the data they are dealing with. With greater levels of automation, the more reliant the operator becomes in the system to provide accurate and trustworthy data. In any field of automation, the data-links must be fail-safe and totally accurate. Otherwise the operator will not trust the system and will feel -uncomfortable working with it. In the ATC world there can be no element of doubt in a controller's mind to the accuracy of the data being handled. Safety depends on it. The question of responsibility in automated systems needs to be looked at closely. If the controller is going to retain the 'control' function, then the responsibility remains with him, or her. It is no good at all suggesting that an error in the computer input is the reason for an airmiss, or that the data that the ATCO was using was corrupt or inaccurate. The legal implications of automated ATC systems need to be identified and responsibilities determined. ATCOs may hold the responsibility, at least in the legal sense, without retaining the means to exercise the responsibility that they possessed when the system was manual. The reliance that will be naturally placed on automated system should be balanced by alternative methods of handling the workload should the system fail and 'manual reversion' takes over. Does 'fail-safe' actually exist and should Air Traffic Services (ATS) authorities ensure that the manual type of ATC procedures are maintained to cope with failures of an automated system?

CONCLUSIONS

Automation is, in many instances, desirable and beneficial. In some cases it can be considered essential. Automation should only be introduced when there is a full awareness of all the associated Human Factor implications. Automation sometimes brings extra tasks, as well as benefits, and many functions cannot be automated entirely. Automated aids should also be trustworthy.

When automation is extended to problem solving and decision making, it affects job satisfaction and the exercise of skills and responsibilities. Computer assistance can alter the nature of the task and new information is of no use unless it is presented in an easily understandable form. There are no equivalents of job satisfaction, stress, status, morale or professional pride in machines.

A fall back system of ATC skills should be maintained, despite the ability to do away completely with some aspects of the ATC task Automation carries implications, both real and perceived, of status and responsibility. As automation evolves, certain automated functions in the system become relatively fixed and inflexible. This reduces the ability of the controller to implement different strategies to solve individual problems. The nature of tile ATC task allows individual controllers to apply differing strategies that result in similar solutions and efficiency of operation.

Automation may limit the freedom of choice for the individual and ATCOs may perceive a loss of control or a reduction in skills to perform their tasks. In efficiency terms, it may be possible to identify and assume that there is only one particular way to complete a task. Though convenient, this will require all ATCOs to be trained in one particular way. If there is no such optimum way, then the machine should not be designed as though there is.

Studies in other safety critical industries, such as nuclear power and the cockpit, have identified undesirable human factors trends within the design and operating philosophy of these system, as a result, the current human factors preference should be to retain the controller as the primary decision maker in future ATC systems, in order to maintain the controller's interest, motivation and skill levels.

There are genuine needs for automation to assist ATCOs, to improve performance and reduce workload, to increase efficiency, to remove non-essential tasks, and to enhance job satisfaction. There is also a need for ATCOs to be involved as an essential part of any future ATC system The man-machine interface needs to be examined closely so that the system fits the man, rather than the other way around. There is a requirement for ATCOs to be involved with evaluations of equipment design and purchase. It is essential that automation work for the benefit of the controller.

IFATCA POLICIES ON AUTOMATION

Automation must improve and enhance the data exchange for controllers. Automated systems must be fail-safe and provide accurate and incorruptible data. These systems must be built with a integrity factor to review and cross-check the information being received.

Automation must assist and support ATCOs in the execution of their duties, to improve performance and reduce workload to remove non-essential tasks, to increase efficiency to enhance not only the job satisfaction of the controller, but also the safety element of the controller's task.

The Human Factors aspects of automation must be fully considered when developing Automated systems and should include the maintenance of essential manual skills and controller awareness.

The controller must remain the key element of the ATC system and must retain the overall control function of the system. Safeguards must be established to ensure that the controller remains an active, rather than a passive, user of an automated system.

The legal aspects of a controller's responsibilities must be clearly identified when working with automated systems.

Additional Policies (1993)

A controller shall not be held liable for incidents that may occur due to the use of inaccurate data if he is unable to check the integrity of the information received.

A controller shall not be held liable for incidents in which a loss of separation occurs due to a resolution advisory issued by an automated system.

REFERENCES

Hopkin, V.D. (1982). Human Factors in ATC. *Nato AGARDograph no. 275.* Paris.
Hopkin, V.D. (1989). *MMI* Problem in designing ATC system. *Proceedings of the LE.E.E.*
ICAO (1994). Human Factors Digest no. I I - Human Factors in CNS/ATM-systems. IC40 *Circular 249*AN1149. Montreal, Canada.
IFATCA (1991). Automation and the ATCO - Human Factors considerations. *Working paper for 30th*

Levesley, J. (1990). Report on NATO ASI Conference - Automation and System Issues in ATC. LJK Guild of Air Traffic Control Officers.

Narbormigh-Hall, C. S. (1987). Automation implications for knowledge retention as a function of operator controllability In: Diaper and Winder (Eds.), *People and Computers 11,* Cambridge University Press.

Thorning, A.G. (1987). Research in Automation for ATC: UK Work and Associated European Projects.

Automation and Free(er) Flight: Exploring the Unexpected

P.G.A.M. Jorna

Netherlands Aerospace Laboratory, Amsterdam

INTRODUCTION

The consistent increase in air travel initiated many activities in the field of automation development. The purpose is to provide means for bypassing bottlenecks in aviation operations. The present route network(s) where aircraft follow each other sequentially, restrict the maximal amount of aircraft. Keeping an overview over traffic is demanding and controllers have to share the work between teams handling separate sectors of the network. This type of task sharing however, increases communications demands creating another limiting factor. Possible solutions are to reduce the overall task load for the controllers, or to open the skies and use more airspace. Technologies are available that would allow a restructuring of air traffic procedures, provided that humans can still work effectively. Two perspectives seem to prevail, a 'ground perspective' where the controller is assisted by software tools, and an airborne 'Free Flight' perspective, where pilots assume some level of responsibility for separation. Automation philosophies, by definition, raise discussion and are susceptible to the risk of taking assumptions with a high 'face validity' for true. Objective and systematic testing of new technologies is therefore mandatory. Examples will be provided of surprises encountered when testing technologies with humans in the loop, underscoring that assumptions continue to be misleading.

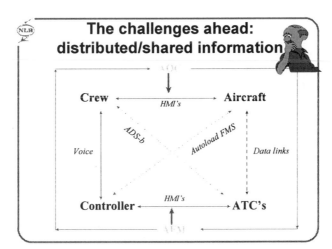

Figure 1. A diagram depicting the overall context, players and technologies in future aviation.

OPTIONS FOR THE AVIATION FUTURE

Airline Operating Centers (AOC) instruct and guide their crews for realizing maximal revenue's at minimal costs. Air Flow Management (AFM) units regulate/ restrict air traffic to manageable levels. Different goals are reconciled for the sake of safety. Presently, information availability is different for air and ground systems, but the digital data links create options for drastic changes. The complexities involved in possible future system configurations are depicted in Figure 1.

Data link can serve ATC communications in a non-voice format. Even more influential is the creation of an auto load connection with the Flight Management System (FMS), allowing direct loading of clearances in the flight computers. Essentially, fully ground controlled flight would be feasible. Alternatively, technologies like Aircraft Dependent Surveillance (ADS-B), enable aircraft to broadcast flight information such as speed, heading and intent to other aircraft. With such information, aircraft can have access to similar data as the air traffic controller. New technologies, provide not only opportunities, but also unknown consequences with respect to responsibilities, work procedures and the potential for human and machine errors.

EVALUATING HUMAN PERFORMANCE

Human performance is depending on a dynamic process comprising multiple steps and mental resources such as data sampling from displays, processing it into task relevant information and executing responses. During the work, an adequate overview (awareness) has to be maintained in order to plan future activities and to detect problems. Individual measurements will therefore not reveal the complexities in human task strategies, so a battery of measurements is required. Examples are:

- Sampling of data: eye tracking, point of gaze, use of specific equipment
- Processing : heart rate (variability), pupil size, event linked measures, embedded dual tasks
- Responses and Actions: internal/ external communications, equipment use, switchology
- Cognition: situation & systems awareness, subjective measures.

Concepts like 'situation awareness' etc. are quantified by combining specific measures like looking at traffic (traffic awareness) or systems (awareness) and processing efforts. (Jorna, in press).

RESULTS FOR ATC AUTOMATION

Software tools can handle conflict detection and even conflict resolution of aircraft route or altitude infringements. The assumed effectiveness of such assistance is based on 'removing' a task from the controller, but alternatively, it can also be assumed that task load will be increased as the controller is still responsible and therefore has to review the adequacy of the advice. If the controller simply accepts, he/she will be essentially 'out of the loop'. This issue was investigated by means of simulations with high and low traffic loads (Hilburn et al. 1995, Hilburn, 1996).

Both psychophysiological measures revealed a systematic effect of traffic load levels as well as reduced workload as a function of automation (heart rate variability increases and pupil size decreases respectively). Performance measures based on the dual task principle were used with the purpose of acquiring an indication of situation awareness, in this case defined as awareness of communication and aircraft status. Occasionally, aircraft would fail to acknowledge the data linked clearances and the controller had to detect these occurrences. Also subjective workload ratings were obtained from the controllers. The results are depicted in Figure 3.

Response times to unconfirmed uplinks, also differentiated between traffic levels and level of automation. With automation, detection improved. So, the results of the objective measures consistently indicate that automation is capable of reducing overall workload with benefits for overall monitoring tasks. However, subjective ratings indicated the opposite, as workload was rated higher with automation. Such perceptions, or assumptions like *more tools must mean more work*, can influence task strategies and automation usage, as demonstrated in an experiment addressing arrival management (Hilburn, this volume).

RESULTS FOR AIRBORNE AUTOMATION

Free flight has been a concept with strong supporters as well as fierce opponents. Many assumptions

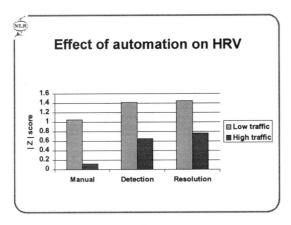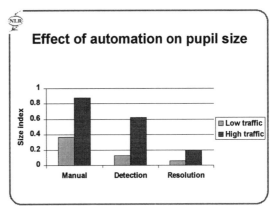

Figure 2. Effects of ATC automation on psychophysiology as compared to manual.

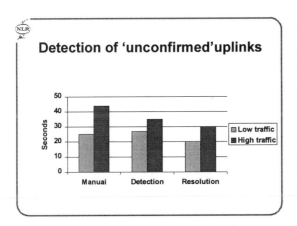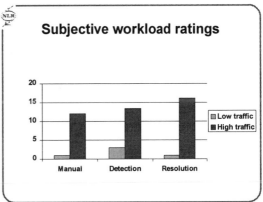

Figure 3. Controller response times to communication status information and subjective ratings.

seem to play a significant role in the discussions. Some view it as the ultimate solution for an outdated ATC system, while others believe that it will never work. Based on careful considerations, an advice was issued recently by the National Research Council to maintain separation responsibility on the ground (Wickens, this volume). Paramount was the complexity of pilot-pilot communication and the effectiveness of 'rules of the road'. However, alternatives have been developed recently using a multi disciplinary and human centered approach (van Gent et. al., 1998). In this approach, ADS-B and other data is used not only to create traffic displays, but also to provide configurable conflict detection and resolution facilities to the flight crew. An essential element is the creation of an *'adaptive aircraft'* responding to other aircraft, either automatically or managed by the crew(s). The automation uses an algorithm that calculates 'escape vectors' from other aircraft in real time. In behavioral terms, the aircraft now has an automatic tendency to avoid others. When all aircraft are equipped, the burden of evasive maneuvers is distributed evenly and fairly, preventing more excessive maneuvering in case of mixed equipage. The crews have access to all the data and can influence the selection of maneuvers. Many configurations can be realized experimentally, using the same basic technology.

A first surprise was encountered after a formal safety analysis by the NLR ATC department using mathematical methods. The results indicated a systematic safety improvement (!) of a factor of ten, in comparison with traditional means for separation. Note that further studies are needed, but the results were counter-intuitive to many. Paramount is the implicit redundancy between automation(s) and crew member(s). The second surprise was that crews quickly adapted to the concept and assessed it as quite promising, but

109

'what will happen if it gets busy?'. Apparently, not all realized that the scenario was not only flown in one of the busiest European sectors, but also contained three times the average traffic as compared to today!

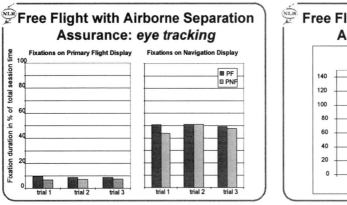
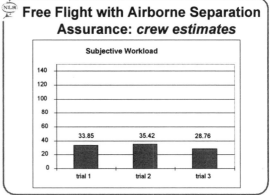

Figure 4. Eye data and subjective ratings for three subsequent sessions with airborne automation.

Eye tracking data revealed that both crewmembers participated in aircraft separation and evenly distributed monitoring of navigation and primary flight displays. Subjective workload ratings were in the low range, and comparable with routine flight. Practice requirements seemed to be low as no apparent learning trends were found. The explanation for these results can be found in a phenomenon related to different perceptions of traffic load. The controller has to monitor all traffic while the crews are only concerned with part of the traffic. A clarification can be provided by the 'sail boat analogy' depicted in Figure 5.

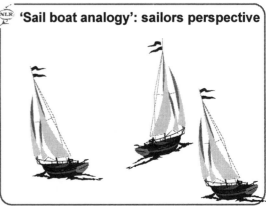

Figure 5. Effect of perspective on human perception of traffic loads and manageability.

If someone doesn't like busy lakes, no sailing will be attempted when observing many, many sailboats from the shore. Suppose that a traditional ATC approach would be applied to this situation, than everybody would have to file a sail plan, the (future) winds should be known precisely to predict courses and everybody would need more than sophisticated auto helms. Many respond to such an approach with amazement, but it is actually pursued in aviation. However, when finally a daring decision has been taken to go sailing, a personal surprise is that traffic is not at all as bad as expected (try it for yourself). Apparently, high traffic loads in free waters are not perceived as such, when being in the middle. Boats that are way ahead, or behind you, are of no interest and the helmsman only needs to concentrate on one or perhaps two other boats. A similar process seemed to work for the flight crews, as they selected a limited display range quite compatible with this sail boat analogy (note that this will not work for high ways with sequential traffic).

DISCUSSION

The studies presented are examples of some unexpected 'lessons learnt' with respect to technology development. They illustrate that subjective idea's can turn out good or bad and that 'face validity' is still a dangerous concept for decision and policy making. Solely relying on subjective user acceptance or assumed merits of new technologies, is a risky strategy without formal and systematic experiments and it can lead to premature blocking of design alternatives. Furthermore, innovative thinking can be put at jeopardy when research is directed by 'believers' in stead of by results building on objective and systematic research. The costs involved in such research is often mentioned as a limiting factor for performing human in the loop experiments with technologies as they are assumed not to add anything to 'real' product development. This assumption is invalid as such experiments not only include the development and integration of Human Machine Interfaces, automation tools and working procedures, but also their integration in the envisaged operational environment. It is in that context that most surprises occur.

REFERENCES

Gent van, R., Hoekstra, J. and Ruigrok, R. (1998). Free Flight with Airborne Separation Assurance. *International conference on Human Computer Interaction in Aeronautics*. Canada.

Hilburn, B., Jorna, P.G.A.M. & Parasuraman, R (1995). The effect of advanced ATC strategic decision aiding automation on mental workload and monitoring performance: an empirical investigation in simulated Dutch airspace. In *Proceedings of the Eighth International Symposium on Aviation Psychology*, April 1995, Columbus, Ohio: Ohio State University.

Hilburn, B. (1996). *The Impact of ATC Decision Aiding Automation on Controller Workload and Human-Machine System Performance*. Dissertation for Ph.D., available as NLR TP 96999.

Jorna P.G.A.M. (In press). Context simulation: an interactive methodology for user centered system design and future operator behavior validation. In N. Sarter & Amelberti (Eds.), *Cognitive engineering in the aviation domain*. Mahwah, NJ: Lawrence Erlbaum Associates..

Wickens, C. (1998). *Automation in Air Traffic Control: The Human Performance Issues*. Paper presented at the Third automation Technology and Human Performance Conference. Norfolk, Virginia.

The Role of the Controller in Future Air Traffic Control: Techniques for Evaluating Human / Machine System Performance

Brian Hilburn

National Aerospace Laboratory NLR
Amsterdam, The Netherlands

INTRODUCTION

The demand for air travel has never been greater. Within Europe, for example, civilian air traffic is expected to grow by nearly 50% over the next seven years. With the busiest airports already saturated, and the introduction of extremely large civilian airliners still some way in the future, accommodating such massive growth promises to be a daunting task. Increasingly, it is being realised that the drive to increase air traffic capacity will ultimately have to address a fundamental redesign of the Air Traffic Control (ATC) system.

One can reasonably argue that ATC represents one of the most complicated human-machine systems in the world, and which can be seen to consist of the following four interrelated elements: interfaces; automation tools; procedures; and operational concepts. At the highest level of description (the operational concept) the system is defined in term of the goals and general rules of the system. At its lowest level (the interface) the display between human and computer is specified. Historically, conventional wisdom has held that advances in ATC system development have to be made incrementally— "evolution not revolution" has been the credo of ATC designers—to displays, tools and procedures supporting the same basic operational concept (i.e., positive air traffic control via radio telephony) that has existed since the advent of the contemporary ATC system some 40 years ago.

Figure 1 shows the hierarchical structure of these four system elements, along with examples of each. Each example corresponds to a research program currently being carried out at the NLR.

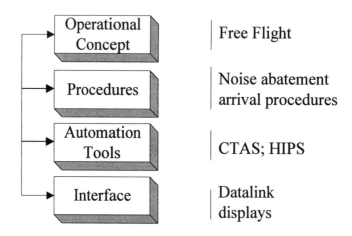

Figure 1. The hierarchy of ATC elements.

These examples can be briefly described as follows:
- *Free Flight*-- a revolutionary operational concept in which aircraft would assume greater authority for route selection and separation assurance
- *CTAS* (Center TRACON Automation System)—a set of three automated decision aiding tools

designed to assist the scheduling and sequencing of arrival traffic;

· *HIPS* (Highly Interactive Problem Solver) an intuitive conflict analysis tool that permits controllers to efficiently de-conflict traffic;

· *Datalink*—a digital replacement for current-day air-ground radio communication.

· *Noise abatement arrival procedures*—in this case, an increase in the glide slope to reduce the arrival noise footprint on densely-populated areas;

Coupled with the attention currently being devoted to developing new ATC systems is the realisation that human error is a causal factor in some 70-80% of all aviation accidents. There is therefore increasingly greater awareness of the need to develop ATC systems that better consider the capabilities and limitations of the human user (i.e., the controller). The following will summarise the results of two experiments, one into a prototype ATC interface, the other into a sweepingly new operational concept, that demonstrate how the NLR uses its battery of human performance measures to evaluate new ATC concepts. These measures include, for example, objective and subjective workload measures, as well as behavioural measures to assess situation awareness and monitoring performance. By way of example, empirical results will highlight the use of objective and subjective workload measures.

EXPERIMENT ONE: "DATALINK" INTERFACES FOR ATC

A recent study conducted at the NLR (Hooijer, 1996) investigated various interface options for digital "datalink" communication interfaces. Digital datalink, currently under development, would enable the controller and pilot to communicate without the use of the traditional radio telephone system. This experiment evaluated three prototype datalink interfaces, which varied in Input Mode (permitting separate versus integrated clearances), and Feedback Mode (providing iconic versus tabular display of aircraft-to-controller communications), as follows:

Interface	Input Mode	Feedback Mode
Baseline	Separate	Tabular
Iconic	Separate	Iconic
Integrated-Iconic	Integrated	Iconic

An additional factor, traffic load, was varied within subject to explore a possible interaction with interface format. Under all three interface conditions, controllers issued aircraft clearances by interacting directly with the flight label of that aircraft. Once the controller clicked on the flight data label, a dialogue box relevant to the active parameter (e.g., altitude) would appear, as depicted in Figure 2.

Heart rate related measures have a long history in the measurement of mental workload (Roscoe, 1978). Heart rate variability (HRV) is a measure of irregularity in the heart's inter-beat-interval. A medium frequency component of HRV (centered around 0.1 Hz), which is associated with short-term blood pressure regulation, has been seen to vary inversely with mental workload. HRV decreased (i.e., indicated workload decreased) with traffic load, $F(1,9)=57.4$, $p<.001$. Datalink interface also had a significant effect on indicated workload—a post-hoc Scheffe test revealed that workload was significantly lower ($p<.01$) for the Integrated-Iconic than for the Baseline interface.

Although the chief purpose of the pupillary response (dilation / constriction) is to accommodate differences in ambient illumination, small but measurable workload effects can be seen superimposed on the larger "light reflex" pupil response. In general, pupil diameter is seen to increase with workload.

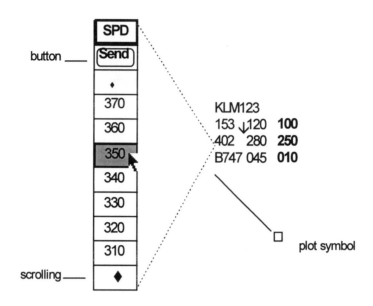

Figure 2. Datalink pop-up dialogue.

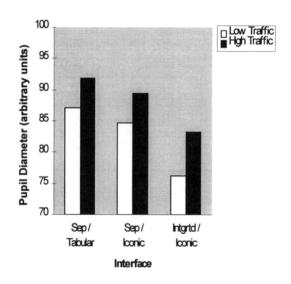

Figure 3. Pupil diameter, by Interface and Traffic.

As with HRV, pupil diameter indicated increased workload, $(F(1,7)= 51.8, p<.001$. As shown in Figure 3, a trend suggested lower indicated workload for the Integrated-Iconic and Iconic displays. The data suggested that the observed workload benefits are more attributable to integrated clearance input than to iconic communication feedback.

EXPERIMENT TWO: "FREE FLIGHT" OPERATIONAL CONCEPT FOR ATC

Although "Free Flight" (FF) is a recent and still loosely-defined concept (RTCA, 1995), research into aspects of the FF concept (e.g., direct routing, self separation) is proceeding quickly. According to a vision of mature FF, aircraft outside of terminal areas would generally be free to fly user-preferred routes, and modify their trajectories en route, with minimal intervention by air traffic control.

Under likely near-term FF scenarios, the controller would continue to play an important (albeit new) role in ATC, especially in the face of unpredictable aircraft behaviour. Rather than strategically controlling air traffic, the "controller" of the future might well fill the role of either a strategic flow manager, and/or a tactical "Separation Assurance Monitor," who would intervene only when losses of separation were imminent. This new role would raise a number of potential human performance problems (e.g., workload extremes, vigilance problems, and reversion-to-manual difficulties).

To understand how profoundly the change to FF could influence controller workload and monitoring, consider the simple diagram of Figure 4, which depicts the principles of controlled and free flight in the en route phase. The diagrams are identical, except that the angle of four of the ten aircraft has been changed under free flight. Notice how much more difficult this makes the task of anticipating traffic conflicts. Under controlled flight (CF), there are a limited number of areas at which conflicts are likely to occur. Indeed, the historical reasons behind the current-day fixed route structure have to do more with human limitations than with technical or procedural concerns. Under FF, on the other hand, assuring separation of the same number of aircraft is now a daunting task for the controller.

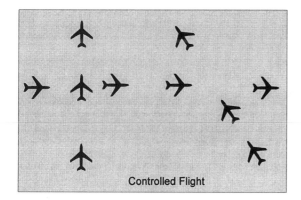 

Figure 4. Controlled versus free flight.

A recent experiment by Endsley, Mogford and Stein (1997) assessed the effect of FF-like scenarios on controller situation awareness and mental workload. Workload in their study was assessed in terms of self-reported subjective workload. The current experiment examined the effect of FF traffic per se, and the withholding of traffic manoeuvre intent information, on the controller (Hilburn, Bakker, Pekela & Parasuraman, 1997). This was done by assessing the performance of currently-active controllers under both conventional (i.e., controlled) and free flight conditions, using the same en route airspace. Two free flight conditions were evaluated: one in which aircraft shared their intentions with ATC before manoeuvring, and one in which aircraft manoeuvred without notifying the controller. In addition to subjective workload ratings, the current study also collected objective (pupil diameter) measures of mental workload, and measures of controllers' conflict prediction performance.

As in Experiment One, a statistically significant difference ($p<.001$) was found between pupil diameter under low and high traffic—pupil diameter was seven percent higher under high traffic than under low traffic. The trend depicted in Figure 5 suggests that free flight in general might reduce workload, and that intent information might provide additional workload benefits, especially under low traffic situations.

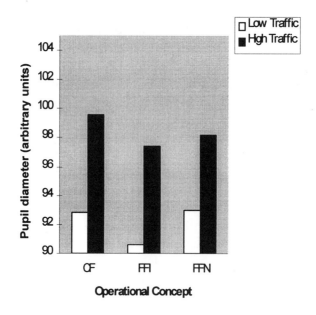

Figure 5. Pupil diameter, by operational concept and traffic.

This experiment also made use of a subjective workload instrument, the Rating Scale for Mental Effort (Zijlstra & van Doorn, 1985). Using this scale, controllers reported significantly higher workload under high traffic conditions, ($p<.0001$). A significant interaction was found between traffic level and control condition, ($p<.05$). This interaction is depicted in Figure 6. A post hoc Newman-Keuls test revealed that, under high traffic, controllers felt significantly more workload under controlled flight than they did under uninformed free flight (FFN), ($p<.05$). Whereas the objective workload measure indicated that aircraft intent under free flight conditions could reduce controller workload, controllers actually reported that, under high traffic, they felt less workload if intent information were withheld under free flight.

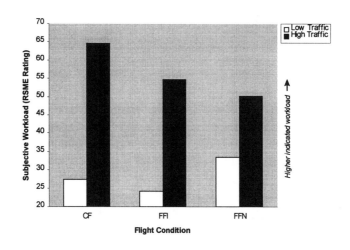

Figure 6. Self-reported workload, by Flight Condition and Traffic Load.

116

DISCUSSION

Over the course of many high fidelity ATC experiments, the NLR has continued to refine its integrated approach to human performance assessment. This experience has provided several lessons. First is the value of objective physiological workload measures, including some (e.g., pupil diameter, or scanning entropy measures) that had fallen out of favour with the greater research community. This is due, in part, to the experimental control that ATC tasks generally afford— such potential artefacts as response demands, or ambient lighting, can be minimised without sacrificing realism.

Second, such techniques by themselves are insufficient. The results of Experiment Two highlight the fact that subjective and objective measures sometimes dissociate. Although the pupil diameter and subjective workload were in essential agreement (i.e., free flight showed workload benefits), controllers reported that, under high traffic, "uninformed" free flight reduced workload. This suggests the importance of both types of measures—objective and subjective-- in evaluating performance in certain settings. Although it may be necessary to demonstrate objective workload gains in ATC developments, system designers cannot overlook the impact that controllers' subjective evaluations can have on the acceptance of new ATC developments. If the introduction of, say, a new ATC tool relies on controllers' willingness to use the tool, then controllers should first recognise the potential benefits of the tool.

Third, strategies play an important part in task performance. In ATC, the low frequency of collisions (or losses-of-separation) can limit their sensitivity as performance criteria, even in the laboratory. Recognising that task performance can be strategy-tolerant (i.e., controllers can develop idiosyncratic ways of operating), greater use should therefore be made of knowledge elicitation techniques (e.g., verbal protocol analysis) to identify individual differences in how controllers use, say, new displays or tools..

Finally, as in many domains of human factors research, the selection of test participants is crucial (Hilburn & Parasuraman, 1997). Previous ATC studies (e.g., Bisseret,1981) have found systematic differences in the performance of experienced and novice controllers. In certain evaluations of ATC concepts, the benefits of using expert participants (e.g., face validity) have to be weighed against such potential costs as: decision making biases, negative transfer of training, and idiosyncratic performance.

REFERENCES

Bisseret, A. (1981). Application of signal detection theory to decision making in supervisory control. *Ergonomics*, *24*(2), 81-94.

Endsley, M.R., Mogford, R.H. & Stein, E. (1997). *Effect of free flight conditions on controller performance, workload and situation awareness: a preliminary investigation of changes in locus of control using existing technology.* Draft FAA technical report. April, 1997.

Hilburn, B. & Parasuraman, R. (1997). *Military Controllers Provide an Especially Appropriate Population for Evaluation of Free Flight and Other Advanced ATM Concepts.* Poster presented at the CEAS 10th European Aerospace Conference on Free Flight. 20-21, October. Amsterdam, The Netherlands: Confederation of European Aerospace Societies.

Hilburn, B., Bakker, M.W.P., Pekela, W. & Parasuraman, R. (1997). The Effect of Free Flight on Air Traffic Controller Mental Workload, Monitoring and System Performance. *Proceedings of the Confederation of European Aerospace Societies (CEAS) 10th European Aerospace Conference*, 20-21 October 1997, Amsterdam, The Netherlands.

Hooijer, J.S. (1996). *Evaluation of a label oriented HMI for tactical communications in Air Traffic Control.* NLR TP 96676L. Amsterdam: National Aerospace Laboratory NLR.

RTCA. (1995). *Report of the RTCA board of director's select committee on free flight.* Washington DC: Author.

Roscoe, A.H. (1978). Stress and workload in pilots. *Aviation, Space & Environmental Medicine, 49*(4), 630-636.

Zijlstra, F.R.H & Doorn, L. van (1985*). The construction of a scale to measure subjective effort.* Technical Report, Delft University of Technology, Department of Philosophy and Social Sciences.

ADAPTIVE AUTOMATION

Adaptive Aiding and Adaptive Task Allocation Enhance Human-Machine Interaction

Raja Parasuraman
Catholic University of America, Washington DC

Mustapha Mouloua
University of Central Florida, Orlando

Brian Hilburn
National Aerospace Lab, Amsterdam

INTRODUCTION

Adaptive automation represents an alternative design approach to the implementation of automation. Computer aiding of the human operator and task allocation between the operator and computer systems are flexible and context-dependent. In contrast, in static automation, provision of computer aiding is pre-determined at the design stage, and task allocation is fixed. Several conceptual and theoretical papers have suggested that adaptive systems can regulate operator workload and enhance performance, while preserving the benefits of static automation (Hancock, Chignell, & Lowenthal, 1985; Parasuraman, Bahri, Deaton, Morrison, & Barnes, 1992; Rouse, 1988). The performance costs of certain forms of automation—over-reliance, reduced situation awareness, skill degradation, etc. (Parasuraman & Riley, 1997)—may also be mitigated. These suggestions have only recently been tested empirically (Hilburn, Jorna, Byrne, & Parasuraman, 1997; Parasuraman, Mouloua, and Molloy, 1996; Scallen, Hancock, & Duley, 1995; see Scerbo, 1996 for a review of earlier work).

Empirical evaluations of adaptive automation have focused primarily on the performance and workload effects of either (1) adaptive aiding of the human operator or (2) adaptive task allocation (ATA), either from the human to the machine (ATA-M), or from the machine to the human (ATA-H). Each of these forms of adaptive automation have been shown to enhance human-system performance, but independently in separate studies. For example, in an early study, Morris and Rouse (1986) showed that adaptive aiding (AA) in the form of target localization support enhanced operator performance in a simulated aerial search task. Benefits of AA in a more complex simulation were reported by Hilburn et al. (1997), who provided air traffic controllers with a decision aid for determining optimal descent trajectories—the Descent Advisor (DA) of the Center Tracon Automation System (CTAS), an automation aid that is currently undergoing field trials at several air traffic control centers (Wickens, Mavor, Parasuraman, & McGee, 1998). Hilburn et al. (1997) found significant benefits for controller workload (as assessed using physiological measures) when the DA was provided adaptively during high traffic loads, compared to when it was available throughout (static automation) or at low traffic loads. With respect to adaptive task allocation from the machine to the human (ATA-H), Parasuraman et al. (1996) showed that temporary return of an automated engine-systems task to manual control benefited subsequent monitoring of the task when it was returned to automated control.

For adaptive systems to be effective, both AA and ATA-H need to be examined jointly in a single work domain. Furthermore, if adaptive systems are designed in a manner typical of "clumsy automation"—i.e., providing aiding or task reallocation when they are least helpful (Wiener, 1989)—then performance may be degraded rather than enhanced (see also Billings & Woods, 1994). One of the drawbacks of some flightdeck automated systems—for example, the Flight Management System (FMS)—is that they often require extensive reprogramming and impose added workload during high task load phases of flight such as final approach and landing, while doing little to regulate workload during the low-workload phase of cruise flight. This suggests the need for linking the provision of aiding to the level of operator workload or performance. Following Hancock et al. (1985), we propose a model for effective combination of adaptive aiding and adaptive task allocation, based on matching adaptation to operator workload. The model is illustrated is Figure 1, which plots operator workload against work period or time. As workload fluctuates due to variations in task load or operator strategies, high workload periods represent good times for AA and poor times for ATA-H; the converse is true for low workload periods. According to this model, if AA and ATA-H are to be combined in a adaptive system, they should be provided at the times indicated in order to regulate operator workload and maximize performance.

We evaluated this model for adaptive automation in a study with pilots performing a flight simulation task under different conditions of AA and ATA-H. We predicted that, compared to a non-adaptive group receiving neither AA nor ATA-H, a workload-matched adaptive group receiving AA at high workload and ATA-H at low workload would show superior performance and reduced workload. In order to confirm that any observed benefits were due to workload-matched adaptation rather than to aiding per se, we also tested a "clumsy automation" group who received aiding at times not matched to their workload, that is, AA at low workload and ATA-H at high workload.

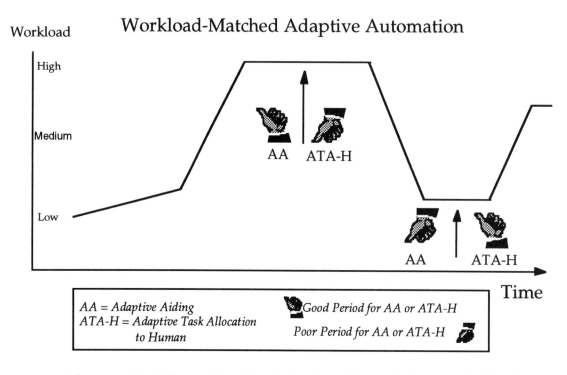

Figure 1. Model for workload-based adaptive aiding and adaptive task allocation.

METHODS

Twenty four pilots (20 male, 4 female) from Embry-Riddle University, 10 of whom were instrument-rated, participated in the study. They ranged in age from 21 to 35 years (mean age 28.5 years). They performed a flight simulation task modified from the Multi-Attribute Task Battery or MAT (Comstock & Arnegard, 1992). The MAT consists of sub-tasks of two-dimensional tracking, engine systems monitoring, and fuel management. The somewhat artificial dial monitoring task of the standard MAT was replaced with a version of the Engine Indicator and Crew Alerting System or EICAS that is typically implemented in current glass-cockpit aircraft. The EICAS display showed digitized versions of analog gauges indicating various engine parameters (EPR, N1, etc.), with regions of safe performance (green), warning (yellow), and danger (red) being marked.

A 60-minute session consisting of a three-phase (each 20 minutes), high-low-high task load profile typical of takeoff/climb, cruise, and approach/landing was simulated by increasing tracking difficulty during the first and last phases of the session. Tracking difficulty was manipulated by varying the bandwidth of the forcing function from .1 (high) to .06 (low) and back to .1 (high) again. The tracking task was performed manually at all times, except when AA was provided. The EICAS task was under automation control throughout the 60-minute session, except when it was returned to manual control. However, pilots were told that the automation was not 100% reliable and that they should monitor the automation for occasional failures. Finally, the fuel management task was performed manually at all times.

AA consisted of a simulated altitude hold for the tracking task. When this was implemented (preceded and followed by clear warning signals on the tracking display), y-axis tracking was automated and participants had only to track in one dimension. ATA-H consisted of a temporary return (10 minutes) of the automated EICAS task (also preceded and followed by clear warning signals on the EICAS display) to pilot control in the middle of the session (25 to 35 minutes).

The 24 pilots were randomly assigned to three groups of 8, two adaptive groups and a nonadaptive control group. For the workload-matched adaptive group, AA—altitude hold automation—was provided to the pilots during the first and last 20-minute phases of the 60-minute session (difficult tracking, high workload), while ATA-H—temporary manual control of the EICAS task—was implemented during the middle phase (easy tracking, low workload). For the "clumsy-automation" adaptive group, these contingencies were reversed; ATA-H was provided during the first and last 20-minute phases, and AA during the middle 20-minute phase. For the control group, neither AA nor ATA-H were provided.

RESULTS AND DISCUSSION

Mean tracking RMS error was significantly lower ($p < .05$) and subjective workload (mean NASA-TLX score) significantly greater ($p < .01$) during the first and last phases compared to the middle phase of the 60-minute session. Figure 2 shows tracking performance and subjective workload for the nonadaptive control group. The results confirm that both tracking performance and subjective workload showed the

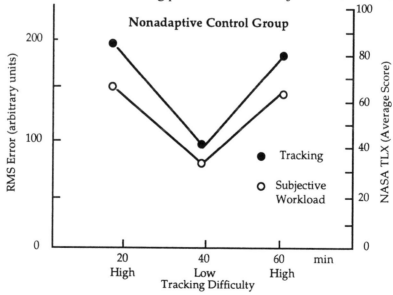

Figure 2. Tracking performance and subjective mental workload as a function of tracking difficulty (forcing function bandwidth) during the 60-minute session.

expected trends across the high-low-high task load profile. Thus, the tracking difficulty manipulation was successful in inducing greater levels of mental workload during the early and late phases of the simulation, as they would be during the takeoff and landing phases of actual flight.

Overall performance on all three sub-tasks was significantly higher ($p<.01$ in each case) in the workload-matched adaptive group than in the other two groups. In addition, subjective workload was significantly lower ($p<.05$) in this group than in the other groups. The performance benefits of workload-matched adaptation were most marked for the EICAS task. The detection rate of automation failures was significantly ($p<.001$) higher for the workload-matched adaptive group than for the other two groups. As Figure 3 shows, the performance benefit of ATA-H persisted beyond the period of manual control (middle 20-minute phase) to the last 20-minute phase when the EICAS task was returned to automated control. This result is consistent with the previous findings of Parasuraman et al. (1996).

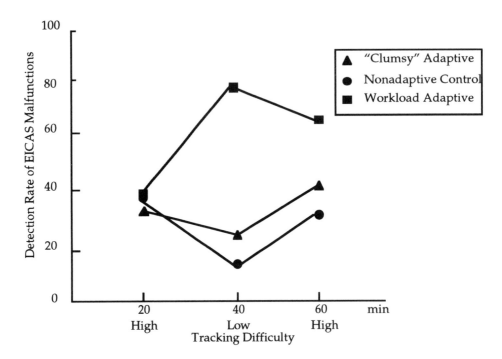

Figure 3. Detection rate of automation failures in the EICAS task for the workload-matched adaptive group, the "clumsy-automation" group, and the nonadaptive control group.

The workload and performance levels of the clumsy-automation adaptive group did not differ significantly from those of the nonadaptive control group. The performance levels for this group were equivalent to or in some cases lower than those of the control group. Figure 3 shows that the monitoring performance benefits of workload-matched adaptation did not accrue to the clumsy-automation group.

These results validate a design approach to adaptive automation involving adaptation matched to operator workload (Hancock et al., 1985; Parasuraman et al., 1992). Under these conditions, adaptive aiding and adaptive task allocation both enhance performance. The results also show, however, that performance benefits are eliminated if adaptive automation is implemented in a clumsy manner.

REFERENCES

Billings, C. E., & Woods, D. D. (1994). Concerns about adaptive automation in aviation systems. In R. Parasuraman & M. Mouloua (Eds.), *Human performance in automated systems: Current research and trends.* (pp. 264-269). Hillsdale, NJ: Erlbaum.

Comstock, J. R. & Arnegard, R. J. (1992). *Multi-attribute task battery.* (Technical Report). Hampton, VA: NASA Langley Research Center.

Hancock, P. A., Chignell, M. H., & Lowenthal, A. (1985). An adaptive human-machine system. *Proceedings of the IEEE Conference on Systems, Man and Cybernetics, 15,* 627-629.

Hilburn, B., Jorna, P. G. A. M., Byrne, E. A., & Parasuraman, R. (1997). The effect of adaptive air traffic control (ATC) decision aiding on controller mental workload. In M. Mouloua and J. Koonce (Eds.) *Human-automation interaction: Research and practice* (pp. 84-91). Mahwah, NJ: Erlbaum.

Morris, N. M., & Rouse, W. B. (1986). *Adaptive aiding for human-computer control: Experimental studies of dynamic task allocation.* (Report No AAMRL-TR-86-005).

Parasuraman, R., Bahri, T., Deaton, J., Morrison, J., & Barnes, M. (1992). *Theory and design of adaptive automation in aviation systems.* (Progress Report No. NAWCADWAR-92033-60). Warminster, PA: Naval Air Warfare Center, Aircraft Division.

Parasuraman, R., Mouloua, M., & Molloy, R. (1996). Effects of adaptive task allocation on monitoring of automated systems. *Human Factors, 38,* 665-679.

Parasuraman, R., & Riley, V. (1997). Humans and automation: Use, misuse, disuse, abuse. *Human Factors, 39,* 230-253.

Rouse, W. B. (1988). Adaptive aiding for human/computer control. *Human Factors, 30,* 431-438.

Scallen, S. F., Hancock, P. A., & Duley, J. A. (1995). Pilot preference for short cycles of automation in adaptive function allocation. *Applied Ergonomics, 26,* 397-403.

Scerbo, M. W. (1996). Theoretical perspectives on adaptive automation. In R. Parasuraman, & M. Mouloua (Eds.) *Automation and human performance: Theory and applications* (pp. 37-63). Mahwah, NJ: Erlbaum.

Wickens, C.D., Mavor, A., Parasuraman, R. & McGee, J. (1998). *The future of air traffic control: Human operators and automation.* Washington DC: National Academy Press.

Wiener, E. L. (1989). *Human factors of advanced technology ("glass cockpit") transport aircraft.* (Report No. 177528). Moffett Field, CA: Ames Research Center.

Does Desire for Control Affect Interactions in an Adaptive Automated Environment?

Cristina Bubb-Lewis
Lucent Technologies

Mark W. Scerbo
Old Dominion University

INTRODUCTION

Adaptive automation refers to dynamic systems which adjust their methods of operation in response to changes in situational demands (Rouse, 1988). In an adaptive automation system, the human and the machine must work together as partners in order to maintain optimal operation of the system (Scerbo, 1994). Because of the close relationship between the human and the system, it seems reasonable that human-machine communication would be critical to adaptive automation. Thus, one goal of the present study was to examine the effects of different communication patterns on performance with an adaptive task.

DESIRE FOR CONTROL

Desirability for Control (DC) refers to one's need to control the environment (Burger and Cooper, 1979). High DC people are described as decisive, assertive, and active while low DC people are described as nonassertive, passive, and indecisive. High-DC participants have been shown to display higher levels of aspiration, have higher expectancies for their performance, and set more realistic expectations than low-DC participants (Burger, 1985). In addition, high-DC participants respond to a challenging task with more effort, persist longer, and perform better than low-DC participants.

Although DC has not been studied with regard to team dynamics, it seems to be relevant. A high-DC person might be less willing to act as a team member in solving problems because of their need to control situations. On the other hand, a low-DC person might rely too heavily on their partner. Either of these effects within a team could have a detrimental impact on the efficiency of the interaction, and consequently could affect human-computer interaction in adaptive automation in a similar manner.

PRESENT STUDY

The current study used a Wizard-of-Oz simulation (Gould, Conti, & Hovanyecz, 1983) to study the effects of communication mode, task complexity, and desire for control on performance with an adaptive task. The focus of this paper will be on the DC results.

The simulation involved a"talking" adaptive computer that helped participants complete computer tasks. Four modes of communication were used that differed in the level of restriction placed on communication between the participant and computer. Two levels of task complexity were used with all participants completing both simple and complex tasks. DC was measured and participants were split into high-DC and low-DC groups for analysis. Dependent measures included task score as well as responses on a participant questionnaire

It was hypothesized that as more restrictions were placed on communication, performance would decrease and computer control would increase. Also, restricting communication was expected to make the interaction less efficient and make it more difficult for participants to complete the tasks, thereby leading to lower scores and increased computer intervention. Task scores were also expected to be higher for simple tasks, but that increases in restriction would lower scores primarily for complex tasks because communication would be more important for the complex tasks. DC was expected to amplify the basic effects described above for communication restriction and task complexity. Specifically, high-DC participants were expected to score higher, but only on the complex tasks because high-DC individuals have been shown to respond better to challenging tasks (Burger, 1985).

METHOD

Participants

Sixty-four university students were prescreened for the DC variable using the DC Scale (Burger & Cooper, 1979). The DC Scale has been found to have adequate internal consistency ($r_{yy} = .80$) and test-retest reliability ($a = .75$), as well as discriminant validity from measures of locus of control ($r = -.19$; Rotter, 1966) and social desirability ($r = .11$; Crowne & Marlowe, 1960). Construct validation evidence was provided by studies on learned helplessness, hypnosis, and illusion of control (Burger & Cooper, 1979).

The mean DC score for females and males was $M = 102.96$, $SD = 12.51$, and $M = 106.17$, $SD = 11.23$, respectively. A t test showed these means to be significantly different ($t = -2.20, p < .05$), therefore separate cutoff scores were used for males and females. The cutoff scores were a DC score less than 93 or greater than 112 for females and a DC score less than 98 or greater than 114 for males. Sixteen participants were assigned to each communication mode in a quasi-matched groups design to reduce the possibility of group differences for DC score. The experimenters were blind to the DC level of each participant.

Apparatus

Two separate rooms were used to run the experiment. The participant's room contained an IBM compatible computer. A mouse was also connected to the participant's computer, but was placed in the experimenter's room by a cable that ran through the wall. A 2-station wired intercom ran between the two rooms and allowed the experimenters to listen to the participants' comments. A signal splitter was connected to the participant's computer and duplicated the images from the participant's monitor on a second monitor in the experimenter's room allowing the experimenters to follow the participant's progress. The experimenter's room contained a separate IBM compatible computer used to generate the computer partner's audio responses. Thus, the speakers connected to this computer were placed in the participant's room.

A Wizard-of-Oz method was used to simulate the computer partner (Gould et al., 1983). The experimenter could hear the participant's questions and responded according to the appropriate communication condition by selecting a prerecorded .WAV file to play to the subject. In addition, an assistant could monitor the participant's actions on the second VDT and used the mouse to take control of the task when necessary.

Each participant was asked to complete a series of tasks using the Expert Travel Planner program by Expert Software. This is a commercial program that allows users to plan trips. The participant used the keyboard to complete the tasks. The mouse was reserved for the computer partner and was used when the computer took over according to the adaptive rules (see below).

Procedure

Participants were tested in a small, sound attenuated room without windows. The experimenter and assistant were in a separate room next door. The participants were told that they were testing a computerized system that would act as a partner in completing the tasks. They were asked to interact with this computer partner and that the computer would act just like a human partner by giving advice or helping them work on the tasks. They were asked to work with the computer to complete the tasks as quickly and accurately as possible. The participants then completed an ice breaking session to show how to speak to the computer and how the computer might take over the task.

During the session, the computer partner answered the participant's questions according to the rules for group assignment. The adaptive rules allowed the computer partner to take over the task if the participant did not speak or press any keys for 30 sec. If the participant spent more than 3 min working on one task, the computer partner would interrupt every 30 sec until the task was completed. Each computer intervention consisted of completing the next step in the task solution and informing the participant what was being done. Participants completed two sessions with simple and complex task sets counterbalanced across sessions. Afterward, they filled out a questionnaire which addressed the ability to communicate with the computer partner, the helpfulness of the computer partner, and enjoyment of the interaction.

Experimental Design

The study used a 4 (communication mode) x 2 (task complexity) x 2 (DC) mixed-model factorial design with participants nested in communication mode and desire for control. There were four levels of communication mode:

1. Context Sensitive Interaction. The participant was permitted to speak normally and the computer partner chose a response from a list that included context sensitive responses. For example, "Use the Facts function on the File menu to find out about Fayetteville's history."

2. Limited Response Interaction. The participant could speak normally and the computer partner chose a response from a limited list that did not include context sensitive responses. For example, "Use the Facts function on the File menu."

3. Limited Human-Computer Interaction. The participant was required to use keywords to formulate utterances and the computer partner chose responses from a limited list that did not include context sensitive responses. For example, participants were told they could use the keyword "Help" plus a menu function to receive information about that function.

4. Control Group. This group was not able to communicate with the computer partner. As in all the other groups, however, the computer partner would take over the tasks according to the adaptive rules.

Task Complexity

Tasks were designed to represent the program's capabilities, divided into simple and complex sets, and then rated for difficulty on a scale of 1 to 7 during a pilot test. The simple tasks had a lower difficulty rating of $M = 2.57$, $SD = .77$, while the complex tasks had a higher difficulty rating of $M = 4.05$, $SD = .92$. In addition, a GOMS analysis (Kieras, 1988) was performed to ensure that the task goals were evenly distributed throughout the simple and complex tasks.

Dependent Measures

The primary dependent measure was the number of tasks completed in the allotted time period. This score was then divided by the minutes on task to obtain tasks per minute. The level of computer control was measured by tabulating the number of times the computer partner took over the tasks from the participant. This number was divided by the minutes on task to obtain the number of interruptions per minute.

RESULTS

The quasi-matched groups assignment of participants in the present study was used so that equivalent distributions of DC scores would be present in each communication mode. An analysis of the DC scores confirmed that there were no significant differences in DC score for communication mode or gender.

Groups who could communicate freely performed better than those where communication was restricted or denied, $F(3,55) = 8.18$, $p < .05$. In fact, restriction of communication lowered performance efficiency by more than 20 percent over unrestricted communication. This large difference between unrestricted and restricted groups was apparent for other dependent measures as well (i.e., task difficulty ratings and computer control).

Participants scored higher on simple tasks than complex tasks, $F(1,55) = 317.17$, $p < .05$. However, the communication mode by task complexity interaction, $F(3,55) = 3.82$, $p < .05$, showed that the unrestricted groups scored higher than the restricted groups on only the simple tasks. This may have been due to the difficulty participants had in formulating questions for the complex tasks, thereby negating the benefits of unrestricted communication on those tasks.

Desire for Control

It was hypothesized that high-DC participants would perform better on complex tasks, but that there would be little difference on the simple tasks. There were, however, no effects of DC on task score at all. Unfortunately, these results do not support Burger's (1985) research on DC and there does not seem to be an explanation for participants approaching these tasks differently from other achievement-related tasks. In the past, researchers studying DC (Burger, 1985; Burger & Cooper, 1979) have used a less powerful median split method and have found significant effects. This suggests that the lack of effects here may be valid.

It is possible that the complex tasks in the present study did not provide enough challenge to illicit greater effort from the high-DC participants. Past research, however, has found DC effects with only minor differences between tasks. For example, Burger (1985) used a proofreading task to examine the responses of high- and low-DC participants. The more challenging condition in his experiment consisted of the proofreading task plus a word counting task. The results showed that high-DC participants proofread more lines in the more challenging condition, but that there were no differences for DC in the less challenging condition. The manipulation used in Burger's (1985) study does not seem particularly strong, yet he still found significant effects due to DC. Given the established differences in difficulty between the task sets used in the present study, the complex tasks should have been challenging enough to reveal a difference in DC.

It should also be noted, however, that participants in the present study did not perceive differences between the simple and complex tasks to be excessive. Therefore, if high-DC subjects did not perceive the complex tasks as challenging, they might not have been motivated to work very hard at them.

It is also possible that the high-DC participants did not view the computer as a threat to their control. If high-DC participants considered the computer to be a tool and not another person to whom they were relinquishing control, they might have been more willing to accept its help. In fact, the participants' ratings of the computer's helpfulness (see below) support this explanation. Perhaps DC does not apply to working with a computer partner in the same way it does with a human. This is an issue that should be explored in more detail.

Participant Opinions

Although there were no performance differences found for DC, there were differences in the way high- and low-DC participants perceived the tasks. Participants rated their ability to communicate with the computer partner on a questionnaire designed to provide additional information about the effects of communication restriction on adaptive task interactions. High-DC participants rated ability to communicate as easier than low-DC participants, $F(1,41) = 5.88, p < .05$. This might be due to the high-DC tendency to want to master a situation. Although they might have been more motivated to do well, it did not improve their performance. The more favorable ratings of the high-DC participants were likely the result of *perceptions* of performance.

None of the independent variables affected helpfulness. Again, this supports the assertion that high-DC participants did not see the computer as a threat to their control because they did not reject its help and rated its helpfulness similar to that of low-DC participants.

Participants were also asked to rate how well they enjoyed interacting with the computer. The results showed an interaction between communication mode and desire for control, $F(3,55) = 3.79, p < .05$. High-DC participants reported more enjoyment than low-DC participants in the limited human-computer interaction condition, but less enjoyment in the control condition. This interaction might be explained by the way high- and low-DC participants perceived the tasks. The limited human-computer interaction group was faced with the most challenging condition (as shown by task score) in which participants were still able to actively communicate with the computer. It is possible that high-DC participants in this group may have enjoyed the interaction more because of the challenge of formulating appropriate utterances to communicate with the computer. By contrast, the control group had no control over their interaction with the computer because they could not communicate in any way. This inability to control when and how the computer would intervene may have caused high-DC participants to rate enjoyment as lower in this group.

CONCLUSIONS

This study was designed to examine the effects of DC on adaptive human-computer interactions. DC has been found to reliably affect achievement-related behavior in human interactions. It was hypothesized that DC would have similar effects on adaptive human-computer interactions. The results, however, did not support Burger's (1985) research on DC. It is possible that DC does not apply to interactions with computers in the same way it does with a human. Future research may need to distinguish between DC for humans and computers.

REFERENCES

Burger, J. (1985). Desire for control and achievement-related behaviors. *Journal of Personality and Social Psychology, 48,* 1520-1533.

Burger, J. & Cooper, H. (1979). The desirability of control. *Motivation and Emotion, 3,* 381-393.

Crowne, D., & Marlowe, D. (1960). A new scale of social desirability independent of psychopathology. *Journal of Consulting Psychology, 24,* 349-354.

Gould, J., Conti, J., & Hovanyecz, T. (1983). Composing letters with a simulated listening typewriter. *Communications of the ACM, 26,* 295-308.

Kieras, D. (1988). Towards a practical GOMS model methodology for user interface design. In M. Helander (Ed.), *Handbook of human-computer interaction* (pp. 135-157). Amsterdam: North Holland.

Rotter, J. (1966). Generalized expectancies for internal versus external control of reinforcement. *Psychological Monographs, 80,* Whole Number 609.

Rouse, W. B. (1988). Adaptive aiding for human/computer control. *Human Factors, 30,* 431-443.

Scerbo, M. W. (1994). Implementing adaptive automation in aviation: The pilot-cockpit team. In M. Mouloua & R. Parasuraman (Eds.), *Human performance in automated systems: Current research and trends* (pp. 249-255). Hillsdale, NJ: Lawrence Erlbaum Assoc.

Adaptive Automation of a Dynamic Control Task Based on Workload Assessment through a Secondary Monitoring Task

David B. Kaber and Jennifer M. Riley
Mississippi State University

INTRODUCTION

Adaptive automation (AA) can be defined as the dynamic allocation of tasks or control functions between human operators and automated control systems over time, based on the state of the human-task-environment system. Research interests in AA are expanding beyond the realm of psychomotor tasks, such as tracking in aircraft piloting, to cognitive tasks involving strategizing for automated system performance, including AA in air traffic control (Hilburn & Jorna, 1997). This interest has been motivated by the success of AA in psychomotor tasks for addressing issues such as increased monitoring and miscalculated trust in automated system reliability, degradation of skills, and attentional resource overload due to static automation. All of these issues may potentially impede human cognition and decision-making under static automation as well; thus, AA may be helpful for cognitive task performance.

Aside from a need to examine AA in the context of different types of tasks, research questions remain as to how AA can be most effectively encouraged in human-machine systems for promoting overall performance. Different mechanisms for triggering dynamic control allocations between human and computer servers have been proposed and some empirically examined in the contexts of monitoring and tracking tasks. These mechanisms include operator performance monitoring, operator workload monitoring, systems activities monitoring for critical events, and system/operator behavior modeling. Of these mechanisms, operator workload (arousal) monitoring has been predominantly examined in experimental settings. Specifically, physiological measures including heart rate variability and electroencephalogram signals have been related to control allocations for increasing or decreasing operator arousal levels (e.g., Byrne & Parasuraman, 1996). However, the relationship of other objective measures of workload, including secondary task performance, to effective facilitation of AA has not been investigated for manipulating operator cognitive workload or indicating the potential for out-of-the-loop performance problems. Potential advantages of secondary task workload methods for AA include reliable indication of resources expended by primary task processing and applicability across a broad range of tasks. Further, such measures directly represent trade-offs in cognitive resources and changes in task performance. With these characteristics in mind, secondary tasks may be useful as effective predictors for AA based on operator primary task workload.

In potential applications of AA based on operator workload, the question can be raised as to who should decide whether, and when, automation should be invoked—the human or the computer. Some have expressed concern that humans may not be the best judges of task allocations because they are limited in decision making by individual perceptions and professionalism and may not be qualified to make control allocation decisions. Others acknowledge that while it may not be optimal for the human to have total control of automation decisions, problems can arise when they are completely removed from the loop. Furthermore, if machine control of task allocations is exclusive and impervious to human override, system safety may be comprised in situations where artificial intelligence regarding task circumstances is not superior to that of the human operator.

The question of who makes allocation decisions needs to be resolved in various contexts, including psychomotor and cognitive task performance. One potential solution may be to design systems for computer decision-making regarding function allocation based on operator workload, but allow the human to invoke automation upon system suggestions. Methods that utilize both human and computer servers in different ways for allocation decisions need to be empirically investigated.

EXPERIMENT

Objective and Subjects

An experiment was conducted to address some of the needs identified above. These included assessing the usefulness of a secondary task measure for capturing and manipulating human workload in a dual-task paradigm by predicting dynamic, manual and automated control allocations in a primary cognitive task. The study was also intended to address the issue of human or automation directed invocation of AA. Specifically, it was to examine the speculated solution to rigid operator or system authority involving human and computer collaboration in allocation decisions (based on operator workload as a trigger to AA). This was to be accomplished by evaluating the effect of human decision-making, using a computer-based assistant, in task control allocations of the primary task on overall system performance and workload quantified by the secondary task.

Ten Mississippi State University students were recruited for the experiment. They ranged in age from 21 to 23 years and possessed 20/20 or corrected to normal visual acuity. The subjects were required to have personal computer (PC) experience.

Tasks and Equipment

Two computer-based tasks were used in this study including a secondary gauge monitoring task and dynamic control task (Multitask©). Both tasks were presented through high resolution, graphics monitors (1024×768 pixels) integrated with 166-MHz Pentium® PCs. Conventional mouse controllers and keyboards were also integrated with the systems.

The gauge monitoring task presented a fixed-scale display with a moving pointer. Subjects were required to monitor pointer movements and detect when a deviation occurred from a central "acceptable" region on the scale to an "unacceptable" region. The subjects were also required to correct for these deviations by depressing keys on a keyboard facilitating upward or downward pointer movements. This task was psychomotor in nature involving subject monitoring diagnosis and action. Performance was recorded as ratio of the number of "unacceptable" pointer deviations detected to the total number of deviations (i.e., the hit-to-signal ratio).

Multitask© presents square targets of different sizes and colors on a "radar scope" display. The targets travel at different speeds towards the center of the display. An operator's task is to select and eliminate the targets by collapsing them before they expire (i.e., reach the center of the display) or collide with one another. Target selection is accomplished by using a mouse controller or keyboard. As targets are collapsed, reward points are added to a total score. If targets expire or collide with each other, penalty points are deducted from a subject's score. Each target has different penalty and reward points associated with it depending upon its size (small, medium and large) and color (red, green and blue). The exact target rewards and penalties are displayed as data tags attached to them. The goals of a subject are to maximize the reward points and minimize the penalty points.

Multitask© is a cognitive task involving operator judgements and projections on the temporal and spatial relationships of targets, as well as interpretation of target characteristics for decision-making and prioritization in processing. The task goals posed to operators are in opposition to each other and require development of a conscious strategy for selecting targets to satisfy both simultaneously. The speed at which targets travel and their distances from the center of the display are also considered to be factors influencing target selection.

The modified version of Multitask© used in this experiment allowed for targets to be collapsed by one-of-two different methods representing levels of automation in Endsley and Kaber's (in press) taxonomy of LOA. These levels included 'Manual Control' and 'Blended Decision-Making', which have been identified as being significantly different in terms of Multitask© performance. Blended Decision-Making performance, as well as Manual Control functioning, was recorded as the ratio of the number of targets collapsed to the total number of targets presented. Since operator cognitive resource demands in the Multitask© simulation as part of Blended Decision-Making and Manual Control performance completely overlapped those of the gauge monitoring task, the latter was anticipated to be a sensitive, objective measure of workload.

Procedures and Data Collection and Analysis

At the onset of the experiment, subjects were trained in the secondary gauge monitoring task. 'Baseline' performance in the task was established as the mean hit-to-signal ratio in the second of two training trials. (This ratio was to later be compared with hit-to-signal ratios observed under test conditions for establishing subject workload.) Gauge monitoring training was followed by practice in the Multitask© simulation using Manual Control and Blended Decision-Making. Subsequently, training in simultaneous performance of both tasks was conducted.

For experimental testing, subjects were divided into two groups based on how primary task allocation decisions were to be made. One group was required to make allocations between manual and automated control modes based on computer assistant mandates. Another group was instructed to consider suggestions from the computer assistant for manual or automated control in making decisions, but the subjects were not required to implement the suggestions. Both the mandates and suggestions were based on computer monitoring of subject workload through the secondary task.

Workload criteria were established for computer assistant mandates or suggestions of manual and automated control of the primary task based on secondary task performance data collected during a pilot study. The coefficient of variation for monitoring performance for four of the ten subjects (19.93) was used as an indicator of the maximum acceptable deviation in test condition workload from baseline workload for the other six subjects during the full investigation. If a subject's performance in the secondary gauge monitoring task under dual-task conditions decreased (i.e., workload increased) from the baseline level by more than 20%, the computer assistant either mandated or suggested that automation be invoked in the primary task (depending upon subject grouping). (The text, "Use automation in the target elimination task" was displayed on the secondary task screen and the computer sounded 3 beeps.) Manual Control was ordered or recommended by the assistant during testing if a subject's workload returned to the baseline level. (The text, "Use manual control in the target elimination task" was displayed on the secondary task screen and the computer sounded 3 beeps.)

During all test trials the hit-to-signal ratio for the secondary task was averaged and recorded at 30-second intervals. Primary task performance was averaged and recorded at 1-min intervals. This data and the observations on subject workload were analyzed through a two-way analysis of variance with subject Group (mandated AA or non-mandated AA) and Level of Automation (LOA) (Manual Control or Blended Decision-Making) as between- and within-subjects predictors, respectively.

RESULTS

Results on primary task performance revealed a significant main effect of LOA ($F(1,4) = 359.11$, $p = .0001$) and the presence of the Group $\times$ LOA interaction ($F(1,4) = 32.97, p = 0.0046$). The Group ($F(1,4) = 1.32, p = 0.3150$) main effect was not significant when assessed in the presence of the variance due to individual differences among subjects within groups. Figure 1 shows a plot of the mean response across subjects for each group at the different LOAs. Orthogonal contrasts controlling for experimentwise error revealed mandated and non-mandated AA group performance under Blended Decision-Making to be significantly superior to Manual Control ($p < 0.05$). Mandated AA group Manual Control was significantly superior ($p < 0.05$) to manual processing by the non-mandated AA group.

Results on workload measured through the secondary monitoring task revealed significant main effects of primary task LOA ($F(1,4) = 11.34, p = 0.0281$) and subject Group ($F(1,4) = 8.56, p = 0.0430$). The two-way interaction of Group × LOA ($F(1,4) = 78.29, p = 0.0009$) was also found to be significant. Figure 2 shows a plot of the mean secondary task workload for each subject group using Manual Control and Blended Decision-Making in the Multitask© simulation. Orthogonal contrasts revealed mandated AA group workload associated with performance of the primary task under Blended Decision-Making to be significantly less ($p < 0.05$) than Manual Control workload. On average, workload for mandated AA subjects during automated control of the primary task was significantly lower ($p < 0.05$) than that experienced by non-mandated AA subjects.

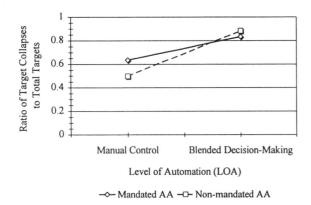

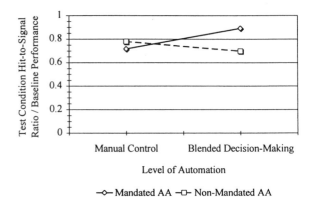

Figure 1. Plot of mean primary task performance.

Figure 2. Plot of mean secondary performance.

DISCUSSION

Subjects who were required to affect task allocation mandates from the computer assistant functioned substantially better under Manual Control than subjects considering computer assistant messages as mere suggestions. Subjects in the mandated AA group may have been able to produce better manual performance than subjects in the non-mandated AA group because they did not devote cognitive resources to evaluating whether the primary task should have been automated. The computer off-loaded this responsibility from them and they demonstrated confidence in the mandates. The non-mandated AA group, however, often differed from the computer assistant in judgement of task allocation opportunities. Subjects may have felt insecure in their own assessments of task allocation upon recognizing that they did not agree with those of the computer assistant or that their performance in the Multitask© was poor. Consequently, it is possible that they devoted significant cognitive resources to continually questioning whether to automate. This activity may have served to significantly distract subjects from task processing. These inferences are in line with the concern that adaptive decision aiding in cognitive tasks may pose increased mental workload for operators due to the need to evaluate a system's advice.

Improved mandated AA group performance in the secondary task during automation of the primary task may be attributed to subjects concentrating on the gauge display to perceive mandates from the computer assistant for task allocations. Non-mandated AA subjects concentrated on the gauge display to a lesser extent because they were not required to implement the computer assistant suggestions, but more importantly because they did not perceive a need to switch from automated control of the primary task to

Manual Control. As a result, mandated AA subjects were more likely to observe "unacceptable" pointer deviations in the secondary task than non-mandated AA subjects, which indicated reduced workload for the former group. Poorer non-mandated AA group performance in gauge monitoring can be attributed to devotion of greater resources to the primary task. Decrease in non-mandated AA subject performance in the secondary task indicated increased primary task workload.

For mandated AA subjects, manual control in the primary task served to significantly increase secondary task workload, as compared to the workload associated with automation. This can be attributed to higher attentional demands being placed on the operator under Manual Control in terms of implementing a target processing strategy, as compared to Blended Decision-Making. Under Manual Control subjects were required to track and collapse targets using the mouse controller; whereas Blended Decision-Making provided for automatic computer target elimination. The result may also be attributed to the observed lack of mandated AA subject concern with allocations to automation from Manual Control and, consequently, decreased operator attention to the gauge display for computer assistant mandates. Under Multitask© Manual Control, the mandated AA group was less likely to observe pointer errors, decreasing secondary task performance and indicating increased subject workload.

CONCLUSIONS

Human workload measurement through comparison of operational secondary task performance against baseline levels established under optimal conditions can be used as a trigger mechanism for AA to cause changes in operator primary task workload. This workload assessment method appears to be sensitive to both the degree of human involvement in a primary cognitive task and human responsibility in decision-making concerning control allocations. Whether completely human manual processing or blended human-computer strategizing accompanied by computer task implementation is employed, objective workload is significantly affected when operators are mandated to use either LOA, as compared to non-mandated use. The responsibility of allocation decisions in a cognitively complex task appears to have an effect on task performance and operator workload; however, this is dependent upon whether the allocation is to manual or automated control. Humans who are mandated in allocation decisions by a computer controller tend to be more vigilant of Manual Control mandates than those for automation. This may occur because operators view the manual processing mandates as directives to work. Lack of vigilance to automation mandates can be attributed to concentration on Manual Control performance.

On the contrary, humans who are not mandated in implementing AA, appear to be more vigilant of suggestions by a computer assistant for automated control allocations, as compared to suggestions for manual functioning. This can be attributed to operators interpreting computer suggestions to use automation as a form of validation of their decision to do so. Lack of non-mandated AA operator vigilance to manual control suggestions is likely due to a lack of a perceived need to use Manual Control. This is a particularly troubling inference, as it suggests that operators may not be aware of out-of-the-loop performance problems or appreciate their associated consequences.

REFERENCES

Byrne, E.A. & Parasuraman, R. (1996). Psychophysiology and adaptive automation. *Biological psychology*, 42(3), 249-268.

Endsley, M.R. & Kaber, D. B. (in press). Level of automation effects on performance, situation awareness and workload in a dynamic control task. *Ergonomics*.

Hilburn, B. & Jorna, P. (1997). The effect of adaptive air traffic control (ATC) decision aiding on controller mental workload. In M. Mouloua & J.M. Koonce (Eds.), *Human-Automation Interaction: Research and Practice* (pp. 84-91). Mahwah, NJ: Lawrence Erlbaum Associates.

Dynamic Function Allocation for Naval Command and Control

Catherine A. Morgan, Catherine A. Cook and Colin Corbridge
DERA Centre for Human Sciences, UK

INTRODUCTION

Dynamic function allocation (DFA) refers to the variable distribution of functions in real time between the system and the human operator(s) to achieve optimal system performance. In the majority of complex man-machine systems, the allocation of functions between the system and operator are decided during the design process, and remain fixed for the life of the system. DFA would enable variable distribution of functions between the system and human according to operational demands and the current human and system resources available. Thus, a dynamic method of allocating functions would enable those tasks that can be performed equally well by both the human and system to be allocated to the entity (human or system) that has the resources available to best deal with the task or part of the task at that particular point in time. DFA has been proposed as a potential design option in future Naval Command and Control systems, for several reasons (Cook et al., 1997). Future threat capabilities mean that system demands will be such that they may exceed human capabilities. Thus, the human operator may be unable to cope with such demands without some form of additional automation. Total system automation in military systems such as Naval Command and Control is unfeasible as the human operator is ultimately accountable for any decision to deploy weapons against a perceived threat. It is therefore essential that the human decision maker is maintained within the decision-making loop and provided with a cohesive set of functions which facilitate the operator's role within the total system, as well as enabling him/her to maintain a high degree of situational awareness. Theoretically DFA can achieve these system requirements. In addition, it has been proposed that DFA may also reduce the longer term adverse consequences of system over-automation such as operator deskilling and reduced job satisfaction (Lockhart et al., 1993). Another advantage of DFA, compared to non-adaptive systems is that it offers a potential solution for optimal manning of complex systems. The UK Ministry of Defence is under increasing pressure to reduce the through life costs of future platforms, with manpower being one of the largest of these costs. Operator workload in Naval Command systems is characterised by long periods of relative inactivity or steady workload followed by bursts of intense activity. However, the requirements for such systems are that manning levels are high enough to cope with the peak demand. It seems reasonable to propose that in situations of intense workload, functions could be offloaded from the operator to the machine, so maintaining the operators workload at an optimal level that is neither too high nor too low. In this way the system could be manned by fewer operators who would not be underloaded during normal situations yet could cope with excessive task demands through the dynamic allocation of tasks to the system. CHS has been investigating whether DFA is a potential option for future Naval Command and Control systems.

The Research Environment

The research testbed is an abstraction of Force/Platform Threat Evaluation Weapon Assignment in which the subject assumes the role of force Anti-Air Warfare Co-ordinator, and is fully described in (Cook et al., 1997). Naval Command and Control comprises three major stages; compilation of the tactical picture; situation assessment and threat prioritisation; allocation of resources to meet the assessed threat. Automation is already being introduced to some degree in the first two stages. Thus these two stages are fully automated within the Naval Testbed. However as the human operator is ultimately accountable for the deployment of weapons it is unrealistic and unfeasible to fully automate the third stage, i.e. allocation of resources. The

134

resource allocation stage of Naval Command and Control was considered to be an appropriate focus for this investigation of DFA. The layout of the interface of the Naval Dynamic Allocation Research Testbed (N-DART) is illustrated in Figure 1. The overall objective of the task is to counter all incoming threats as efficiently as possible to ensure own force safety. As tactical picture compilation and threat evaluation are performed automatically, the subject's role for these two sub-tasks is to monitor the situation presented by the CMS through the Label Plan Display and the Threat Evaluation List. The resource allocation subtask can either be performed manually by the subject or allocated to the Combat Management System (CMS) - an underlying expert system that can assign weapons to selected threats, (DFA). In addition, a communications channel provides subjects with early warning information about threats or neutral aircraft that are beyond Own Ships sensor range. Subjects are required to input this early warning information to update the tactical picture accordingly. The communications sub-task can also be performed manually or automated (DFA). Although the simulation was designed to be performed by non-Naval subjects, it incorporates sufficient realism that more detailed research can be conducted with subjects with domain expertise.

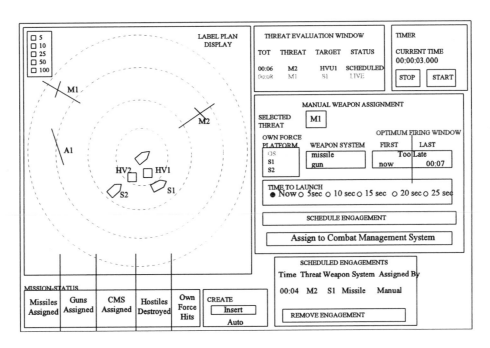

Figure 1. Naval Dynamic Allocation Research Testbed

CASE STUDY

Rationale

The main aim of the case study was to investigate whether an explicit mode of DFA is beneficial in a complex task environment resembling Naval Command and Control, in terms of overall performance and situation awareness, compared to a non-adaptive control condition. A further aim of the case study was to compare the performance of Subject Matter Experts (SME) and non-naval subjects, and the strategies they used when performing the task.

Subjects

Two SMEs and two non-experts participated in the case study. The SMEs were both officers in the Royal Navy and with extensive Anti-Air Warfare experience, aged between 50-55 years old. The non-expert subjects were matched as closely as possible for age and had no experience or knowledge of Naval Command and Control. None of the participants had any previous experience of the task.

Task

Performance was assessed using the Naval Dynamic Allocation Research Testbed (N-DART) (Figure 1). Three training scenarios were developed plus two experimental scenarios. Training scenarios 1 and 2 introduced subjects to the resource allocation task and communications task respectively, with the two sub-tasks being combined in the third training scenario. The two experimental scenarios were based on realistic demands and lasted 30 minutes. The experimental scenarios were matched as closely as possible in terms of level of difficulty, with each scenario containing a total of 44 threats, 12 hostile early warning messages and 2 unknown early warning messages. Own Force consisted of Own Ship, plus two other armed ships and two high value units. The threats were sequenced in raids, with a total of 9 raids in each scenario. The number of threats in each raid varied from 3-6.

Performance Measures

A range of objective performance measures were obtained for both the weapon assignment task and communications task, such as number of hits sustained to Own Force and number of threats removed. Feedback from subjects was used to gain information about subjects' situation awareness, and the strategies used when performing the task.

Experimental Design

All subjects had to successfully complete a two and a half hour training programme, before being able to participate in the study. A within-subject matched pairs experimental design was used. Therefore, all subjects performed scenario A followed by scenario B. One SME and one non-SME performed the task manually throughout scenario A, and with DFA in scenario B, and vice versa for the other SME and non-SME. Subjects were instructed that their objective was to defend Own Force, by removing threats as quickly as possible using the minimum of resources. Throughout each trial subjects were required to give verbal reports on the tactical situation, and the strategies they were using when performing the task. More detailed feedback on the task and strategies employed when performing the task, were obtained from subjects in a post-trial debriefing session.

SUMMARY OF RESULTS

Performance Data

As only four subjects participated in the case study, the numbers were considered too small for any statistical analyses to be performed. An indication of each subject's overall performance on the task was gained from the total number of hits sustained to Own Force and the percentage of manually assigned weapons that were successful in destroying incoming missiles. The performance measures obtained for each scenario, together with number of threats each subject assigned to the CMS, were compared. The performance data showed that in both cases the SMEs performed much better on the task when able to allocate the weapon assignment task to the CMS (DFA), compared to when they were required to perform the task manually throughout the experiment. Fewer hits were sustained to Own Force platforms, and a much higher percentage of manually assigned weapons were successful in eliminating incoming threats. However, only one of the non-SMEs assigned threats to the CMS in the DFA condition. Although this non-SME assigned considerably more threats to the CMS than the SMEs, the subject's overall performance on the task was also enhanced by DFA. A much higher percentage of manually assigned weapons successfully destroyed incoming threats in the DFA condition compared to the fully manual condition. There were no performance differences between the fully manual condition and explicit control condition for the non-SME that chose not to assign threats to the CMS.

Subjective data

The information obtained from subjects' verbal reports and post-trial feedback indicates that being able to offload tasks to the system when overloaded not only helped to reduce their workload to a level that was more manageable, but also helped to enhance their awareness of the tactical situation. The two SMEs and one non-naval subject reported that in the fully manual condition under high levels of task demand, they did not have enough time to gather all the information available from the data sources, to evaluate which weapon should be assigned to each threat. As a result they repeatedly assigned the same weapon to each threat even though it wasn't always the most appropriate response. However, being able to offload threats to the CMS in the DFA condition when task demands were high was seen as a great benefit. Subjects reported that as they had more time to decide which weapons to manually assign, they were able to gain and assimilate tactical information from more data sources. As a result the subjects felt that under high task demands the manual weapon assignment decisions made in the DFA condition were more effective than in the non-adaptive condition, as they were more aware of the tactical situation. Interestingly the non-Naval subject that chose not to use automation in the DFA condition reported that he felt that he was able to deal with the threats effectively without any aid from the CMS. He also felt he could perform the task better than the CMS. Consequently he was unwilling to hand over responsibility to the system.

DISCUSSION

The performance data obtained from this case study suggests that compared to a fully manual system, a system incorporating an operator controlled (explicit) method of DFA can enhance overall system performance when task demands are high. These findings are in accordance with previous studies which also found that, compared to a non-adaptive manual system explicit DFA improved overall performance in a multiple task environment when task demands were high (Tattersall & Morgan, 1997). Subjective data also suggest that, compared to a non-adaptive fully manual system, explicit DFA can reduce operator's workload, and enhance their situation awareness. Both SMEs and the non-SME that used automation, reported that they were more proficient at dealing with the incoming threats, and had more time to assess the tactical situation and assimilate all the information presented. This also resulted in their feeling less stressed in the DFA condition compared to the fully manual system. Thus the performance data together with the subjective data suggest that offloading some of the threats to the CMS enabled the operator to analyse his options in-depth and select the most appropriate weapon system to assign to a threat, so increasing the likelihood that the assigned weapon would successfully hit an incoming threat. Further, subjective data relating to the strategies used by subjects when allocating either the weapon assignment and/or communications task to the CMS, indicates that DFA is a method of aiding an operator in a way that supports their existing decision making process. The information collected from the non-naval subject that chose not to allocate threats to the CMS, highlights two previously identified drawbacks of explicit DFA (Reiger & Greenstein, 1982). Firstly operators may not be aware of any performance decrements on a task, and secondly the operator may be reluctant to hand over control to the system, especially in circumstances where the situation is life threatening. This raises important issues relating to operator trust, which may effect the acceptability of a dynamic method of allocating functions in complex systems. In summary, the results of this study indicate that if used appropriately explicit DFA can be an effective, non-intrusive way of aiding an operator that complements the way that they work. However, although this case study provides some evidence to suggest that compared to a fully manual system, DFA can enhance overall system effectiveness, more extensive empirical evidence is necessary. Further, although this study showed apparent benefits of explicit DFA compared to fully manual performance, it is as yet unclear how system performance with DFA compares to full automation. In addition, research is needed to investigate other modes of DFA, such as the various system controlled (implicit) methods of dynamically allocating functions and a comparisons between implicit and explicit modes. Again, studies are planned to investigate these issues. This research to investigate DFA is still in its initial stages and more extensive research is necessary to evaluate fully the potential of various methods of implementing DFA before it can be considered for inclusion in operational Naval Command and Control systems. However, our initial findings are promising and suggests further work in this area is warranted..

© Crown Copyright (1998) - DERA
Published with the permission of DERA on behalf of the Controller HMSO
Any views expressed in this paper are those of the authors and do not necessarily represent those of DERA.

REFERENCES

Cook, C.A., Corbridge, C., Morgan, C.A., & Tattersall, A.J. (1997). Developing Dynamic Function Allocation for Future Naval Systems. *Proceedings for the Human Factors and Ergonomics Society 41st Annual Meeting*, 1047-1054.

Lockhart, J.M., Strub, M.H., Hawley, J.K., & Tapia, L.A. (1993). Automation and supervisory control: A perspective on human performance, training, and performance aiding. *Proceedings of the Human Factors and Ergonomics Society 37th Annual Meeting*, 1211-1215.

Reiger, C.A., & Greenstein, J.S. (1982). The allocation of tasks between the human and computer in automated systems. *Proceedings of the IEEE 1982 International Conference on Cybernetics and Society*, 204-208.

Tattersall, A.J., & Morgan, C.A. (1997). The Function and Effectiveness of Dynamic Task Allocation. *Engineering Psychology and Cognitive Ergonomics, Vol. 2.* (pp. 247-257). Ashgate Publishing Ltd.

Behavioral, Subjective, and Psychophysiological Correlates of Various Schedules of Short-Cycle Automation

Gerald A. Hadley   **Lawrence J. Prinzel III** [1]   **Frederick G. Freeman**   **Peter J. Mikulka**
Old Dominion University

INTRODUCTION

The presence of automation has pervaded almost every aspect of modern life. The use of automation has made work easier, more efficient, and safer, as well as giving us more leisure time. However, research has also shown that the increase in automation has not been without costs. These costs have resulted in increased interest in advanced automation concepts. One of these concepts is automation that is adaptive or flexible in nature and can respond to changing situational demands (Hancock & Chignell, 1987; Scerbo, 1996).

One challenge facing those seeking to implement adaptive automation technology concerns how changes among modes of operation will be accomplished (Scerbo, 1996). This question requires research into the various issues surrounding the use of adaptive automation. These issues include the frequency that task allocations should be made, when automation should be invoked, and how adaptive automation changes the nature of the human-automation interaction.

Although many proposals have been made concerning adaptive automation, few studies have been directed at determining how adaptive automation will impact human performance. One such study is by Scallen, Hancock, and Duley (1995) who reasoned that situations could exist with adaptive automation wherein a certain threshold for triggering an automation change is exceeded. This would result in an allocation in automation mode. However, the automation change prompted by the system may result in a further increase in workload leading again to another automation change. Such a situation could then lead to a number of automation changes within a short period of time. Therefore, these researchers examined how various short-cycle schedules of automation affects correlates of mental workload. They reported that performance was significantly better but workload was higher under a 15-sec schedule of automation than 60-sec. The results were interpreted in terms of a micro-tradeoff; that is, participants performed better under the 15-sec schedule at the expense of working harder.

The present study was an attempt to further the findings reported by Scallen, Hancock, and Duley, (1995). Vidulich and Wickens (1986) noted that greater effort can result in higher subjective ratings, but greater effort can also improve performance. This is a workload dissociation because increased subjective ratings and degraded performance together is what is usually indicative of increased workload. Such a dissociation may prove troublesome for system designers seeking to develop adaptive automation systems. Therefore, we sought to examine how other measures of workload are impacted by schedules of short-cycle automation. Specifically, we gathered EEG and ERP measures under the 15, 30 and 60-sec schedules of automation investigated by Scallen, Hancock, and Duley. In addition, because a concern voiced by many has been directed at how quickly and effectively operators can transition from automation to manual control when needed to do so, another goal of the study was to examine the impact of manual reversions on these correlates of workload.

METHOD

Participants

Nine graduate students served as participants for this experiment. The ages of the participants ranged from 18-35. All participants had normal or corrected-to-normal vision.

Experimental Variables

Tracking performance was measured by root-mean-squared-error (RMSE), and subjective workload was assessed using the NASA-TLX (Hart & Staveland, 1988). Relative power of theta, beta, and alpha at each cortical site was measured as was the accuracy in the counting of the high tones. Event-related potentials were analyzed using base-peak amplitude and latency. Return-to-manual deficits were determined by analyzing RMSE five seconds after a manual task reversion.

Experimental Tasks

Participants performed a modified version of the NASA-Multi-Attribute Task (MAT) battery (Comstock & Arnegard, 1992). Only the compensatory tracking task was used for the present study. The auditory oddball secondary task consisted of high (1200 Hz) and low (960 Hz) tones. The frequency of the tone presentation was once per second, and the probability of a high tone was .10 which was randomly assigned for presentation. Therefore, over a ten-minute trial there were 60 high tone signals and 540 low tone signals. The tones were gated to provide a rise and fall of .10 shaping a square wave signal. The tones were presented to both the participant's ears through stereo head phones.

Physiological Recording and Analysis

EEG. The EEG was recorded from sites Fz, Cz, and Pz. Measures of eye blinks (VEOG) and horizontal eye movement (HEOG) were also taken from electrodes above and below the right eye, and at the inner and outer canthus, respectively. A ground site was used located midway between Fpz and Fz. Each site was referenced to the right mastoid.

The EEG was routed through the SynAmps amplifier which converts it to a digital signal. The total EEG power from the bands of theta, alpha, and beta for each of the three sites was measured. The EEG frequency bands were set as follows: alpha (8-13 Hz), beta (13-22 Hz), and theta (4-8 Hz).

ERP. The Neuroscan SynAmps amplifier system was used for ERP acquisition and analyses. The software package for gathering ERPs is the Acquire386 SCAN software version 3.00. Data was acquired based upon assigned bit numbers placed in the data record from the MAT computer. The signal was gathered with 500 sweeps and points in the time domain with an A/D rate of 500. All corrections and artifactual reduction were done off line. All electrodes had an impedance of below 5 KOhms.

The continuous EEG data file was analyzed to reduce ocular artifact through VEOG and HEOG electrodes. These channels were assigned weights according to a sweep duration of 40ms and minimum sweep criteria of 20. The continuous EEG data file was then transformed into an EEG epoch file based upon a setting of 500 points per data file. The epoch file was then baseline corrected in the range -100 to 0 from the onset of the high tone. ERPs were then acquired through a sorting procedure based upon the assigned bit numbers in the data file. The signal was then further filtered with a low pass frequency of 62.5 and a low pass slope of 24 dB/oct. The high pass frequency was set at 5.00 with a high pass slope of 24 dB/oct. All filtering was performed in the time domain. All EEG was referenced to a common average and was smoothed by the SCAN software.

Experimental Procedure

The participants' scalps were prepared with rubbing alcohol and electrolyte gel. A reference electrode was then attached to their right mastoid by means of electrode tape. ECI Electro-Gel conductive gel was then placed in the reference electrode with a blunt-tip hypodermic needle. Electrode gel was also placed into each of the three electrode sites (Fz, Cz, Pz), the two VEOG and the two HEOG sites, and the ground site. Using the blunt-tip hypodermic needle, the scalp was then slightly abraded to bring the impedance level of the sites, relative to the ground, to less than 5KOhms. The participant was then hooked up to the SynAmps headbox connector, headphones placed on their heads, and exposed to the three 10-minute experimental tasks. Participants were allowed to practice for five minutes, or until they had reached asymptotic level for the two tasks. After each of the three experimental trials, participants were asked to complete the NASA-TLX.

RESULTS

RMSE

A within-subjects analysis of variance (ANOVA) was conducted on the tracking performance relative to the three different automation cycles. A main effect was found for cycle, F (2,16) = 4.18, $p <$ .05. A Student-Newman-Keuls (SNK) post-hoc analysis of the means revealed that subjects performed better (lower error scores) under the 15-second automation cycle ($M = 10.863$) as opposed to the 30 and 60-second cycles ($M = 13.670$ & 13.955, respectively). A second ANOVA examined return-to-manual deficits and a main effect for cycle was also found, F (2,16) = 9.43, $p <$ .05. A SNK post-hoc analysis of the means indicated that the return-to-manual deficits were significantly greater under the 15-second cycle ($M = 23.255$). The 30 second ($M = 19.299$), and 60 second cycle ($M = 17.753$) were not found to be different from one another.

Subjective workload

A within-subjects ANOVA was conducted on the NASA-TLX scores as a function of automation cycle. A main effect for cycle was found, F (2, 26) = 4.57, $p <$ .05. A SNK post-hoc analysis revealed that only under the 60-sec cycle did participants report significantly lower workload overall ($M = 36.22$).

Electroencephalogram

An ANOVA was performed on the relative power of the alpha, beta, and theta bands as a function of automation cycle and no significant differences were found. A similar analysis was conducted on the ERPs to the high tone presentations and, while visible components of the ERP waveform were present, they were not significantly different from one another, F (2, 15) = 2.27, $p >$ .05. However, an analysis of the ERP during task allocations revealed a significant decrease in P3 amplitude during the 15-sec condition, F (2, 15) = 3.75, $p <$ .05, compared to the 30- and 60-sec conditions. Figure 1 presents the composite ERP amplitude

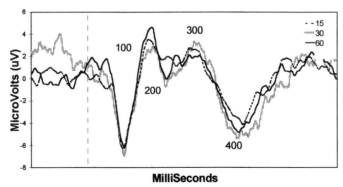

Figure 1. Composite Event-Related Potential

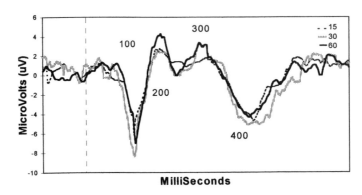

Figure 2. Return-to-Manual Event-Related Potential

for the 15-, 30-, and 60-sec conditions. Figure 2 shows the ERP gathered to high tones five minutes after a manual reversion.

DISCUSSION

The present study reports on behavioral, subjective, and psychophysiological correlates of various schedules of short-cycle automation. We attempted to examine the impact that 15-, 30-, or 60-second allocations in automation modes have on performance and workload. The study was an off-shoot of previous research by Scallen, Hancock, and Duley (1995) who found that shorter cycles of automation results in better performance but higher reported subjective workload scores.

Our findings were that performance was indeed better and subjective workload was higher under the 15-sec schedule of automation as reported by Hancock, Scallen, and Duley (1995). Also, although performance was better, there were greater return-to-manual deficits under the shorter cycle of automation.

The study was also designed to supplement the behavioral and subjective measures by examining psychophysiological measures of workload under short-cycle automation. We found that EEG power was not found to be significantly different across the three automation schedules. Also, although components emerged in the ERP waveform, overall base-peak amplitude or latency was not found to be significant. However, if you examine the ERP around the task allocations, you see a significant decrease in P3 amplitude and an increase in P3 latency in the 15-sec condition.

In conclusion, these results suggest that workload demands were highest during the automation transitions. The participants were able to regulate the increased workload demands by working harder leading to an increase in their subjective impressions of workload during the 15-sec condition. The increase in effort effectuated into better performance overall during this condition. However, participants may have found it difficult to sustain their performance after manual reversions leading to a higher return-to-manual deficit. Furthermore, although the subjective measures were not sensitive to the increased workload demands that occur during task allocations, the results provide that the amplitude and latency of the ERP was diagnostic of the workload increase.

Taken together, although brief periods of manual reversions may ameliorate some of the "out-of-the-loop" problems associated with automation use, the increase in workload demands with schedules of short-cycle automation may outweigh overall performance benefits. Therefore, system designers should safeguard against the possibility of such short cycling of automation in task situations where return-to-manual deficits will negatively impact human performance. In addition, these results support the use of psychophysiological measures in the design and evaluation of adaptively automated systems (for a more detailed discussion, see Byrne & Parasuraman, 1996). Future research should focus on the impact that other schedules of automation may have on correlates of workload as well as other psychophysiological measures that could serve in the development of advanced automation technologies.

ACKNOWLEDGMENTS

1. Lawrence J. Prinzel III served as principal co-author on paper.

The study was supported by the NASA Summer Scholars and NASA Graduate Student Researchers programs (Grant #: NGT-1-52123). The authors would like to acknowledge the considerable assistance of Dr. Alan T. Pope and Dr. Raymond Comstock at the NASA Langley Research Center. We would also like to thank Dr. Mark W. Scerbo for his help with data interpretation.

REFERENCES

Byrne, E., & Parasuraman, R. (1996). Psychophysiology and adaptive automation. *Biological Psychology, 42*, 249-268.

Comstock, Jr., J.R. & Arnegard, R.J. (1991). Multi-attribute Task Battery: Applications in Pilot Workload and Strategic Behavior Research. *Proceedings of the Sixth International Symposium on Aviation Psychology, 2*, 1118-1123.

Hancock, P.A., & Chignell, M.H. (1987). Adaptive control in human-machine systems. In P.A. Hancock (Ed.), *Human factors psychology* (pp. 305-345). North Holland: Elsevier Science Publishers.

Hart, S.G., & Staveland, L.E. (1988). Development of NASA-TLX (Task Load Index): Results of empirical and theoretical research. In P.A. Hancock & N. Meshkati (Eds.), *Human Mental Workload* (pp. 138-183). Amsterdam: North-Holland.

Scallen, S.F., Hancock, P.A., & Duley, J.A. (1995). Pilot performance and preference for short cycles of automation in adaptive function allocation. *Applied Ergonomics, 26*, 397-403.

Scerbo, M.W. (1996). Theoretical perspectives on adaptive automation. In R. Parasuraman & M. Mouloua (pp. 37-64), *Automation and human performance: Theory and applications*. Mahwah, NJ: Lawrence Erlbaum Associates.

Vidulich, M.A., & Wickens, C.D. (1986). Causes of dissociations between subjective workload measures and performance: Caveats for the use of subjective assessments. *Applied Ergonomics, 17*, 291-296.

Skill Acquisition with Human and Computer Teammates: Performance with Team KR

Kristin Krahl   Jamie L. LoVerde   Mark W. Scerbo
Old Dominion University

INTRODUCTION

Many issues involving skill acquisition in teams are tenuous and need to be studied experimentally (e.g., practice strategies, types of feedback, personality and motivational differences). One such issue is whether knowledge of results (KR) should be given as individual scores for individual members, or as an overall team score to all members, or whether both types of KR are necessary. At present, there is little research on this topic, thus, the purpose of the present study was to address this need in the literature.

Often times, team members must tradeoff between optimizing team performance or maximizing individual performance. Salas, Dickinson, Converse, & Tannenbaum (1992) suggest that performance should benefit from those aspects of a task to which feedback is given (i.e., individual feedback facilitates individual performance, team feedback facilitates team performance). For example, individual feedback may cause members to concentrate on their own tasks and neglect their team duties thereby potentially reducing overall team performance. Consequently, Jentsch, Navarro, Braun, and Bowers (1994) and Jentsch, Tait, Navarro, and Bowers (1995) performed studies to observe the effects of different forms of feedback on team performance.

Based on results of Saavedra, Earley, and Van Dyne (1993), Jentsch et al. (1994) had teammates work toward a common goal on a task requiring reciprocal interdependence, i.e., one member's output became another member's input and vice versa. They accomplished this by using a tracking task, referred to as Team Track (Jentsch, Bowers, Compton, Navarro, & Tait, 1996), where members relied upon their partner to guide them about how to move their joystick/mouse. Each participant was assigned to control either the horizontal or vertical position of the cursor. Participants saw the target along only their partner's axis and, therefore, they required positional cues from their partner. Those subjects controlling the horizontal direction of the cursor either received individual, team, or no feedback; those in the vertical condition received no feedback. Their goal was to minimize team tracking error.

Overall, the results (Jentsch et al., 1994) showed that team members receiving individual feedback performed significantly better than their partners who received no feedback, however, those receiving individual feedback optimized their individual performance at the expense of their partner. This finding suggests that individual feedback causes team members to neglect their team duty (i.e., provide directional cues) while concentrating on their own performance. On the other hand, those receiving team feedback did not perform significantly better than those members receiving no feedback. Members receiving team feedback could not effectively use this information to enhance performance; team feedback provides aggregated scores which does not allow for the decomposition of strengths and weaknesses of each team member. The authors suggested that because individuals did not know their own levels of performance it inhibited them from taking remedial action. Therefore, team feedback may only be effective if members are able to derive information about their own performance.

One approach to reducing the effects found in the above study is to provide both individual and team feedback to members. Usually, members on a team must perform both individual and team duties, thus Matsui, Kakuyama, and Onglatco (1987) suggested providing both forms of feedback in order to maximize team performance. They believe it is necessary to provide individual feedback to those members whose performance is below the team average. On the other hand, team feedback is useful to those teams whose performance is below average as a whole even though individual members are meeting their performance goals.

Previous research (Krahl & Scerbo, 1997) showed that high levels of performance can be achieved in an adaptive environment with human and computer teammates given both individual and team KR. Therefore, the goal of the present study was to extend this line of research and examine performance in the same team conditions when only team KR is available. Based on previous research (Matsui, Kakuyama, & Onglatco, 1987), it was expected that the performance of team members who were provided with only team KR would suffer compared to those who received both forms of KR.

METHOD

Participants

Thirty-six undergraduate students from Old Dominion University participated in this study and received extra credit for their participation.

Task

Participants performed a dual-axis pursuit tracking task presented on a VDT using an IBM-compatible personal computer. The target was a red 5x5 mm square that moved around the computer screen in the pattern of a circle 150 mm in diameter. Participants used a joystick to position a blue 5x5 mm cursor on top of the red target.

Participants completed 20 blocks of 5 trials, whereby once around the circle constituted a single trial. In all conditions, each participant began by controlling movement along either the horizontal or vertical axis, thereby sharing control over the blue cursor. However, on some trials they were able to take control of their partner's axis and assume total control over the blue cursor. Either partner could request control on any trial, but the computer would only relinquish control after evaluating each player's performance on the previous trial. If a participant's performance was worse than their partner's, they continued to share control on the next trial. If a participant sought control and had outperformed their partner, they were given full control on the next trial. At the end of that trial, control would revert back to both players for the subsequent trial. Players were instructed to achieve their lowest "team" score which was a composite of performance along both axes. Team scores were presented after each block. Information about who had control was presented on every trial.

Procedure

Upon arrival, the participants were seated in front of the computer and shown how to use the joystick. Participants were allowed to practice the task just once before beginning the experimental session.

Participants were randomly assigned to one of four experimental conditions. In the human-human team condition (HHT), twelve participants were assigned at random to pairs and completed the experimental task together as a team. Another twenty-four participants were assigned at random to each of the human-computer teams. In these conditions, participants controlled the movement of the blue cursor along one axis and the computer controlled the other axis. Assignment of participants to the horizontal and vertical axis conditions was counterbalanced.

In the human-computer teams, the computer's control over the cursor was driven by human performance data. The horizontal and vertical movements of the cursor were recorded from another set of participants as they practiced the task. Data from these records were used to create stimulus sets with two different skill levels. For the novice human-computer team (NHCT) condition, data were drawn from trials with RMSEs greater than 13 on the X axis ($M = 21$, sd = 6.7) and 9.5 on the Y axis ($M = 15$, sd = 5.1). For the expert human-computer team (EHCT) condition, data were drawn from trials with RMSEs lower than 11 on the X axis ($M = 8.3$, sd = 4.6) and 13 on the Y axis ($M = 6.5$, sd = 3.4) .

RESULTS

The present study used a 4 group condition (HHT, NHCT, EHCT, and CT) by 2 axis (horizontal and vertical) by 2 operator (human and computer) by 20 blocks by 5 trials mixed design. The between-subjects factors were group, operator, and axis and the within-subjects factors were blocks and trials. The CT condition is data from Krahl and Scerbo (1997) and is included in the analysis for comparison purposes. It represents a control condition in which a participant tracked along *both* axes for entire session. Performance was measured using participants' RMSE scores along both horizontal and vertical vectors. A preliminary analysis showed that participants in each group performed better over trials but at different rates within each block. Therefore, to facilitate the analyses, data were collapsed across trials and reanalyzed.

Significant main effects of RMSEX were found for group, $F(3, 20) = 9.62$, $p < .001$, and blocks, $F(19, 380) = 5.30$, $p < .001$. The group and block interaction was also significant for RMSEX, $F(57, 380) = 1.77$, $p < .001$. In addition, there were significant main effects of RMSEY for group, $F(3, 20) = 3.49$, $p < .05$, and blocks, $F(19, 380) = 8.04$, $p < .001$. Moreover, the group and block interaction was significant for RMSEY, $F(57, 380) = 1.69$, $p < .01$.

The results from all four groups are plotted in Figure 1. The data in the figure are for RMSEX, but are representative of scores for the Y axis as well. The information presented for novice and expert represents only the human participants' performance. Overall, error scores for all groups declined over the session. Subsequent Student Newman-Keuls (SNK) tests revealed that those in the novice computer and human teams performed significantly better across the blocks; they began poorly and performed better throughout the experiment. However, those in the expert computer and control conditions performed approximately the same across blocks; they began moderately well and performed consistently throughout the experiment, except in Block 14 which appears to be an anomaly. Also, focusing only on Block 1, SNK tests showed that performance in the novice condition was significantly worse than performance in both the expert computer and control conditions (see Figure 1). On the other hand, performance in the novice

Figure 1

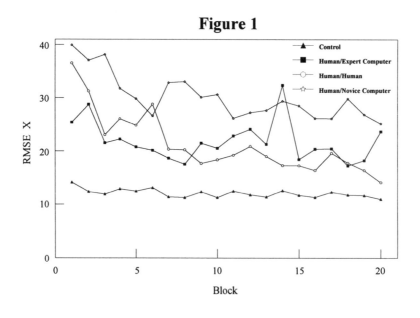

Figure 1. Mean RMSEX scores for each team over blocks of trials.

condition only differed from performance in the control condition by Block 20. Thus, overall, groups began the experiment at significantly different performance levels in the beginning of the experiment but attained comparable levels of performance by the end of the session.

The results from a previous study (Krahl & Scerbo, 1997) are shown in Figure 2. Similar to the findings in the current study, overall error scores declined over the session and SNK tests revealed that those in the novice computer and human team conditions improved over the blocks. As one can see from viewing both figures, the pattern of performance for each group in the current study is similar to that in the previous study (i.e., performance lowest for novice computer condition), however, the RMSE scores for the current study were higher overall.

Figure 2

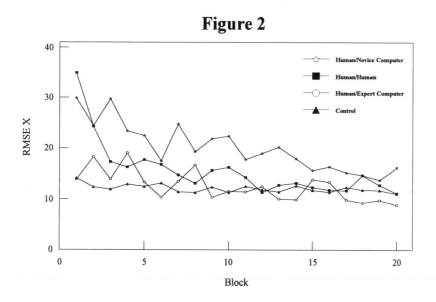

Figure 2. Mean RMSEX scores for each team over blocks (Krahl & Scerbo, 1997).

DISCUSSION

The present study compared the effects of team KR with team and individual KR. The team compositions were the same as those used by Krahl and Scerbo (1997). Participants in this study, however were given only team KR, whereas those in the previous study were presented with both individual and team KR.

The primary difference between the two studies concerns the overall level of performance. Overall RMSE scores were higher in the present study than what was observed by Krahl and Scerbo (1997). Thus, it appears that the absence of individual KR results in performance that is worse than what might be obtained when both forms of KR are available.

With the exception of the elevated RMSE scores, the results of the present study were quite similar to those of Krahl and Scerbo (1997). In general, performance improved over blocks. These effects, however depended upon group assignment. Those paired with another human or computer of novice-level skill started poorly but showed improvement over the session. Those paired with a computer of expert-level skill began at a much better level of performance and maintained that level throughout the session. Moreover, in both studies those who worked with the expert computer performed at a level comparable to the individuals who tracked both axes throughout the session.

The comparison between the present study and previous one suggests that differences in performance may be tied to the types of KR provided. Team KR alone appears to be less beneficial to team performance then both types of KR for all pairings. According to Jentsch et al. (1994, 1996) and Matsui et al. (1987), this effect is the result of individual team members being unable to extract usable information concerning their own performance from a team score and modify their efforts accordingly to bolster their performance.

Collectively, the results of Krahl and Scerbo (1997) and the present study suggest that, in order to optimize team performance, both individual and team feedback must be presented. Offering only team KR does not provide enough information to team members to enable them to optimize their performance. By contrast, offering both types of feedback may permit team members to determine whether increasing individual effort is necessary to improve overall team performance.

REFERENCES

Jentsch, F., Bowers, C., Compton, D., Navarro, G., & Tait, T. (1996). Team-Track: A tool for investigating tracking performance in teams. *Behavior Research Methods, Instruments, & Computers, 28,* 411-417.

Jentsch, F. G., Navarro, G., Braun, C., & Bowers, C. (1994). Automated feedback in teams: Trading individual for team performance. *2nd Mid-Atlantic Human Factors Conference, 69-73.*

Jentsch, F. G., Tait, T., Navarro, G., Bowers, C. (1995). Differential effects of feedback as a function of task distribution in teams. *Proceedings of the Human Factors and Ergonomics Society 39th Annual Meeting,* 1273-1277.

Krahl, K., & Scerbo, M.W. (1997). Performance on an adaptive automated team tracking task with human and computer teammates. *Proceedings of the Human Factors Society 41st Annual Meeting,* 551-555.

Matsui, T., Kakuyama, T., & Onglatco, M.L. (1987). Effects of goals and feedback on performance in groups. *Journal of Applied Psychology, 72,* 407-415.

Saavedra, R., Earley, P.C., & Dyne, L.V. (1993). Complex interdependence in task-performing groups. *Journal of Applied Psychology, 78,* 61-72.

Salas, E., Dickinson, T.L., Converse, S.A., & Tannenbaum, S.I. (1992). Toward an understanding of team performance and training. In R. W. Swezey & E. Salas (Eds.), *Teams: Their Training and Performance* (pp. 3-29). Norwood: Ablex.

The Effects of Task Partitioning and Computer Skill on Engagement and Performance with an Adaptive, Biocybernetic System

Todd M. Eischeid   Mark W. Scerbo   Frederick G. Freeman
Old Dominion University

INTRODUCTION

Automated systems that can shoulder changing responsibilities are generally considered adaptive. (Hancock & Chignell, 1987; Morrison, Gluckman & Deaton, 1991; Rouse, 1976). Adaptive automation allows the level or mode of automation or the number of systems currently automated to be modified in real time in response to situational changes (Scerbo, 1996). Parasuraman, Bahri, Deaton, Morrison, and Barnes (1992) state that adaptive automation produces an ideal coupling of the level of automation to the level of operator workload. Moreover, adaptive automation represents an attempt to maintain an optimal level of operator engagement.

One adaptive system, a biocybernetic closed-loop system developed at NASA (Pope, Bogart, & Bartolome, 1995; Pope, Comstock, Bartolome, Bogart, & Burdette, 1994), uses biofeedback to avoid hazardous states of operator awareness. The closed-loop system used by Pope et al. (1994; 1995) is truly adaptive and changes tasks in real time. The criterion for the changes in automation is an index of operator engagement [20 beta/(alpha+theta)], derived from relative powers of various bandwidths of EEG activity. The present study also used this index to determine operator engagement.

The Electronic Teammate

The idea of a computer as a teammate has been explored by Scerbo, Ceplenski, Krahl, & Eischeid (1996), who compared the performance of human-human and human-computer teams. In that study, a pursuit tracking task was partitioned into horizontal and vertical axes, and control of each axis was allocated to a member of the team. In addition, the skill level of the computer was manipulated. Participants paired with an expert computer outperformed all other participants, showing that the skill level of one's teammate affected performance. The human-human team exhibited the most improvement over time, but never did reach the high level of performance of participants paired with an expert computer. Scerbo et al. (1996) concluded that partitioning may be a desirable strategy for adaptive automation.

It is important to note that the tracking task used by Scerbo et al. (1996) was not adaptive in nature. The present study addressed this gap by utilizing the biocybernetic closed-loop system.

The Present Study

Prior to the present study, task partitioning had not been tested in the closed-loop environment. The tracking task of the MAT was partitioned into horizontal and vertical axes similar to Scerbo et al. (1996). Control of one axis was allocated to or taken away from the operator (by the biocybernetic system) depending upon level of engagement and feedback conditions (positive or negative). In addition, the computer demonstrated either expert- or novice-level performance.

Since task partitioning presumably relieves the operator of performing the entire task, overall better performance should have been observed in the partitioned mode. However, computer skill level was predicted to interact with task mode, and participants in the expert condition, partitioned mode were expected to exhibit significantly better performance than those in the novice condition, partitioned mode. This would show the benefits of having an expert-level computer as a teammate and would coincide with Scerbo et al.

(1996).

Negative feedback situations attempt to maintain the operator at an optimal level of engagement, and thus better performance should have been observed in negative feedback. Further, feedback was expected to interact with skill level such that any performance benefits produced by negative feedback situations would be nullified by having a novice-level computer for a teammate. Better performance was, however, expected in negative feedback than positive feedback in the expert-level condition.

METHOD

Participants

The participants for this experiment were 24 undergraduate psychology students, consisting of 6 males and 18 females and ranging in age from 19-45 years.

Task

A modified version of the compensatory tracking task from the MAT Battery (Comstock & Arnegard, 1992) was used that allowed three modes of operation: fully manual (participant controlled both axes), partitioned (participant and computer each controlled an axis), and fully automatic (computer controlled both axes). Only one axis was allocated during a given automation switch. In addition, the computer demonstrated either expert-level (M Root Mean Squared Error = 13.28, SD = 3.21), or novice-level (M Root Mean Squared Error = 40.09, SD = 2.07) performance. The criterion used to invoke a change in the automation status of the tracking task was level of operator engagement as defined by the EEG engagement index.

Experimental Design

The present study used a 3 task mode (automatic, partitioned, and manual) by 2 feedback condition (positive, negative) by 2 computer skill level (expert, novice) design. Task mode and feedback were within-subject conditions, and computer skill level was a between-subject condition.

The dependent variables were mean index of engagement, number of task mode allocations, time between allocations, and tracking root mean squared error (RMSE).

RESULTS

Task Allocations & Engagement Index

An ANOVA of the task allocations showed a significant main effect for feedback, $F(1, 22)=4.86$. There were significantly more task allocations in the negative feedback condition (M=56.84) than in the positive feedback condition (M=53.68).

An ANOVA of the engagement index revealed a significant feedback by task mode interaction, $F(2, 44) = 172.41$. An analysis of simple effects revealed that feedback conditions were significantly different in automatic mode, $F(1, 22) = 177.62$; and also in manual mode, $F(1, 22) = 125.97$.

Time Between Task Allocations

In addition, an ANOVA of the mean time between task mode allocations revealed a main effect for feedback, $F(1, 22) = 5.45$. There was significantly more time between task mode allocations in positive feedback ($M = 6.13$) than in negative feedback ($M = 5.67$). There was also a significant main effect for task mode, $F(2, 44) = 76.20$. More time was spent in both automatic ($M = 7.18$) and manual ($M = 6.89$) modes than partitioned mode ($M = 3.62$).

Finally, there was a significant feedback by task mode interaction, $F(2, 44) = 5.24$. An analysis of simple effects showed there was significantly more time between allocations in positive feedback ($M = 8.16$), than in negative feedback ($M = 6.21$), only for the automatic mode $F(1, 22) = 11.29$.

Tracking RMSE

For the tracking analyses, RMSE for partitioned mode was calculated for only the axis the participant controlled. For manual mode, RMSE was calculated using the same axis the participant controlled during the partitioned mode. A significant main effect for axis condition, $F(1, 20) = 5.53$, showed that participants who controlled the vertical axis performed better ($M = 8.15$) than those who controlled the horizontal axis ($M = 10.85$). Also, a significant main effect for skill, $F(1, 20) = 8.73$, showed that participants paired with an expert computer ($M = 7.80$) performed better than those paired with a novice computer ($M = 11.20$). There was also a skill by task mode interaction, $F(1, 20) = 82.82$. An analysis of simple effects showed that there was a significant difference between skill level only in the partitioned mode, $F(1, 20) = 22.64$; participants in the expert condition ($M = 8.62$) performed better in the partitioned mode than did participants in the novice condition ($M = 14.71$).

DISCUSSION

Engagement

For engagement, the interaction between type of feedback and task mode with engagement shows that the biocybernetic system was responding just as expected in a multi-level situation. Other patterns in the data reveal that positive and negative feedback do differ in another manner, implying that some vital information may be lost when averaging the indices across experimental conditions. For example, an analysis of the range of values of the engagement index around the baseline show that it was more narrow in negative feedback, as compared to positive feedback. These results are exactly what cybernetic theory would predict and further serve to validate the system developed by Pope et al. (1994; 1995).

Skill

The skill level of an adaptive computer teammate was found to have a considerable effect on participants' tracking performance. Specifically, participants paired with an expert computer outperformed those paired with a novice computer, even though *all* participants had complete control of both axes in the manual mode and control of one axis in the partitioned mode. This effect of computer skill level coincides with the results of Scerbo et al. (1996). However, unlike that study, in which the tracking task was not adaptive, the present research shows that the effect of computer skill level on performance similarly applies in an adaptive environment. Modeling adaptive systems after expert-level performance may be the best method to help human operators achieve the best performance.

Regarding the interactive effects of skill and feedback on performance, participants paired with an expert-level computer, in negative feedback were expected to have the best tracking performance. However, this was not observed in the present study. There was, however, a marginal, overall main effect of feedback on performance, showing overall performance to be better in negative feedback. This may show that negative feedback applied to regulation of operator engagement may elicit slightly better performance.

Task Partitioning

When the human portion of RMSE in partitioned mode was compared to RMSE for the same axis in manual mode, a performance advantage of partitioned mode was not evident. Indeed, *overall* performance in partitioned mode was worse than in manual mode. However, an advantage of being paired with an expert-level computer was evident in the partitioned mode, when the human and computer were sharing the task. When participants fully controlled the task (manual mode), there were no performance differences between skill conditions.

The analysis of mean time between task mode allocations showed there was significantly less time spent in the partitioned mode overall than in the fully automatic or fully manual conditions. This result was unexpected and may be indicative of an overly narrow range around the mean engagement ("0.2 SD) used for partitioned mode. Figure 1 illustrates how the range around the mean engagement is related to system behavior. Any differences among the task modes, particularly in negative feedback, should show more mean time spent in partitioned mode, simply because the biocybernetic system attempts to maintain operator engagement close to the baseline mean. Recently, Scallen, Hancock, and Duley (1995) tested the effects of automation cycles on performance, and suggested imposing a damping factor, or governor, on an adaptive system which sets a minimum frequency with which changes in automation may occur. For the system used in the present study, a wider range around the baseline mean may be needed in order to increase the time between allocations. A follow-up to this study is addressing that issue.

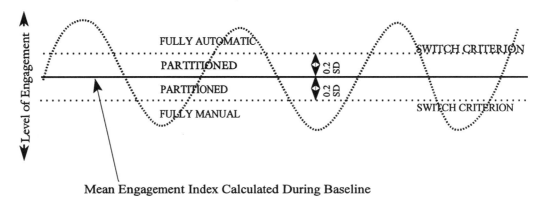

Figure 1. Example of Ideal System Behavior in Negative Feedback

Conclusions

The present study indicates a definite performance benefit for having a highly-skilled computer as a teammate. Thus, adaptive systems utilizing performance modeling should exhibit expert-level behavior to best aid the human operator.

Although the effects of task partitioning were examined in the present study, the extremely short periods that the task remained in partitioned mode prevented a valid assessment of this. A corrective measure may include widening the range around the baseline mean to limit switching frequency. A follow-up study is addressing that issue.

ACKNOWLEDGEMENTS

I would like to acknowledge my advisor, Dr. Mark Scerbo, and also Dr. Ray Comstock and Dr. Alan Pope at NASA Langley Research Center for their help with this research.

REFERENCES

Comstock, J. R., Jr., & Arnegard, R. J. (1992). *The Multi-Attribute Task Battery for human operator workload and strategic behavior research.* National Aeronautics and Space Administration Technical Memorandum No. 104174.

Hancock, P. A., & Chignell, M. H. (1987). Adaptive control in human-machine systems. In P. A. Hancock (Ed.), *Human Factors Psychology* (pp. 305-345). North Holland: Elsevier.

Morrison, J. G., Gluckman, J. P., & Deaton, J. E. (1991). *Program plan for the adaptive function allocation for intelligent cockpits (AFAIC) program* (Final Report No. NADC-91028-60). Warminster, PA: Naval Air Development Center.

Parasuraman, R., Bahri, T., Deaton, A. E., Morrison, J. G., & Barnes, M. (1992). *Theory and design of adaptive automation in aviation systems.* (Progress Report No. NAWCADWAR-92033-60). Warminster, PA: Naval Air Warfare Center, Aircraft Division.

Pope, A. T., Bogart, E. H., & Bartolome, D. S. (1995). Biocybernetic system evaluates indices of operator engagement in automated task. *Biological Psychology, 40,* 187-195.

Pope, A. T., Comstock, J. R., Bartolome, D. S., Bogart, E. H., & Burdette, D. W. (1994). Biocybernetic system validates index of operator engagement in automated task. In M. Mouloua & R. Parasuraman (Eds.), *Human performance in automated systems: Current research and trends* (pp. 300-306). Hillsdale, NJ: Lawrence Erlbaum Associates.

Rouse, W. B. (1976). Adaptive allocation of decision making responsibility between supervisor and computer. In T. B. Sheridan & G. Johannsen (Eds.), *Monitoring Behavior and Supervisory Control* (pp. 295-306). New York: Plenum.

Scallen, S. F., Hancock, P. A., & Duley, J. A. (1995). Pilot performance and preference for short cycles of automation in adaptive function allocation. *Applied Ergonomics, 26,* 397-403.

Scerbo, M. W. (1996). Theoretical perspectives on adaptive automation. In R. Parasuraman and M. Mouloua (Eds.), *Automation and human performance: Theory and applications* (pp. 37-63). Mahwah, NJ: Lawrence Erlbaum Associates.

Scerbo, M. W., Ceplenski, P. J., Krahl, K., & Eischeid, T. M. (1996). Performance on a team tracking task with human and computer teammates. *Proceedings of the Human Factors and Ergonomics Society 40th Annual Meeting,* 1271.

Situation-Adaptive Autonomy: Trading Control of Authority in Human-Machine Systems

Toshiyuki Inagaki
University of Tsukuba

INTRODUCTION

How tasks should be allocated to the human and the computer in the supervisory control is one of the central issues in research on human-machine systems. *Human-centered automation* (Billings, 1991; Woods, 1989) assumes that human must be at the locus of control, and that the computer should be subordinate to the human. Does this mean, among various levels of autonomy listed in Table 1, the autonomy with levels 6 or higher must be denied? Do we have to assume that the level of autonomy must stay within the range of 1 through 5 at all times in every occasion?

Table 1   Levels of Autonomy (Sheridan, 1992)

1.	The computer offers no assistance, human must do it all.
2.	The computer offers a complete set of action alternatives, and
3.	narrows the selection down to a few, or
4.	suggests one, and
5.	executes that suggestion if the human approves, or
6.	allows the human a restricted time to veto before automatic execution, or
7.	executes automatically, then necessarily informs human, or
8.	informs him after execution only if he asks, or
9.	informs him after execution if it, the computer, decides to.
10.	The computer decides everything and acts autonomously, ignoring the human.

By using mathematical models, Inagaki (1991, 1993, 1995) has proposed the concept of *situation-adaptive autonomy* in which (i) the autonomy changes its level dynamically depending on the situation, and (ii) the autonomy with levels 6 or higher may be adopted for attaining system safety. It is proven that conditions exist for which the computer should be allowed to perform safety-control actions even when the human has given the computer no explicit directive to do so.

This paper discusses how authority can be traded between human and automation in a situation-adaptive manner. The following two problems are taken as examples: (i) Rejected takeoff problem in aviation, in which a human pilot must decide, upon an engine failure during the takeoff run, whether to continue the climb out or to abort the takeoff, and must take necessary control actions within a few seconds. (ii) Process control problem, where a human operator is requested to perform specified control tasks, and to take an appropriate action against a malfunction in the process.

REJECTED TAKEOFF (RTO) IN AVIATION

Suppose an engine failure warning is set off while an aircraft is making its takeoff run. The pilot must decide whether to continue the takeoff (GO) or to abort it (NO-GO). The GO/NO-GO decision is made basically depending on whether the airspeed is less than a predesignated speed, V1. The standard procedure for the GO/NO-GO decision, upon an engine failure, is specified as follows:

(1) If the airspeed is less than V1, then reject the takeoff.

(2) If V1 has already been achieved, then continue the takeoff.

Roughly speaking, V1 has the following two meanings: (i) the maximum speed at which the human pilot must *initiate* RTO actions (instead of just *deciding* whether to go or not) to stop safely within the remaining field length, and (ii) the minimum speed to continue an engine out takeoff and to attain a height of 35 feet at the end of the runway.

According to statistics, (i) one RTO occurs per 3,000 takeoffs, (ii) one out of 1,000 RTOs ends with an overrun accident/incident, (iii) 58% of the accidents/incidents are due to RTOs which were made at an airspeed greater than Vl, and (iv) 80% of the RTO accidents could be avoidable by continuing the takeoff or by correct stop techniques. The statistics show that human errors can happen in GO/NO-GO decision making under time-criticality. Some arguments about whether to automate GO/NO-GO decision making reflect this fact.

Inagaki (1997a, 1997b) has proven mathematically that GO/NO-GO decision making should neither be fully automated nor be left always to a human pilot for assuring takeoff safety. The formal proof has been given based on a probability model where the following assumptions are made:

(1) An engine failure warning, which is given to a human pilot, is not always correct.

(2) Delay is inevitable before a human pilot initiates actions, because he/she responds to a warning in the following way: (i) He/she firstly recognizes the warning, (ii) analyzes the situation and decides whether to go or not, and (iii) finally takes control actions based on his/her decision.

(3) A human pilot's understanding of a given situation is not always correct: He/she may regard a correct warning as being false, or vice versa.

(4) A human pilot may fail to judge either that the warning is correct or that the warning is incorrect. Two policies are distinguished for this case: (i) Trustful Policy (TP), in which the warning is trusted and the engine is assumed failed, and (ii) Distrustful Policy (DP), in which the warning is distrusted and the engine is assumed working.

(5) Erroneous decision, or delay in making decision or initiating actions can cause cost the size of which varies depending on the situation.

The conditional expected cost, given an engine failure warning, is denoted as L_{AS} when a GO/NO-GO decision is made by an automated system (AS), and is denoted as L_{TP} or L_{DP} when the decision is made by a human pilot with the TP policy or the DP policy, respectively. Three phases are distinguished as shown in Figure 1:

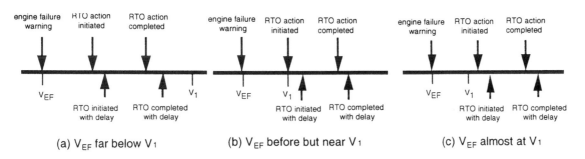

Figure 1. Engine Failure and Rejected Takeoff

Phase 1: Suppose an engine failure warning is given at V_{EF} which is far below V1 (Fig. 1a). We have $L_{DP} \leq L_{TP} \leq L_{AS}$, which implies GO/NO-GO decision should not be automated in this phase.

Phase 2: If a warning appears before but near V1 (Fig. 1b), then we have $L_{DP} < L_{TP}$. There is no fixed order relation between L_{AS} and L_{TP}, or between L_{AS} and L_{DP}.

155

Phase 3: Suppose an engine failure warning is given almost at V1 (Fig. 1c). We categorize the cases into two classes: (i) Phase 3a, in which no human pilot can initiate RTO by V1 but the automated system can, and (ii) Phase 3b in which neither a human pilot nor the automated system can initiate RTO by V1. For Phase 3a, we have $L_{DP} < L_{TP}$, but no fixed order relation exists between L_{AS} and L_{TP}, or between L_{AS} and L_{DP} On the other hand, for Phase 3b, we have $L_{AS} \leq L_{DP} \leq L_{TP}$, which implies that the automation should have authority for decision and control. The less effective performance of a human pilot in Phase 3b is due to the possibility that he/she may fail to recognize properly that it is too late to reject the takeoff. This type of misunderstanding can happen because the engine failure has occurred "before V1." Note that the automation is *never* used in Phase 3b to abort the takeoff. No pilot would accept situations where the automation rejects the takeoff while he/she tries to continue the takeoff.

The above results proves that authority for decision and control must be traded dynamically between humans and automation depending on the situation, and that the authority may be given to the automation when time-criticality is extremely high.

Further analyses in (Inagaki, 1997a, 1997b) have shown the following:

(1) The DP policy corresponds to *Go-Mindedness*. The Go-Mindedness is effective to attain takeoff safety in Phases 2 and 3a, even under higher degree of situational uncertainty.

(2) If we still wish a human pilot to be in authority at all times in every occasion, then human-interface of a warning should be altered. An "engine failure" message gives no direct information on whether to go or not, because the message implies GO in some cases, and NO-GO in other cases. An alternative interface of a warning may be to give an explicit sign of GO or NO-GO to a human pilot. Showing GO message enables a human pilot to exhibit as good performance as the automation in Phase 3b: viz., $L_{AS} = L_{DP} = L$.

(3) Improvement of warning interface cannot solve the problem completely. Even under the new interface, order relation between L_{AS} and $L_{TP} = L_{DP}$ is still situation-dependent in Phase 3a.

PROCESS CONTROL

The microworld SCARLETT (Supervisory Control And Responses to LEaks: TARA at Tsukuba) has been developed in (Inagaki, Moray, Itoh, 1997). One of tasks for the operator is to control manually the temperature of the central heating system for the apartment at the bottom right in Figure 2 which depicts the control panel through which the operator communicates with SCARLETT.

The operator is given another task of fault management when pipe faults occur in the plant. Pipe faults are classified into two classes: (i) A pipe leak where a portion of flow quantity is lost, and (ii) a pipe break where the flow quantity is lost completely at the site of the break. The observable phenomena differ depending on the size or the location of the fault, and the operating condition (such as the pump rate). Without having sound knowledge of the controlled process, it is easy for the human to make mistakes in fault detection and location.

Upon detecting a leak or a break, the computer gives a red light which flashes at the site of the leak or the break, and gives a message in the Message Window, such as: "Leak! Send repair crew," or "Pipe break! Shut the plant down." However, the warning message may not be always correct: It says "Leak!" when "Pipe break" is actually the case, and vice versa. Three levels of reliability are distinguished regarding the warning messages: (i) Completely reliable messages, where every one of them is correct, (ii) highly reliable messages, where 90 % of them are correct, and (iii) less reliable messages, where 70 % of them are correct.

Two levels are distinguished for control mode of the fault management task: (i) Manual mode (M-mode) in which the computer gives warning messages or suggests control actions, but it performs no action without explicit commands by the human, and (ii) Situation-Adaptive Autonomy mode (SAA-mode) which includes levels of autonomy where the computer may send repair crew or shut down the plant when it think it appropriate to do so, even if it receives no directive by the human.

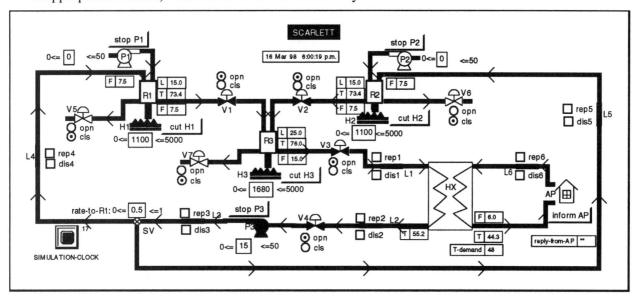

Figure 2. Control panel of SCARLETT

The experiment has a 2 x 2 x 3 x 2 factorial design with two within-subjects factors (Mode of Control and Type of Fault) and two between-subjects factors (Reliability of Warning and Order of Presentation), where the order of presentation refers to the order in which subjects are exposed to two types of control modes: (i) Half of the subjects receive M-mode before SAA-mode and (ii) half of the subjects receive SAA-mode before M-mode.

Thirty graduate and undergraduate students in computer science participated in the experiment. The experiment consists of the following sessions: Guidance, training, pre-test, data collection, and post-test sessions. Several performance measures are recorded, including: (i) RMS error around the target temperature, (ii) subjective rating of trust in automation (Lee & Moray, 1992), (iii) subjective rating of self-confidence (Lee & Moray, 1992) in the ability to control the process manually by himself/herself, (iv) lost amount of water due to pipe break, (v) number of accidents, where the accident is defined as one of the following events: (1) A reservoir becomes dried out while heated, (2) a pump runs with no incoming fluid flow, (3) the fluid coming out of the pipe floods the floor, or (4) temperature of fluid to the apartment complex is too high or too low.

ANOVA has been carried out on the performance measures. Some of the results are:

1. Accidents: There are two significant main effects, reliability of warning ($F(2,24) = 9.825$, $p = 0.0008$), and mode of control ($F(1,2) = 438.86$, $p = 0.0023$). The control mode effect shows that there are far more accidents overall in M-mode than in SAA-mode, which justifies the use of automation to reduce accidents.

2. False Shut-downs: There is a significant main effect of reliability of warning ($F(2,24) = 25.56$, $p < 0.0001$). The probability of false shut-downs decreases as reliability of warning increases. There is an interaction of reliability of warning x mode of control ($F(2,24) = 4.800$, $p = 0.0176$), which proves that the human should be retained as controllers to avoid unnecessary shut-downs.

3. Loss of fluid during faults: There are two significant effects. The first is the main effect of the type of the fault ($F(1,2) = 2160.22$, $p = 0.00046$), which is straightforward. The second is a three-way interaction of reliability of warning x mode of control x type of fault ($F(2,23) = 4.630$, $p = 0.020$), which implies that for small faults, and especially when the information provided by the warnings is reliable,

157

humans are more efficient. For severe problems, SAA is more effective in managing fluid loss during faults, whether or not the displayed information is reliable.

The experiment has also shown the need for autonomy which comes between levels 6 and 7 (say, level 6.5) in Table 1. Level 6.5 autonomy is defined as: "The computer executes automatically after telling human what it will do. No veto is allowed for the human." This type of autonomy may be effective to avoid *automation surprises* (Sarter & Woods, 1995; Wickens, 1994) induced by safety-related actions by automation under high time-criticality.

CONCLUSION

This paper has shown, through a mathematical analysis and an experiment, how authority can be traded between human and automation in a situation-adaptive manner. The situation-adaptive autonomy is closely related to the *adaptive automation:* See (Scerbo, 1996) for a survey. We need further research for validation of those concepts, because effectiveness may be dependent on task context.

ACKNOWLEDGEMENTS

This work has been done jointly with Neville Moray of the University of Surrey and Makoto Itoh of the University of Tsukuba. This work has been supported by the Center for TARA (Tsukuba Advanced Research Alliance) at the University of Tsukuba, Grants-in-Aid 08650458 and 09650437 of the Japanese Ministry of Education, and the First Toyota High-Tech Research Grant Program.

REFERENCES

Billings, C.E. (1991). Human-Centered Aircraft Automation. NASA TM-103885.

Inagaki, T. & Johannsen, G. (1991). Human-computer interaction and cooperation for supervisory control of large-complex systems. In Pichler & Moreno Diaz (Eds.), *Computer Aided System Theory*, LNCS 585, Springer-Verlag, 281-294.

Inagaki, T. (1993). Situation-adaptive degree of automation for system safety. *Proc. 2nd IEEE Int. Workshop on Robot and Human Communication*, 231-236.

Inagaki, T. (1995). Situation-adaptive responsibility allocation for human-centered automation. *Trans. SICK of Japan, 31*(3), 292-298.

Inagaki, T. (1997a). To go or not to go: Decision under time criticality and situation-adaptive autonomy for takeoff safety, *Proc. IASTED Int. Conf. Appl. Modeling & Simul,* 144-147.

Inagaki,T. & Itoh, M. (1997b) Situation-adaptive autonomy: The potential for improving takeoff safety, *Proc. 6th IEEE Int. Workshop on Robot and Human Communication*, 302-307.

Inagaki, T., Moray, N., & Itoh, M. (1997) Trust and time-criticality: Their effects on the situation-adaptive autonomy, *Proc. Int. Symp. AIR-IHAS*, 93-103.

Lee, J. & N. Moray (1992). Trust, control strategies and allocation of function in human-machine systems. *Ergonomics, 35*(10), 1243-1270.

Sarter, N. & Woods, D.D. (1995). Autonomy, authority, and observability: Properties of advanced automation and their impact on human-machine coordination, *Proc. IFAC MMS*, 149-152.

Scerbo, M.W. (1996). Theoretical perspectives on adaptive automation. In R. Parasuraman & M. Mouloua (Eds.), *Automation and human performance: Theory and applications* (pp. 37-63). Mahwah,NJ: Erlbaum.

Sheridan, T. (1992). *Telerobotics, automation, and human supervisory control.* MIT Press.

Wickens, C.D. (1994). Designing for situation awareness and trust in automation. *Proc. IFAC Integrated Systems Engineering*, 77-82.

Woods, D. (1989). The effects of automation on human's role: Experience from non-aviation industries. In Norman and Orlady (Eds.), *Flight Deck Automation: Promises and Realities.* NASA CP- 10036, 61 -85.

SITUATION AWARENESS

The Global Implicit Measurement of Situation Awareness: Implications for Design and Adaptive Interface Technologies

Bart J. Brickman, Lawrence J. Hettinger, and Dean K. Stautberg
Logicon Technical Services, Inc.
Dayton, Ohio

Michael W. Haas and Michael A. Vidulich
Human Effectiveness Directorate
Air Force Research Laboratory
Wright-Patterson Air Force Base, Ohio

Robert L. Shaw
F.C.I. Associates
Beavercreek, Ohio

INTRODUCTION

The purpose of this paper is to describe the initial results of a novel approach to develop a reliable, objective methodology to assess crewmember Situation Awareness (SA) during simulated air-to-air combat missions. The Global Implicit Measurement (GIM) technique was developed at the US Air Force Research Laboratory's Synthesized Immersion Research Environment (SIRE) facility at Wright-Patterson Air Force Base, Ohio as part of a program of research devoted to developing virtually-augmented, multisensory, adaptive crewstation technologies for use in future tactical aircraft.

The GIM was designed to incorporate the strengths of traditional implicit measurement techniques (e.g., non-intrusive, performance-based, objective, precisely measurable) while limiting the weaknesses typically associated with implicit measures and other methods such as memory probe techniques (e.g., intrusiveness, reliance on imperfect memory, few observations, limited scope and generalizability) (see Brickman, et al. 1995 for a description of GIM).

In order to ensure that it benefits from a high level of operational validity, the first step in the GIM procedure is to provide a high level of specificity to the nature of pilot tasks involved in air combat. To this end, a detailed task analysis of the general characteristics of a typical mission is conducted in association with subject matter experts (USAF fighter pilots). The results of the task analysis provide a detailed description of the mission parsed into mission phases. The tasks associated with each phase are further segmented according to three major sub-headings: aviation, navigation, and communication. Specific mission tasks are then listed under each subheading to reflect the changing goals and sub-goals of the pilot throughout the mission. During an Air Base Defense mission, for example, a transition from a CAP phase to an intercept phase represents a radical change in the pilot's immediate goals, while the overall goal of bomber interception and return to safe air space remains. The task analysis is used to create a highly detailed set of rules of engagement, similar in nature to rules of engagement encountered by fighter pilots in an operational context. The rules derived from the task analysis place constraints on both mission performance parameters as well as the engagement of hostile targets. Each of the "rules" of engagement is treated as an individual implicit probe and assessed at the frame rate of the simulation.

As an example, the rules of engagement specify a radar mode for each mission segment (i.e., "track while scan" as opposed to "single target track"). If the pilot configures the radar correctly for that portion of the mission, the resulting score on that implicit metric would reflect adequate SA (i.e., a numerical value of "1"). If the pilot sets the radar to any other mode, the score on that item would reflect inadequate SA (i.e., a numerical value of "0"). In addition, the implicit probes can be weighted in order to yield a more realistic metric of SA based upon the relative importance of each probe as judged by subject matter experts. Thus, the GIM implicit probes yield a comparison between actual versus ideal performance and are used to determine the degree to which a pilot succeeds in adhering to each segment's specified performance constraints.

For any time period during the mission (mission phase, or segment within a phase), a proportion score can be calculated for each of the implicit probes. This proportion consists of the sum for each measure, divided by the number of observations. The individual implicit probe scores may also be combined, yielding a composite SA score for that segment. Each of the segment scores may then be combined to yield overall SA phase and mission scores.

In summary, the Global Implicit Measurement technique affords comparisons of actual versus ideal performance in very complex task domains. It provides metrics of situation awareness that are sensitive to the changing goals and sub-goals associated with good mission performance, both within mission phases and across entire missions. The following experiment describes the first attempt to apply the Global Implicit Measurement approach to a complex simulation of air-to-air combat.

METHOD

Subjects

Eighteen military pilots from the United States, France, and Great Britain served as subjects. Their total flight experience ranged from 1720 to 4808 hours with a mean of 2980 hours.

Apparatus

An F-16 fiberglass cockpit was configured with six head-down displays and F-16C throttle and sidestick controller. Each trial was flown with the crewstation in either a *conventional* mode (consisting of somewhat standard F-15/F-16 type displays and controls) or a *candidate* mode (consisting of virtually augmented displays and controls). Out-the-window visuals were projected onto an 8'x 8'x8' room and provided a 240 x 120 degree field-of-view.

Procedure

Each pilot was trained in both crewstations prior to data collection. Each of the subjects performed half of the trials using the conventional and half using the candidate configuration. The task required the primary fighter to intercept and destroy four computer-controlled bombers and return to safe airspace. Two human controlled hostile fighters defended the bombers.

RESULTS

The results of the GIM scoring technique are presented for the initial Combat Air Patrol mission phase and are summarized by two measures of performance. First, the overall average GIM SA score collapsed across subjects and crewstations is presented in Figure 1. Next, a comparison of overall average GIM SA scores for the Conventional vs. the Candidate cockpit configurations is presented in Figure 2.

The overall CAP phase GIM SA scores are presented in Figure 1. An examination of the figure reveals a well-defined pattern of increasing SA early in the mission, followed by a relatively flat period of high SA scores from approximately frame 82 to 200. The average scores begin to drop rapidly beginning about frame 200 (e.g., 50 seconds into the mission), followed by a "plateau" at a somewhat lower level. There is a fairly gradual decline until around frame 450, when there is a fairly rapid rise, followed by a more gradual climb until the end of the CAP phase.

An analysis of the individual missions indicates that the initial rise in SA scores was due to an "initial cockpit setup" period, during which the pilots checked the displays and flight conditions and configured the crewstation as required by the rules of engagement. Following this initial period the pilots only had to concentrate on flying straight-and-level on the outbound leg of the CAP pattern at the required altitude and airspeed. The GIM SA scores were very high throughout this portion of the mission. This is not surprising as the CAP represents a relatively easy portion of the mission.

The sharp drop in SA scores following this plateau appears to be almost entirely due to the subjects losing awareness of the vertical coverage of the radar. The rules of engagement specified radar coverage from the ground to 50,000 MSL at the border of the threat sector at all times while in the CAP. It is apparent that the subjects simply placed the radar elevation control on one setting and did not raise the elevation to provide the prescribed coverage as the distance to the threat sector boundary decreased on the outbound leg of the CAP.

Following this drop in SA scores there is another brief plateau until the fighter reached the 30-NM turn point at the far end of the CAP pattern (about frame 300). At this point SA scores began to decrease gradually due to occasional late turns and deviations from the prescribed altitude and airspeed. This also is not surprising, since pilot workload increases substantially during such difficult turns. It may also reflect pilot attention being diverted to assessing radar contacts, which often began to appear near the turn point or during the turn itself.

The drop in scores generally ceased at about frame 400, as the turn from the outbound to inbound leg of the CAP was completed and workload decreased. This was followed (about frame 450) by another rapid rise in scores as distance to the threat sector boundary increased during the inbound leg of the CAP, followed by a gradual rise in SA scores after the turn from inbound to outbound legs of the CAP.

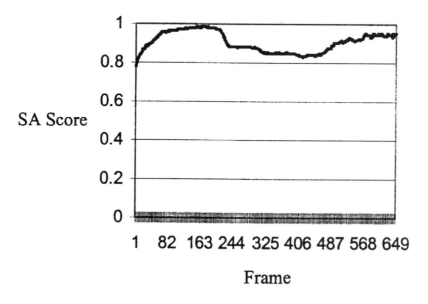

Figure 1. Overall CAP Phase Mean GIM Situation Awareness Scores

The CAP phase typically ended during the second outbound leg of the CAP pattern when threat aircraft began to cross the outer boundary of the threat sector, requiring interception by the primary fighter.

162

A comparison of the mean GIM SA scores by crewstation configuration is presented in Figure 2. An examination of Figure 2 reveals only minor differences between the crewstations (512.7 for the candidate vs. 507.9 for the conventional). The lack of differences in GIM SA scores between the two crewstation designs is probably due the nature of the requirements placed on the pilots during the CAP phase of the mission. This phase required little more than basic instrument flying skills for which pilots tended to use the HUD almost exclusively. Except for one minor difference, both HUD formats were essentially identical during this period, so the lack of significant differences in performance is not surprising. Greater differences would be expected during the more dynamic segments of the mission such as the intercept, weapons deployment, and defensive maneuvering phases.

The candidate crewstation did begin to exhibit a marked superiority over the conventional crewstation configuration near the end of the CAP Phase, as the primary fighter transitioned from the CAP to the intercept phase. At this point in the data analysis, we are unsure of the cause of this phenomenon. The analysis of the remaining mission segments is underway in order to further examine this result.

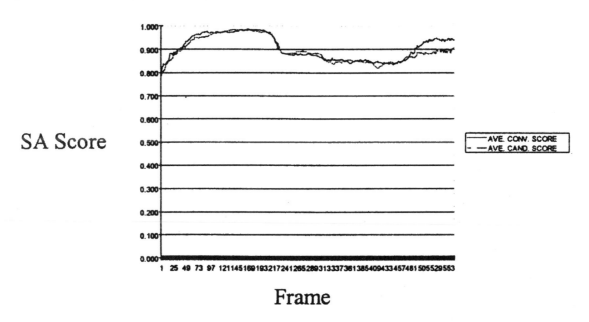

Figure 2. CAP Phase mean GIM Situation Awareness Scores by Crewstation.

DISCUSSION

The initial results of the Global Implicit Measurement approach presented here are limited in scope, reflecting the use of the GIM scoring algorithms only on the first phase of a simulated mission. As such, any inferences based on these results must remain guarded. We are, however, encouraged by the initial results. The GIM SA scores appeared to accurately track the changing demands on the pilot throughout the first phase of the air defense mission. The CAP Phase GIM SA scores initially rose as the pilot setup the crewstation to the proper configuration as required by the rules of engagement then fell as the pilots' workload increased during the initial turn and inbound leg of the CAP. Finally, the GIM SA scores increased during the outbound leg. This pattern of results was corroborated by the variability in SA scores among the subjects.

Unfortunately, the analysis thus far has yielded little differences in GIM SA scores between the two alternative crewstation configurations as shown in Figure 2. This is most probably due to the nature of the tasks demanded in the CAP Phase. The CAP phase required little more than instrument flying skills and the

163

primary instrument (Head-Up display) was nearly identical in both conventional and candidate crewstations. However, as the pilots begin to transition from the CAP Phase to the intercept phase, there does appear to be an increasing dissociation in GIM SA scores between the two crewstations. This will be examined in more detail with further development of the GIM technique.

REFERENCES

Brickman, B.J., Hettinger, L.J., Roe, M.M., Stautberg, D.K., Vidulich, M.A., Haas, M.W., & Shaw, R.L. (1995). An assessment of situation awareness in an air combat simulation: The Global Implicit Measurement Approach. In D.J. Garland, & M.R. Endsley (Eds.), *Experimental Analysis and Measurement of Situation Awareness: Proceedings of an International Conference* (pp. 339 - 344). Daytona Beach, FL: Embry-Riddle Aeronautical University Press.

Level of Automation Effects on Telerobot Performance and Human Operator Situation Awareness and Subjective Workload

David B. Kaber
Mississippi State University

Emrah Omal
Lucent Technologies

Mica R. Endsley
SA Technologies

INTRODUCTION

Recent literature (Endsley and Kaber (in press); Endsley and Kiris, 1995; Milgram et al., 1995) on advances in automation for complex, dynamic systems in general, and telerobots (remote controlled robotic manipulators) specifically, has presented taxonomies of levels of automation (LOAs) detailing control function allocation schemes that may improve systems performance over that resulting from traditional automation. (Traditional automation is considered to be the implementation of technology based on its capabilities, but lacking in consideration of the effects of its application on a human operator.) The taxonomies have also been put forth as vehicles by which to address out-of-the-loop performance and safety problems along with their underlying causes. Root causes to performance problems include operator vigilance decrements and complacency, which may both contribute to problems with operator situation awareness (SA). Most of the taxonomies offer intermediary LOAs falling somewhere between manual control and full automation. These levels are intended to maintain both human and computer involvement in active systems control for improving operator SA (critical to failure recovery and system safety) and increasing system performance (through computer data processing). The taxonomies present levels by identifying or describing the roles that the human operator and computer are to play in controlling a system.

Endsley and Kiris (1995) presented a taxonomy of LOA developed in the context of the use of expert systems to supplement human decision-making for automated systems control. They identified five functions that either a human operator or expert system could play including suggest, concur, veto, decide, and act. They offered five LOAs by structuring allocation of these roles to both servers ranging from "Manually" to "Full Automation." Endsley and Kiris (1995) empirically assessed the effect of the LOAs in this taxonomy on out-of-the-loop performance problems and SA in a simulated automobile navigation task, revealing it to be more significant under fully automated conditions than under intermediate LOAs. They found that using lower LOAs, which maintain human operator involvement in active control, was beneficial to SA and subjects were better able to perform tasks manually when needed.

Endsley and Kaber (in press) present a taxonomy of LOAs developed by allocating to a human and/or computer generic control system functions including monitoring, generating, selecting and implementing based on the capabilities of each server to perform the functions. Endsley and Kaber (in press) formulated 10 LOAs feasible for use in the context of teleoperation. The levels have been assessed as to their effect on human-machine system performance, and operator SA and workload in a dynamic control task. They have also been studied as to their potential for facilitating a smooth transition (in terms of performance) between normal operations and simulated automation failures. Endsley and Kaber (in press) found human-machine system performance to be enhanced by automation that provided computer aiding or control in the implementation aspect of the task. With respect to performance during failure modes, the authors found human control to be significantly superior when preceded by functioning at LOAs involving the operator in the implementation aspect of the task, as compared to being preceded by higher LOAs. Improved SA and lower levels of overall task demand corresponded with higher LOAs.

Another taxonomy of LOA has been developed by Milgram et al. (1995) in the context of telerobot control. They structured five LOAs by considering the different roles a human operator could play in controlling a semi-automated robotic manipulator including decision-maker and direct controller. Their levels ranged from "Manual Teleoperation" to "Autonomous Robotics" including intermediate levels of "Telepresence," "Director/Agent Control" and "Supervisory Control." Milgram et al. (1995) have not empirically assessed the LOAs in their taxonomy for an effect on telerobot performance.

Criticism has been made of these and other LOA taxonomies that they may be incomplete in different senses and may have limited applicability to specific types of systems for improving performance and abating the negative consequences of out-of-the-loop performance. To date, a limited amount of experimental work has been conducted to investigate the usefulness of intermediary LOAs for enhancing specific task performance or examining the effects of LOAs on operator SA and mental workload. The studies of Endsley and Kiris (1995) and Endsley and Kaber (in press) demonstrate that traditional, full automation of a system or task may not be advantageous if joint human-machine performance is to be optimized. They also offer support to the usefulness of intermediate LOAs presented in general taxonomies of control for specific tasks and functions in order to keep human operator SA at higher levels and allow them to perform critical functions during failures.

The purpose of the present research was to further examine the benefits of intermediary LOAs, specifically in a high-fidelity simulation supporting generalizability of results to a real-world application. Further, it was intended to demonstrate the usefulness of LOAs in the context of a specific application. This was accomplished by assessing the impact of LOA on telerobot performance under both normal operating conditions and failure modes, and its effect on operator SA and subjective workload.

EXPERIMENT

An experiment was conducted in which Endsley and Kaber's (in press) LOA taxonomy was further explored using a high-fidelity simulation of a telerobot performing safety tests on plutonium storage containers. The simulation required subjects to interact with a computer in controlling a Fanuc® S-800 robotic arm through a graphical user interface designed by Kaber et al. (1997). In the simulation, the robotic arm was used to remove storage containers from a staging rack one-at-a-time. Each container was placed on a pedestal adjacent to the staging rack and its lid unbolted. The lid and packing materials were then removed from the storage container along with a plutonium containment vessel (CV). The CV was then placed at a weigh station followed by a leak testing station. Subsequently, the CV was re-packed in the storage container. Both the packing materials and lid were then returned to the container and the lid was bolted. Finally, the inspected container was returned to the staging rack. The operator's goal in the simulation was to safely and efficiently handle the plutonium containers through the inspection process.

Equipment, Subjects and Design

The telerobot task was simulated on a Silicon Graphics® (SGI) Indigo² workstation using a 21-in graphics monitor operating under 1600×1280 resolution. The system was integrated with a mouse, standard keyboard and SpaceBall® controller. An Apple Macintosh, configured with a 15-in graphics monitor, a standard keyboard, and a mouse, was also used in the study to electronically present Situation Awareness Global Assessment Technique (SAGAT) queries (Endsley, 1988) and NASA-Task Load Index (TLX) demand-factor ranking and rating surveys. The SAGAT and the TLX were intended to establish operator SA and overall perceived workload, respectively, in performing the task.

Ten subjects (six males and four females) were recruited for participation in this study from the

Texas Tech University student population. All subjects had normal or corrected to 20/20 visual acuity, full-color vision, and computer and mouse experience. They were all right-hand dominant.

The subjects served as replications in a repeated measures investigation examining the effect of LOA as an independent variable in performance during both normal operation of the simulation and the simulated failures. As well, they served as repeated observations of the LOA effect on SA and mental workload. The simulated system was programmed to allow for the use of five LOAs in Endsley and Kaber's (in press) taxonomy including: (1) 'Action Support', (2) 'Batch Processing', (3) 'Decision Support', (4) 'Supervisory Control', and (5) 'Full Automation.' These LOAs represented a range of automated system control function allocations from complete human control (Action Support) to computer control (Full Automation). Action Support required subjects to generate a CV processing plan and control the telerobot in implementing the plan. Batch Processing was identical to Action Support, except implementation was fully automated. Decision Support involved joint human and computer task strategizing, human selection of an "optimal" processing plan and automated implementation. Supervisory Control provided automation of all control functions except monitoring and permitted human intervention in the loop for potential error correction. Under Full Automation the human was an observer of the system. The dependent variables recorded during the experiment included time-to-container completion under normal operating conditions, time-to-system-recovery during automation failures, and operator SA and perceived workload. Operator SA was measured using SAGAT queries regarding the 3 levels of SA proposed by Endsley (1988).

Procedures

This study was conducted across five consecutive days within a single week. On the first day, subjects were trained in how to control the simulated robotic arm under direct teleoperation (Action Support) using the SpaceBall® without any interruptions (e.g., automation failures). They were also required to practice Action Support with three simulation freezes at random points in time for administering the SAGAT queries. When a freeze happened, the display screen of the SGI workstation was blanked and subjects responded to the queries electronically using the Macintosh®. The first day of the study was concluded by subjects being tested on their control of the robot under direct teleoperation. During the test period, all performance measures were recorded. Three simulation freezes were conducted to capture operator SA. NASA-TLX rating forms were administered during each of the SAGAT freezes.

The second through the fifth days of the experiment involved training subjects in control of the telerobot under the other LOAs (i.e., Batch Processing, Decision Support, Supervisory Control and Full Automation) and testing their performance under normal conditions and failures, as well as assessing SA and workload. During training, three simulation freezes were used for SAGAT queries and two automation failures occurred at random times. A system failure caused, a "pop-up" message box to be displayed on the SGI monitor indicating the type of error encountered (e.g., "robot stuck"). Subjects were required to depress keys on the keyboard corresponding to the first letter of various error messages. Once the operator responded to an error appropriately, the LOA of the system shifted to direct teleoperation. The system remained under this level until the operator successfully completed an action sequence (e.g., picking-up an object with the arm), which caused performance to resume under the test LOA.

Data Collection and Analysis

Time-to-container completion during normal functioning was averaged across the test periods (excluding the time spent in failure modes using Action Support to control the robot) for 50 trials (5 LOAs × 10 subjects). Time-to-system recovery during automation failures was recorded for the two automation failures simulated during 40 of the 50 test trials. Observations were not collected on the ten trials involving subject control of the robot using Action Support because the LOA for normal operations did not differ from that used after a system error.

SA was recorded as the percentage of correct responses to the SAGAT queries. The queries were posed to subjects during the three simulation freezes across all 50 test trials yielding 150 observations (3 Freezes × 5 LOAs × 10 Subjects) on each. The percentage of correct responses were averaged across queries targeted at a particular level of SA (Level 1 SA, Level 2 SA and Level 3 SA). These data served as composites of operator perception and comprehension of system information, as well as future system state predictions. During each SAGAT freeze an observation was recorded on workload.

RESULTS AND DISCUSSION

An analysis of variance (ANOVA) conducted on time-to-container completion under normal functioning of the telerobot revealed LOA to be significant in its effect, $F(4,9) = 53.85, p < 0.001$. In general, performance improved (i.e., task time decreased) as the LOA increased. Tukey's honestly significant difference (HSD) test demonstrated Action Support to be significantly lower ($p < 0.05$) in terms of performance than all other levels, and Supervisory Control and Full Automation to be significantly higher ($p < 0.05$) than Batch Processing.

This analysis reveals the benefit of computer programmed motion control over the telerobot. Direct teleoperation (Action Support) required human involvement in implementation of the task (motion path control) and produced the lowest performance. This can be attributed in part to difficulty subjects had in controlling the telerobot using the SpaceBall®. (Extensive training in the use of the SpaceBall® for direct teleoperation was provided; however, an effect of the controller configuration appeared to remain.) The superiority of Supervisory Control and Full Automation to Batch Processing can be attributed to the roles that the human and computer maintained during the teleoperation. Under Batch Processing, subjects were required to generate a container-processing plan and develop specific move sequences to that plan. These sequences were implemented by the computer. Although the system allowed for operator advanced planning of telerobot moves, subjects seldom took advantage of this capability often waiting for processing of a particular sequence to finish to ensure its safety. Consequently, Batch Processing never managed to produce performance equivalent to complete computer control under Full Automation or Supervisory Control.

An ANOVA was conducted on the data recorded during simulated automation failures in which subjects were required to perform direct teleoperation (Action Support) to return the system to a higher LOA. Results revealed a significant effect ($F(3,9) = 3.11, p < 0.05$) of the LOA preceding a failure on time-to-system recovery from the failure. In general, recovery time increased with LOA. Tukey's HSD test on the LOA effect revealed Full Automation to significantly differ ($p < 0.05$) from Batch Processing and Supervisory Control, but not Decision Support.

According to Endsley and Kaber's (in press) taxonomy, the definition of Supervisory Control involves Full Automation along with human process intervention, when needed, through a shift in the LOA to Decision Support. During the experiment, subjects frequently intervened in systems control. Decision Support required a greater degree of human involvement in the simulation including jointly generating processing plans with the computer and selecting plans. The interventions may have increased subject awareness of system states prior to a failure promoting faster recovery times, as compared to Full Automation. Under normal conditions, Supervisory Control only required the human to monitor the robot while the computer maintained all other functions off-loading system responsibilities including planning and decision-making from operators to the computer. These responsibilities included evaluating computer processing plans for safety and efficiency, which may have prevented subjects using Decision Support from monitoring system states and, consequently, relating such information to a failure condition for efficient recovery, as compared to Supervisory Control.

Batch Processing was related to Decision Support in that the human and computer jointly performed monitoring of the system and the computer implemented the process plan. However, Batch Processing limited the generating and selecting functions to the human, whereas, Decision Support permitted both human and computer responsibility for task planning. This difference in the function allocation schemes explains the observed difference between the two LOAs in terms of time-to-recovery in the simulation. Batch Processing involved subjects in the simulation to a greater extent than Decision Support possibly promoting heightened states of subject system awareness for efficient recovery.

In summary, these results are supportive of lower level automation (Batch Processing) maintaining human involvement in the control loop during normal system functioning. This reduced the time-to-recovery during failure modes.

ANOVA results on the average percent correct responses to SAGAT queries covering the three levels of SA revealed a significant effect of LOA ($F(4,9) = 3.4$, $p < 0.05$) only on Level 3 (system state projection). The mean percentage of correct responses to Level 3 SA queries decreased from low to intermediate LOA and increased slightly from intermediate to higher LOAs. According to Tukey's HSD test, Action Support was significantly superior to Batch Processing in terms of the percentage of correct responses. Action Support varied from Batch Processing in that the latter stripped subjects of the capability to manually control the robot using the SpaceBall®. This difference between the two LOAs was associated with a degradation in, for example, operator ability to predict the next move sequence in the telerobot task.

ANOVA results revealed LOA to be significant in its effect on the NASA-TLX, $F(4,9) = 22.538$, $p < 0.001$. The general trend of perceived workload decreased at progressively higher LOAs. Tukey's HSD test on the overall workload scores revealed Action Support to significantly differ ($p < 0.05$) from all other levels and the same to be true for Batch Processing. This result reflects the reduction of human involvement in the active system control loop. As subjects were progressively removed from plan implementation (Batch Processing), limited in their capability to plan move sequences (Decision Support), and reduced to the status of system monitor (Supervisory Control) or observer (Full Automation), workload significantly decreased.

CONCLUSIONS

This study reviewed LOA taxonomies presented in the literature. Arguments both for and against the use of general taxonomies were levied. In particular, a potential lack of applicability of many LOA taxonomies to real-world tasks was raised. This issue was addressed through further empirical exploration of Endsley and Kaber's (in press) LOA taxonomy demonstrating the applicability of specific LOAs to a simulated telerobot task. The study demonstrated that higher LOAs enhance performance during normal operating conditions through computer processing. This was accompanied by lower levels of subjective workload observed at the same levels. The experiment revealed intermediate LOAs to promote higher operator SA and enhance human manual performance during system failure modes as compared to high level automation. This effect was also attributed to maintaining operator involvement in the system control loop during normal operations.

From an automation design perspective, this research validates LOA as an alternate approach to traditional automation. It provides detailed guidance to teleoperation/telerobotic systems designers in allocating system responsibility to a human and computer for safe performance in hazardous operations.

REFERENCES

Endsley, M.R. (1988). Design and evaluation for situation awareness enhancement. In *Proceedings of the Human Factors Society 32nd Annual Meeting* (pp. 97-101). Santa Monica, California: Human Factors and Ergonomics Society.

Endsley, M.R. & Kaber, D.B. (in press). Level of automation effects on performance, situation awareness and workload in a dynamic control task. *Ergonomics*.

Endsley, M.R. & Kiris, E.O. (1995). The out-of-the-loop performance problem and level of control in automation. *Human Factors*, 37(2), 381-394.

Kaber, D.B., Onal, E. & Endsley, M.R. (1997). Design and development of a comprehensive user interface for teleoperator control in nuclear applications. In *Proceedings of the 7ᵗʰ Topical Meeting on Robotics and Remote Systems* (pp. 947-954). LaGrange Park, Illinois: American Nuclear Society.

Milgram, P., Rastogi, A. & Gordski, J.J. (July, 1995). Telerobotic control using augmented reality. In *Proceedings of the 4ᵗʰ IEEE International Workshop on Robot and Human Communication (RO-MAN '95)*. IEEE.

A Role Theory Approach to Strategizing for Team Situation Awareness

Alexis A. Fink and Debra A. Major
Old Dominion University

Although definitions vary, the following elements are considered key in defining situation awareness (SA): an awareness of surroundings and objectives, comprehension of their meaning, an ability to observe, process and take action on the basis of relevant information, an awareness of the dynamic nature of the situation and one's current temporal place within the dynamic system, and a planful anticipation of future events (Endsley, 1995; Sarter & Woods, 1991; Shrestha, Prince, Baker, & Salas, 1995). As this definition suggests, situation awareness is not equally relevant and important across performance contexts. SA is most critical to task performance when operating in a context that is complex, dynamic, risky, cognitively demanding, and time-constrained (Sarter & Woods, 1991). The environment must be constantly scanned for relevant information, which must be processed, prioritized and incorporated in planning. In this type of context critical cues may be subtle, meaningful only in combination, or fleeting (Gaba & Howard, 1995). The complexity, criticality, and volume of information presented in the SA context often precludes a single individual working alone from effectively operating in that environment; it is simply too demanding. Therefore, teams are often required to ensure safe, effective decision-making, and appropriate action in this context. In general, teams are defined as "distinguishable sets of individuals who interact interdependently and adaptively to achieve specified, shared and valued objectives" (Morgan, Glickman, Woodard, Blaiwes, & Salas, 1986, p. 3).

TEAMS AND TEAM SITUATION AWARENESS

With few exceptions, theoretical and empirical work to date has focused primarily upon SA at the individual level with emphasis on cognitive processes. The shared understanding approach to team SA (e.g., Stout et al., 1996/1997) is more complex and comprehensive than others that have been suggested in the literature. According to this perspective, team SA is defined as a function of individual team member SA and the degree of shared understanding among team members. Thus, the same level of team SA may exist either because individual team members are maintaining a high degree of individual level situation awareness, or because there is a great degree of shared understanding among team members. However, Stout et al. contend that the largest incremental increases in team SA will be due to increased understanding among team members. Shared understanding is hypothesized to be the direct result of both shared mental models and team process behavior. Although Stout et al. (1996/1997) provide a more thorough treatment of the role of mental models in contributing to shared understanding and team situation awareness, the importance of team processes is explicitly acknowledged.

Stout et al. (1996/1997) argue that one way shared mental models among team members develop is as the result of explicit strategizing and that these mental models then allow for the maintenance of SA under dynamic circumstances in which explicit strategizing is not possible. A key point in this argument is that explicit strategizing among team members must occur in order for an accurate and effective shared mental model to exist. We take the position that while mental models are cognitive constructs, the process by which they become shared among team members is necessarily social. Team members must interact with one another to develop a shared understanding. Thus, the focus of the current effort is to explore the interactive processes by which the shared understanding that is so critical to team SA is developed.

AN APPLICATION OF ROLE THEORY

The potential relevance of role theory to the development of shared understanding is evident in the definition of a role. In general, a role is an expected pattern or set of behaviors (Biddle, 1979). "By definition, expectations are beliefs or cognitions held by individuals. Therefore, roles exist in the minds of people," (Ilgen & Hollenbeck, 1991, p.169). According to Stout et al. (1996/1997) shared mental models are cognitive structures that provide a team with shared expectations for behavior and task performance. Thus, both roles and mental models are cognitive and center around shared expectations.

In their work on the relationship between shared understanding and team SA, Stout et al. (1996/1997) concentrated on the content of shared mental models. In particular, they argued that the accuracy of declarative, procedural, and strategic mental models determined the level of individual SA, one contributor to team SA. Moreover, shared understanding among team members, the other determinant of team SA, is the result of the commonalties of these three mental models across team members. While this work focuses on *what* expectations need to be shared, role theory concentrates on *how* expectations are shared, suggesting that these perspectives may be complementary.

The Content of Shared Expectations

In attempting to explicate the basis of team SA, Stout et al. (1996/1997) argued for the relevance of three types of mental models: declarative, procedural, and strategic. According to Stout, Cannon-Bowers, and Salas (1994), declarative models are comprised of facts, rules, and relationships and include knowledge of relevant systems, task goals, the relation among system components, equipment/hardware, position, team members. Procedural mental models contain knowledge about the timing and sequencing of activities, essentially the steps or order in which tasks must be accomplished for goal attainment (Converse & Kahler, 1992; Stout et al., 1994). Drawing on declarative and procedural mental models, strategic mental models "are comprised of information that is the basis of problem solving, such as action plans to meet specific goals, knowledge of the context in which procedures should be implemented, actions to be taken if a proposed solution fails, and how to respond if necessary information is absent" (Converse & Kahler, 1992, p. 6).

Strategic mental models are thought to be dynamic and continuously updated based on task parameters, team interaction, and situational demands, in contrast to declarative and procedural mental models which are more static (Stout et al., 1994). In essence, strategic mental models foster the capability of understanding cue/action sequences, cue patterns, and their significance, team resources and capabilities, and appropriate task strategies *as they evolve* in a dynamic context (Cannon-Bowers, Tannenbaum, Salas, & Volpe, 1995). For this reason, Stout et al. (1996/1997) contend that of the three types of mental models, strategic mental models are most critical to team SA and must be shared among team members.

The content distinction between declarative and procedural mental models versus strategic mental models is similar to the distinction made between jobs and roles. In general, a job is comprised of a specific set of tasks and activities, such that the knowledge, skills, and abilities required for performance are identifiable. Ilgen and Hollenbeck (1991) argue that jobs have four defining features: they are created by prime beneficiaries (i.e., certain task elements were grouped for a specific purpose by individuals in positions of authority with a vested interested in the performance of these tasks); they are objective in the sense that they can be documented by a formal, written description and individuals are able to come to ready consensus about what the job is; they are bureaucratic in that they exist independent of the job incumbent; and they are quasi-static, staying relatively constant over time.

Although jobs are defined by a set of established task elements, it is common for a jobholder to be expected to perform additional tasks, known as emergent task elements. Because jobs are performed in an environment that is subjective, personal and dynamic with a number of diverse constituencies, prime beneficiaries cannot pre-determine all the necessary task elements. As Ilgen and Hollenbeck (1991) explain, "to make jobs work in their environment, an extra set or collection of task elements needs to be added to those that originally constituted the job. It is important to note that this does not expand the universe of task elements that we have chosen to call jobs and have defined as finite. Rather, it adds some elements from another domain onto the job that did not exist until there was the necessity of enacting the job in a particular work setting. These additional task elements are specified by a variety of social sources, not the least of which is the incumbent," (p. 174). Thus, a role contains the established task elements of the job plus emergent task elements. It is important to emphasize that roles are defined in terms of expected behaviors and are seen as existing in the shared perceptions of people.

As noted by Stout et al. (1996/1997), teams require SA in environments where the opportunity for explicit strategizing is likely to be limited. Thus, teams need to engage in the active negotiation of developing shared strategic mental models prior to encountering the SA context. Like roles, more complete, useful, and compatible strategic mental models should develop with time and experience. To the extent that circumstances and team membership are stable, shared understanding is likely to be greater. Interestingly, the benefits of successful role development are very similar to proposed benefits of shared understanding in team SA, namely the capacity to interpret the cues in the environment in such a way to be able to understand and anticipate required behaviors, and act in a timely and effective manner. In the next section, we apply the role literature to elucidating the barriers to the development of shared understanding.

Role Conflict and Role Ambiguity

There are two different types of failure to develop shared expectations within a role set. The first condition, *role ambiguity*, exists when there is uncertainty and a lack of clarity regarding the expectations of a single role. Cook, Hepworth, Wall, and War (1981) point out that the ambiguity may exist regarding the behavioral expectations of a given role and/or the consequences resulting from certain behaviors. *Role conflict* results when the expectations or demands faced by a focal person are incompatible. Often these conflicting demands stem from the differing expectations of members of the role set.

There is some debate regarding whether or not the role ambiguity and role conflict constructs are empirically distinct (Ilgen & Hollenbeck, 1991). In a meta-analysis, Jackson and Schuler (1985) estimated a correlation of .42 between the two variables. As Ilgen and Hollenbeck (1991) point out, examination of Jackson and Schuler's (1985) meta-analysis also reveals that the causes and consequences of role conflict and role ambiguity appear to be highly related (i.e., a correlation of .87). For present purposes, both role conflict and role ambiguity are indicators of a lack of shared understanding and expectations. Jackson and Schuler's (1985) meta-analysis demonstrated that a similar set of conditions was associated with both role ambiguity and conflict. While they examined a variety of organizational and task factors, this review will concentrate on those likely to be most relevant to shared understanding in the context of team SA. The meta-analytic results showed that feedback from others, participation, leader initiating structure, and leader consideration were all negatively related to the experience of role ambiguity and role conflict.

The role theory and SA literatures both demonstrate that shared understanding is associated with positive outcomes. Research shows that role clarity and absence of conflict are positively related to commitment, involvement, satisfaction, and performance. The experience of role ambiguity and conflict, on the other hand, is positively related to tension and anxiety. Stout et al. (1996/1997) commented on the positive linkages between shared mental models and implicit coordination in teams, performance under stress, and anticipation of needs. Stout et al. were only able to find weak support for a relationship between planning and explaining and the development of shared mental models. The role theory literature reviewed here clearly suggests that planning and explicit strategizing are critical to shared understanding. We propose that planning content informed by role theory will result in more effective strategizing.

Explicit Strategizing

The complexity of SA environments requires that teams engage in strategy development prior to encountering the environment. For instance, aircrews engage in extensive mission briefs prior to manning the aircraft. The purpose of these strategy sessions is to make explicit the expectations regarding conduct in the particular SA environment about to be encountered. This is intuitively reasonable; explicitly addressing who is to do what, *beyond* standard job descriptions is a critical process. Role theory is useful in prescribing the timing, content, and climate of a strategy session.

Timing. When to strategize is a relatively straightforward matter. In depth and effective strategizing must occur prior to encountering the SA context. Effective strategizing requires time that will not be available to the team once they enter the SA context. The role theory literature has demonstrated that role development occurs through negotiation. This process is likely to be compromised if it ensues while the team is concurrently attending to other performance demands. Although a certain amount of strategizing may become necessary in crisis situations, having clear roles previously established will minimize coordination demands so that greater effort and attention can be devoted to addressing the problem.

Content. Acknowledging the distinction between jobs and roles may be an effective way to determine the issues that a team must be trained to address during a strategy session. Recall that jobs are comprised of a specific set of tasks and activities that are well defined and mutually agreed upon. A role, on the other hand, is a social construction that may be specific to the individual role holder, his or her role set, and the particular setting. It is likely to be unnecessary to emphasize jobs during a strategy session. Team members should already be quite competent when it comes to the specific expertise of their individual jobs. In other words, provided that they have been adequately trained, team members do not need others to tell them how to do their jobs. Team members do require a clear set of role expectations, however. This may include information on how one team member can share his or her expertise to help another, the form and frequency of team interactions, and expected behavioral contingencies associated with the specific characteristics of the situation encountered. Thus, the content focus of an explicit strategizing session should be on role making (i.e., negotiating mutually agreeable role expectations among team members) and the resolution of role conflict and ambiguity.

Contingency plans and the roles to be adopted under various conditions are particularly important. Since the SA context is dynamic by definition, change is to be expected and flexibility is a must. Team members, however, should be adequately prepared so that they can anticipate these demands without experiencing role conflict or ambiguity. For instance, if the manner in which team members are expected to communicate differs between standard and emergency conditions (e.g., the use of simple commands or code words), these expectations should be made explicit during the strategy session. Similarly, if the team is expected to behave "democratically" at times, while responding to centralized authority at others, these contingencies should be made clear in advance.

Climate. The role theory literature demonstrates that effective role development involves the role holder, as well as members of the role set. Thus, strategizing regarding roles hinges on the active participation of all team members. Though this notion may sound simple, establishing a climate in which team members of varying expertise and hierarchical levels feel free to participate is likely to prove challenging.

Team leaders can begin to foster an open climate by clearly articulating the importance and value of team member participation in role negotiation. Furthermore, the leader can serve as a role model, demonstrating active participation, respectful listening, and effective negotiation techniques. An open climate for role negotiation is critical in order to make explicit the expectations and assumptions held by all members of the role set. Through this articulation it becomes possible to identify potential sources of role conflict and ambiguity and resolve them before they detract from the team's ability to achieve and maintain SA. Moreover, since team members have actively and openly participated in establishing their own roles and those of their teammates, they are likely to experience greater commitment to fulfilling those roles for the team.

REFERENCES

Biddle, B. J. (1979). *Role theory: Expectations, identities and behaviors.* New York: Academic Press.

Cannon-Bowers, J. A., Tannenbaum, S. I., Salas, E., & Volpe, C. E. (1995). Defining team competencies and establishing team training requirements. In R. Guzzo & E. Salas (Eds.), *Team effectiveness and decision making in organizations* (pp. 333-380). San Francisco, CA: Jossey Bass.

Converse, S. A., & Kahler, S. E. (1992). *Knowledge acquisition and the measurement of shared mental models.* Unpublished manuscript, Naval Training Systems Center, Orlando, FL.

Cook, J. D., Hepworth, S. J., Wall, T. D., & War, P. B. (1981). *The experience of work.* London: Academic Press.

Endsley, M. R. (1995). Toward a theory of situation awareness in dynamic systems. *Human Factors, 37*(1), 32-64.

Gaba, D. M., & Howard, S. K. (1995). Situation awareness in anesthesiology. *Human Factors, 37*(1), 20-31.

Ilgen, D. R., & Hollenbeck, J. R. (1991). The structure of work: Job design and roles. In M. D. Dunnette & L. M. Hough (Eds.), *Handbook of Industrial and Organizational Psychology* (Second ed. vol. 2), (pp. 165-207). Palo Alto, CA: Consulting Psychologists Press.

Jackson, S. E., & Schuler R. S. (1985). A meta-analysis and conceptual critique of research on role ambiguity and role conflict in work settings. *Organizational Behavior and Human Decision Processes, 36,* 16-78.

Morgan, B. B., Jr., Glickman, A. S., Woodard, E. A., Blaiwes, A., & Salas, E. (1986). *Measurement of team behaviors in a Navy environment* (NTSC Report, No. 86-014). Orlando, FL: Naval Training Systems Center.

Sarter, N. B., & Woods, D. D. (1991). Situation awareness: A critical but ill-defined phenomenon. *The International Journal of Aviation Psychology, 1*(1), 45-57.

Shrestha, L. B., Prince, C., Baker, D. P., & Salas, E. (1995). Understanding situation awareness: Concepts, methods and training. *Human/Technology Interaction in Complex Systems, 7,* 45-83.

Stout, R. J., Cannon-Bowers, J. A., & Salas, E. (1994). The role of shared mental models in developing shared situational awareness. In R. D. Gilson, D. J. Garland, & J. M. Koonce (Eds.), *Situational awareness in complex systems* (pp. 297-304). Daytona Beach, FL: Embry-Riddle Aeronautical University Press.

Stout, R. J., Cannon-Bowers, J. A., & Salas, E. (1996/1997). The role of shared mental models in developing team situational awareness: Implications for training. *Training Research Journal, 2,* 85-116.

MONITORING

AND

VIGILANCE

The Effects of Search Differences for the Presence and Absence of Features on Vigilance Performance and Mental Workload

Victoria S. Schoenfeld and Mark W. Scerbo
Old Dominion University

Automation has been introduced into systems to increase human performance and decrease human errors (Wiener, 1988). Although there are many examples of successful applications, automation has also increased opportunities for other types of human error (Woods, 1996). For example, automated technology assumes responsibility for tasks and often leaves workers the role of monitoring the system. Unfortunately, as Warm (1993) indicates, humans are not well suited for these less active, more supervisory roles.

Research on vigilance, or the ability to sustain attention and react to changes in stimuli over time, has shown that performance declines within only a short time after monitoring begins (e.g., Mackworth, 1948; Warm, 1984). This diminished ability to detect critical stimuli quickly and reliably over time has become known as the vigilance decrement (Davies & Parasuraman, 1982). As the use of automation continues to increase, more and more individuals are being forced to assume this new monitoring role (Warm, 1984); therefore, investigations regarding monitoring activities and ways to improve it are essential.

When working in an automated environment, individuals must be able to notice and react to any unusual changes in the system indicators. These differences might include changes in the size of a signal (Scerbo, Greenwald, & Sawin, 1993), the magnitude of a signal (Mackworth, 1948), or the duration of a signal (Warm & Alluisi, 1971). Other times the critical signal might consist of the presence or absence of some stimulus or stimulus feature.

Differences in searching for the presence and absence of stimulus features have not yet been investigated over time; however, Treisman and her colleagues (Treisman & Gelade, 1980; Treisman & Souther, 1985) have examined search differences under alerted conditions. According to feature integration theory, if a unique separable feature is present only in the target and not in the surrounding stimuli, then this feature will appear to pop out of the display and is quickly detected via preattentive or parallel processing. If, on the other hand, this distinguishing feature is not present in the target but does exist in the surrounding stimuli, the observer must use more deliberate, serial processing to detect the target.

One objective of the present study was to examine how searching for the presence and absence of features affects performance over time. Fisk and Schneider (1981) showed that vigilance performance was unaffected when critical signals were detected with automatic processing. On the other hand, the typical vigilance decrement was observed in the context of effortful, controlled processing. If the results of Treisman and her colleagues generalize to vigilance tasks, one would expect that searching for the presence or absence of features might interact with time on task. Specifically, a decrement should only occur in the feature absence condition where effortful, serial processing is required. By contrast, Treisman's findings were obtained in test sessions of relatively brief duration. Thus, it is possible that search differences for the presence and absence of features might only exist under these alerted conditions. In this case, one would expect to see a decline in performance over an extended watch keeping session regardless of critical signal type.

Treisman and Souther (1985) have also shown that searching for the presence and absence of features is moderated by the number of distractors on the display. Because searching for the presence of features can be done with parallel processing, the number of nontargets in the display has relatively little impact on performance. However, when looking for the absence of features, search times are positively related to the number of distractors because this type of search requires focused attention. To date, few investigators have examined how the number of distractors in a display affect vigilance performance. There is, however, research surrounding the number of displays. For example, several investigators have shown that overall performance suffers when observers must monitor more than one display (Grubb, Warm, Dember, & Berch, 1995; Jerison, 1963). For example, Grubb et al. (1995) required observers to monitor a simulated aircraft instrument panel containing four displays. Observers were required to monitor either one, two, or four displays. The results showed that performance suffered as the number of displays to be monitored increased. Further, a decrement was only found in the two and four display conditions. These results suggest that greater decrements might also be expected over the course of a vigil as the number of nontargets in the display increases. On the other hand, performance might also be moderated by the type of search. Specifically, if searching for the presence of features is a truly parallel activity, then one would expect the number of distractors to have little impact on performance. By contrast, searching for the absence of features should be adversely affected by the number of distractors. A second goal of this study was to examine this idea.

The monitoring required in automated tasks has also been found to affect mental workload (Warm, Dember, & Hancock, 1996). Mental workload can be thought of as the information processing demands placed on an individual during a task (O'Donnell & Eggemeier, 1986) and can be assessed with subjective measures like the NASA Task Load Index (TLX; Hart & Staveland, 1988). Many studies (e.g., Scerbo, Greenwald, & Sawin, 1992; Warm, Dember, Nelson, Grubb, & Davies, 1993) have shown that vigilance tasks produce high workload ratings. For example, Scerbo et al. (1992) used the TLX to examine subjective workload during a 40-min, successive discrimination task. In one condition, participants were provided with the appearance of some control over the task. In the other condition, no means of control was provided. Overall workload levels, particularly ratings of frustration, were significantly higher after the vigil, regardless of the level of perceived control. A third goal of the present study was to investigate workload ratings before and after the monitoring task. Based on previous research (e.g., Scerbo et al., 1992), overall workload scores were expected to be higher after the vigil, with ratings especially high in the feature absence conditions due to the more demanding type of attention required.

METHOD

Participants

Eighty undergraduate women and men from Old Dominion University participated in this study. Participants ranged in age from 18 to 40 years. All had normal or corrected to normal vision. Each participant received course credit to participate.

Apparatus

An IBM 386 personal computer with a SVGA monitor presented the stimuli and recorded responses.

Design

A 2 feature (presence or absence) x 2 display size (small or large) x 4 monitoring period (10-min periods) experimental design was used. Feature type and display size were manipulated between participants creating four treatment conditions, while monitoring period was manipulated within participants.

Vigilance Task

A computerized, successive discrimination task was used to present the stimuli. Every participant performed the task over four consecutive 10-min periods. A high event rate of 30 events per minute was used. Two features conditions (presence and absence of a line) were manipulated. In the presence conditions, participants had to detect a line intersecting the bottom of a circle in an array of circles without lines. In the absence conditions, participants were required to detect a plain circle in an array of circles with lines. All lines were 4 mm in length. One critical signal appeared at random intervals within each minute and occurred equally often in each spatial location.

Two display size conditions (small and large) were also used. The small display size condition consisted of two circles located 75 mm apart in the center of the computer screen. These stimuli remained on the screen for 300 ms. The large display size conditions consisted of five circles arranged in a circular fashion around the center of the screen, with circles located at the 12, 3, 5, 7, and 9 O'clock positions. These stimuli remained on the screen for 400 ms. Pilot testing indicated that stimulus times in the large display size conditions needed to be 100 ms longer to produce performance comparable to the small display size conditions. Circles in all conditions were 14 mm in diameter.

Procedure

Participants were randomly assigned to one of the four conditions and given a brief explanation regarding the nature of the experiment. They were then given a shuffled deck of standard playing cards and asked to sort the cards into four piles according to suit at a constant rate of one per 1.5 s established by a prerecorded audio tape. Afterward, the participants used the NASA TLX to rate the subjective mental workload they experienced during the card-sorting task.

Upon completion of the TLX, the participants were seated in front of a computer in a sound-attenuated room. They were familiarized with the stimuli and then participated an 8-min practice session that duplicated the conditions of the appropriate experimental session. An A' score of 0.7 was required to participate in the experimental session. Participants who scored below this cutoff, which included 4 in the large absence condition and 2 in the large presence condition, were replaced. Participants then completed the 40-min vigil. After the experimental session, participants completed a second TLX to rate the amount of mental workload experienced during the vigil. Participants were then debriefed and allowed to ask questions regarding the experiment.

RESULTS

A 2 feature type (presence or absence) x 2 display size (small or large) x 4 monitoring period (10-min periods) mixed-model factorial analysis of variance with participants nested in feature type and display size was performed on the performance measures (A' and B'') and workload scores. An a priori alpha level of 0.05 was set for all effects. Student Newman-Keuls post hoc tests were performed for all significant main effects and tests of simple main effects for all significant interaction effects.

Perceptual Sensitivity

A significant period effect was found, $F(3, 228) = 9.71$, and a post hoc test revealed that mean A' scores at period 1 ($M = 0.931$), period 2 ($M = 0.928$), and period 3 ($M = 0.910$) were significantly greater than at period 4 ($M = 0.887$).

Significant effects were also observed for feature type, $F(1, 76) = 64.22$, display size, $F(1, 76) = 50.65$, and the interaction between feature type and display size, $F(1, 76) = 32.68$. This interaction is illustrated in Figure 1. As display size increased from two to five, the difference between the presence and absence conditions became more pronounced. An analysis of simple effects showed that the difference between the small display size condition ($M = 0.942$) and large size display condition ($M = 0.793$) was only significant in the feature absence condition, $F(1,76) = 82.35$.

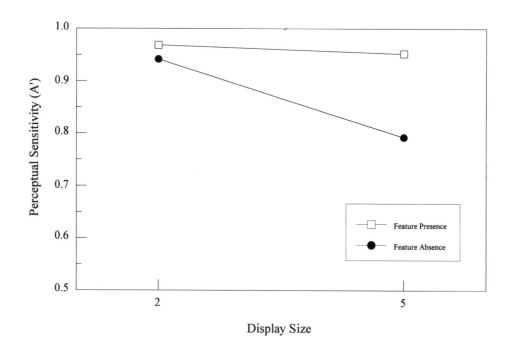

Figure1. Mean A′ scores for feature types as a function of display size.

Response Criterion

A significant period effect was found for response criterion, F (3, 228) = 6.34. A post hoc test revealed that the mean B″ score at period 1 (M = 0.245) was significantly lower than at period 2 (M = 0.481), period 3 (M = 0.525), and period 4 (M = 0.560). Also, a significant display size effect was found, F (1, 76) = 11.12, with mean B″ scores lower in the small display size condition (M = 0.282) than in the large display size condition (M = 0.623).

Mental Workload

A significant main effect for task workload was observed, F (1, 76) = 85.42. Overall workload ratings were found to be higher after the vigil (M = 61.29) than after the card-sorting task (M = 35.44). A two-way interaction between feature type and task approached significance (p = .07), suggesting that the magnitude of the increase in workload scores was dependent upon the feature being detected. Workload ratings increased slightly after searches for feature presence (from M = 38.75 to M = 59.48); however, the ratings nearly doubled after searches for feature absence (from M = 32.13 to M = 63.10).

DISCUSSION

The present study was designed to investigate how searching for the presence and absence of features in different display sizes impacts vigilance performance and mental workload. Traditional vigilance research (Grubb et al., 1995; Jerison, 1963) has shown greater decrements in performance when more complex displays are monitored, suggesting that searching among more displays results in decreased performance. This effect receives partial support from the present study. Performance was found to be significantly higher in the small as compared to large display size condition, but the decline in performance over time was comparable across display size. Observers also adopted a more conservative search strategy over time. A change in response criteria often accompanies a change in perceptual sensitivity in vigilance studies because observers become more familiar with the actual probability of the occurrence of a critical signal and are less willing to respond when they are unsure (Davies & Parasuraman, 1982).

Although these results agree with predictions made from vigilance research, an interaction between feature type and display size was also found. This interaction appears to support the predictions of Treisman and her colleagues (Treisman & Gelade, 1980; Treisman & Souther, 1985). Mean A' scores in the feature presence conditions were similar in the two display sizes, suggesting the use of parallel processing. In the feature absence conditions, however, mean A' scores were significantly higher in the small display size condition, indicating the use of a more deliberate type of processing that is negatively impacted by the number of distractors on the display.

A three-way interaction between feature type, display size, and monitoring period was not found. Instead, performance declined over time in a similar manner in every condition. These results suggest that Treisman's predictions do not generalize to vigilance situations, but only hold up under alerted conditions. Another possibility is that the absence of a significant interaction may have been due to the differences in exposure times of the large and small display size conditions. By using longer exposure times in the large display size condition, differences in performance over time may have been eliminated. This possibility is currently being investigated in a study in which the initial levels of difficulty (as indicated by mean A' scores) have been equated.

The current study also examined subjective mental workload ratings before and after the vigil. In accordance with other vigilance studies (e.g., Scerbo et al., 1992), workload scores were found to be higher after the vigil. Further, there was some evidence that the magnitude of this increase was depended upon signal type. Although workload ratings increased somewhat in the feature presence conditions, the postvigil ratings were nearly twice as high in the feature absence conditions. These results suggest that although observers in both feature conditions found the vigilance task to be demanding, those searching for the absence of a line experienced higher levels of mental workload.

Results from the present study suggest that monitoring for the presence or absence of features in automated systems can have dramatically different effects on both performance and mental workload. When designing automated systems, care should be taken to ensure that critical information is represented by the presence of features instead of their absence. Operators who must look for the absence of features run a greater risk of missing critical information. Moreover, they may perceive that activity as more demanding.

REFERENCES

Davies, D. R., & Parasuraman, R. (1982). *The psychology of vigilance.* London: Academic Press.

Fisk, A. D., & Schneider, W. (1981). Control and automatic processing during tasks requiring sustained attention: A new approach to vigilance. *Human Factors, 23,* 737-750.

Grubb, P. L., Warm, J. S., Dember, W. N., & Berch, D. B. (1995). Effects of multiple-signal discrimination on vigilance performance and perceived workload. *Proceedings of the Human Factors and Ergonomics Society 39th Annual Meeting,* 1360-1364.

Hart, S. G., & Staveland, L. E. (1988). Development of NASA-TLX (Task Load Index): Results of empirical and theoretical research. In P. A. Hancock & N. Meshkati (Eds.), *Human mental workload* (pp. 139-183). Amsterdam: North-Holland.

Jerison, H. J. (1963). On the decrement function in human vigilance. In D. N. Buckner & J. J. McGrath

(Eds.), *Vigilance: A symposium* (pp. 199-216). New York: McGraw-Hill.

Mackworth, N. H. (1948). The breakdown of vigilance during prolonged visual search. *Quarterly Journal of Experimental Psychology, 1,* 6-21.

O'Donnell, R. D., & Eggemeier, F. T. (1986). Workload assessment methodology. In K. R. Boff, L. Kaufman, & J. P. Thomas (Eds.), *Handbook of perception and human performance, volume II: Cognitive processes and performance* (pp. 42/1-42/49). New York: John Wiley & Sons.

Scerbo, M. W., Greenwald, C. Q., & Sawin, D. A. (1992). Vigilance: It's boring, it's difficult, and I can't do anything about it. *Proceedings of the Human Factors and Ergonomics Society, USA, 36,* 1508-1512.

Scerbo, M. W., Greenwald, C. Q., & Sawin, D. A. (1993). The effects of subject-controlled pacing and task type on sustained attention and workload. *The Journal of General Psychology, 120,* 293-307.

Treisman, A. M., & Gelade, G. (1980). A feature-integration theory of attention. *Cognitive Psychology, 12,* 97-136.

Treisman, A., & Souther, J. (1985). Search asymmetry: A diagnostic for preattentive processing of separable features. *Journal of Experimental Psychology: General, 114,* 285-310.

Warm, J. S. (1984). An introduction to vigilance. In J. S. Warm (Ed.), *Sustained attention in human performance* (pp. 1-14). New York: John Wiley & Sons.

Warm, J. S. (1993). Vigilance and target detection. In B. M. Huey & C. D. Wickens (Eds.), *Workload transition: Implications for individual and team performance* (pp. 139-169). Washington, D.C.: National Academy Press.

Warm, J. S., & Alluisi, E. A. (1971). Influence of temporal uncertainty and sensory modality of signals on watchkeeping performance. *Journal of Experimental Psychology, 87,* 303-308.

Warm, J.S., Dember, W.N., & Hancock, P.A. (1996). Vigilance and workload in automated systems. In R. Parasuraman, & M. Mouloua (Eds.), *Automation and human performance: Theory and applications* (pp. 183-200). Mahwah, NJ: Erlbaum.

Warm, J. S., Dember, W. N., Nelson, T., Grubb, P. L., & Davies, D. R. (1993). Perceived workload in cognitive vigilance tasks. *Proceedings of the First Mid-Atlantic Human Factors Conference, 1,* 137-142.

Wiener, E.L. (1988). Cockpit automation. In E.L. Wiener & D.C. Nagel (Eds.), *Human factors in aviation* (pp. 433-461). San Diego, CA: Academic Press.

Woods, D.D. (1996). Decomposing automation: Apparent simplicity, real complexity. In R. Parasuraman, & M. Mouloua (Eds.), *Automation and human performance: Theory and applications* (pp. 3-17). Mahwah, NJ: Erlbaum.

Intraclass and Interclass Transfer of Training for Vigilance

James L. Szalma   Louis C. Miller   Edward M. Hitchcock   Joel S. Warm   William N. Dember
University of Cincinnati

INTRODUCTION

Studies on training for vigilance reveal that feedback in the form of knowledge of results (KR) enhances performance efficiency, and that these effects *transfer* to subsequent conditions where KR is withdrawn (Davies & Parasuraman, 1982). A central issue in transfer of training is whether transfer results from general, nonspecific factors, such warm-up or learning-to-learn, from factors specific to the particular task in question, or some combination thereof (Underwood, 1966). Wiener (1967) argued that the problem of training for vigilance may not be especially complex, since such training is primarily general in character. This view was challenged by Becker, Warm, and Dember (1994), who used what Davies and Parasuraman (1982) describe as simultaneous- (comparative judgment) or successive- (memory based absolute judgment) type vigilance tasks. They found strong evidence for specific transfer in both the simultaneous and successive formats, but little evidence for general transfer. Accordingly, Becker et al. suggested that training for vigilance is *task-type* specific.

Becker et al. tested specific transfer by training observers on one task and testing them on the *identical* task, making use of the maximum conditions for assessing specific transfer (Underwood, 1966). While an approach of this sort can provide strong evidence for such transfer when the training task *per se* is the primary dimension of interest, it does not provide compelling evidence for specific transfer when the training task is intended to represent a *class* of tasks, since category-specific effects are then potentially confounded with display-specific effects. Thus, the limits of the specific component of transfer cannot be determined from their experiment.

The present investigation was designed to disengage category-specific and display-specific components in training for vigilance. Toward that end, a SIMULTANEOUS-TYPE TASK featuring vernier discriminations was employed as a *criterion task*, and comparisons were made of transfer effects which resulted from (a) training on the Criterion task itself, (b) training on another *SIMULTANEOUS* task featuring discriminations of spatial distance, and (c) training on a *SUCCESSIVE* task which also featured spatial distance discriminations. If transfer of training in vigilance is indeed task-type specific, we would expect transfer to the Criterion task to be greater from another simultaneous task than from a successive task.

METHOD

Ninety-six undergraduates, 48 men and 48 women, served as observers to fulfill a course requirement. They ranged in age from 17 to 34 years, with a mean of 21 years. All participants had normal or corrected-to-normal vision and were free of any known hearing impairment.

Three task conditions (Simultaneous/Criterion, Simultaneous/Alternate, and Successive) were combined factorially with two levels of KR (KR, No-KR) to produce six experimental groups. Sixteen observers were assigned at random to each group with the restriction that the groups were equated for sex. The groups were defined by the nature of their KR-training/transfer experience: (1) *Simultaneous/Criterion/KR*--observers received KR training on the SIMULTANEOUS/Criterion task and were then tested on *that* task; (2) *Simultaneous/Alternate/KR*--observers received training on the alternate SIMULTANEOUS task and were then tested on the Criterion task; (3) *Successive/KR*--observers were trained on a SUCCESSIVE task and then were tested on the Criterion task. The observers in the other three groups did not receive KR-training. They were tested on the CRITERION task after initial experience with either the Criterion task itself, the SIMULTANEOUS/Alternate task, or the SUCCESSIVE task. The No-KR groups served as controls for the effects of KR as well as for any effects associated with task exposure itself, such as changes in display format.

Observers experienced a 24-min *training* vigil divided into three continuous 8-min periods and a 32-min *test* vigil divided into four continuous 8-min periods. KR was provided during training and withdrawn during the test vigil. In both phases of the study, observers monitored 200 msec presentations of stimuli presented on a video display terminal (VDT). All stimuli were green in color and were presented against a gray background. The *Criterion* task consisted of a *DOUGHNUT-LIKE CIRCULAR DISPLAY*. The diameter of the "doughnut" was 2 cm, while that of its inner opening was 1 cm. The ring of the doughnut was 0.5 cm thick. The circular display was flanked (0.3 cm) on both sides by two 1.1 x 0.1 cm horizontal lines which were normally aligned with each other along the horizontal diameter of the doughnut or along chords 0.2 cm above or below that diameter. Critical signals for detection were cases in which the lines were misaligned (one line was 0.2 cm above or below the horizontal diameter of the doughnut). The *Simultaneous/Alternate* task consisted of a *LARGE CENTERING DISK* 1.4 cm in diameter flanked by two dots 0.3 cm in diameter arrayed along a horizontal vector which passed through the center of the disk. Normally, the dots were positioned so that they *both* were *either* 1.5 cm or 1.0 cm away from the disk. Critical signals were cases in which one of the dots, left or right, was 1.5 cm away and the other was 1.0 cm from the disk. The *Successive* task made use of a display *identical* to that employed in the Simultaneous/Alternate task, except that only a single flanking dot was used. The dot appeared either to the left or right of the disk. Normally, it was positioned 0.9 cm away; critical signals were cases in which the dot was 0.3 cm farther away from the disk than *usual*. The three tasks were equated for discrimination difficulty under alerted conditions.

During training, observers who received KR heard a recorded female voice (50 dBA at the ear) announce "correct," "false alarm," and "miss," to indicate correct detections, false alarms, and misses, respectively. In order to control for accessory auditory stimulation, observers in the control groups heard the word "save" which was announced in the female voice after each response. A 15-min interval separated the training and transfer phases of the study.

Observers were tested individually in an Industrial Acoustics sound-attenuation chamber. They were seated in front of the VDT, which was mounted on a table at eye level with a viewing distance of 70 cm. Ambient illumination in the chamber was 0.74 cd/m^2 provided by a 25W soft white incandescent bulb housed in a parabolic reflector positioned above and behind the seated observer so as to minimize glare on the VDT.

Stimuli for inspection in all experimental conditions occurred once every 2 sec (event rate = 30/min); critical signal probability was .04. Stimulus presentations and the delivery of KR or response acknowledgement were orchestrated by a Macintosh IIci computer in conjunction with a color monitor and SuperLab(v1.5)software. Observers indicated the detection of critical signals by pressing the spacebar on the computer keyboard. Responses occurring within 1.3 sec after the appearance of a critical signal were recorded automatically by the computer as correct detections. All other responses were tallied as false alarms.

RESULTS

For each observer, percentages of correct responses and false alarms during the training and transfer phases of the study were used to compute signal detection theory measures of sensitivity (d'; See, Howe, Warm, & Dember, 1995) and response bias (c; See, Warm, Dember, & Howe, 1997).

Perceptual Sensitivity

Analysis of variance of the d' data during *training* indicated that KR generally enhanced perceptual sensitivity, M(KR)=3.07, M(NoKR)=2.55, $F(1,90) = 13.45, p < .001$. This effect was similar for the three tasks throughout the training session since all interactions involving KR, task, and periods of watch were not significant, $p>.05$. Overall sensitivity did not vary significantly with periods of watch ($p>.05$), but it did with task, $F(2,90) = 12.03, p < .001$. Newman-Keuls tests ($\alpha=.05$ for each comparison) indicated that the d' scores were significantly lower in the Simultaneous/Alternate task ($M= 2.32$) than in the Criterion ($M=3.04$) and Successive ($M=3.08$) tasks. The latter two did not differ significantly from each other.

In order to assess the effects of KR-training on subsequent perceptual sensitivity in the *transfer* phase of the study, comparisons were first made of the sensitivity scores attained by observers in the three No-KR groups. These values were quite similar: While perceptual sensitivity declined significantly over time in these groups (M's periods 1 through 4 = 2.93, 2.73, 2.53, 2.59, respectively), $F(3,135) = 6.75, p < .001$, there were no significant overall differences among the groups and there was no significant Groups x Periods interaction ($p > .05$ in each case). Given that equivalence, the data for the Criterion/No-KR group were used as the baseline against which to assess the transfer effects to the Criterion task of KR-training on that task, the Simultaneous/Alternate task, and the Successive task.

Mean d' scores during transfer for the three KR groups and the Criterion/No-KR group are presented in Table 1. During the transfer phase, the scores for observers who received KR-training on the Simultaneous/Alternate task were similar to those who were originally trained on the Criterion task itself, and the scores for both of these groups were consistently higher than those for the Criterion/No-KR control group. The scores for observers originally trained on the Successive task and tested on the Criterion task fell between those of the control group and the two other groups. In all cases, sensitivity declined over the course of the watch.

Table 1

Mean Perceptual Sensitivity Scores on the Criterion Task During Transfer

Groups	Periods of Watch				
	1	2	3	4	*M*
Criterion/KR	3.43	3.34	3.17	3.06	3.25
SIM/Alternate/KR	3.64	3.22	3.08	2.91	3.21
SUCC/KR	3.08	2.84	3.03	2.67	2.91
Criterion/NoKR	2.93	2.65	2.47	2.45	2.62
M	3.27	3.01	2.94	2.77	

Analysis of variance of the transfer d' data revealed significant main effects for Groups, $F(3,60) = 3.34, p < .05$, and for Periods, $F(3,180) = 11.12, p < .001$. The interaction between these factors was not significant ($p>.05$). Subsequent Newman-Keuls tests ($\alpha=.05$ for each comparison) showed that the Criterion/KR and the Simultaneous/KR groups did not differ significantly from each other and that the d' scores of *both* of these groups significantly exceeded those of the Criterion/No-KR control group. In contrast, the transfer performance of the Successive/KR group failed to differ significantly from that of the control group. Thus, in terms of perceptual sensitivity, the data provide strong evidence for intra-task or specific transfer but no evidence for inter-task or general transfer.

Response Bias

The principal finding for response bias during *training* was that the effects of KR were not uniform across tasks. Analysis of variance revealed a significant main effect for KR, $F(1,90) = 54.24, p < .001$, and

185

a significant Task x KR interaction, $F(2,90) = 4.24$, $p < .05$. Tests of the simple effects of KR within each task indicated that observers in both the Criterion (MKR= .83, MNoKR= .12) and the Simultaneous/Alternate (MKR= .93, MNoKR= .22) tasks were significantly more conservative when KR was present than when it was not, $F(1,90) > 29$, $p<.001$ in each case. In the Successive task, KR had no significant effect on response bias (MKR= .63, MNoKR= .38), $F(1,90) = 3.51$, $p>.05$.

The absence of a KR effect for the Successive Task during training mandated that the assessment of transfer effects be restricted to the Criterion and Simultaneous/Alternate tasks. In order to assess the effects of KR-training on subsequent response bias in the *transfer* phase of the study, comparisons were first made of the c-scores attained by observers in the NoKR groups. Analysis of variance of these data revealed a significant main effect for Periods (M's periods 1 through 4 = .10, .24, .36, .47, respectively) $F(3,90) =$ 8.38, $p < .001$, but no significant overall task difference, and no significant Periods x Task interaction, $p >$ .05 in each case. Hence, the data of the Criterion/No-KR group were again employed as the baseline against which to assess transfer effects in the Criterion/KR and Simultaneous/ alternate-KR groups.

The c-scores on the Criterion task during transfer for observers who received KR-training on the Simultaneous task (M=.87) were similar to those who received KR-training on the Criterion task itself (M=.93), and the level of conservatism in both groups was greater than that in the control group (M=.32). The level of conservatism increased over time in all three groups (M's periods 1 through 4 = .49, .67, .80, .88, respectively). These impressions were supported by analysis of variance which indicated that there were significant main effects for groups, $F(2,45) = 10.94$, $p < .001$ and for periods, $F(3,135) = 16.51$, $p < .001$. The interaction between these factors was not significant, $F<1$. Subsequent Newman-Keuls tests (α=.05 for each comparison) revealed that c-scores in the Simultaneous/Criterion/NoKR group were significantly lower than those in the Simultaneous/Criterion/KR and the Simultaneous/Alternate/KR groups. The latter two did not differ significantly from each other.

DISCUSSION

As in earlier experiments (c.f. Becker, Warm, Dember, & Hancock, 1995), KR-training was found to enhance performance in terms of perceptual sensitivity. In addition, the training effects transferred from one task to another, but successful transfer depended upon the types of tasks involved. During the transfer phase of the study, perceptual sensitivity on the Criterion task was identical for observers trained initially on that task and those trained on another simultaneous task having a different display architecture. Moreover, the scores for both groups significantly exceeded those for the Criterion control group who were not afforded training with information feedback. In contrast to the Simultaneous/Alternate group, transfer scores on the Criterion task for observers whose initial training was on the Successive task did not differ significantly from those of controls who experienced the Criterion task without KR-training. Thus, the transfer effects noted in regard to perceptual sensitivity did not support Wiener's (1967) claim that training in vigilance is primarily general in character, nor did they support the argument that such effects might be display-specific. Instead, they sustained the conclusion reached by Becker et al. (1994) that perceptual learning in vigilance is task-type specific. The similarity in the sensitivity scores of the Criterion/KR and Simultaneous/Alternate/KR groups during transfer is especially noteworthy, given the fact that the level of perceptual sensitivity associated with the Simultaneous/Alternate task during training was lower than that associated with the Criterion task, and the fact that transfer is often weakened by differences in the level of difficulty between training and test tasks (Holding, 1987).

KR-training also affected observers' response strategies, inducing those in the Criterion and Simultaneous/Alternate groups to adopt a more cautious mode of responding. This effect also transferred across task boundaries: The level of conservatism evident in the transfer phase of the study was identical for observers trained initially on the Criterion task and those trained on the Simultaneous/Alternate task. Unfortunately, the degree to which specific and general factors contributed to transfer in this case could not

186

be assessed because KR-training had no impact on response bias with the Successive task.

The dissimilarity in results obtained with the Simultaneous/Alternate and Successive tasks on both TSD indices is dramatic, given the fact that except for the type of discrimination required, these tasks were more alike in terms of general display architecture than either was to the Criterion task. Consequently, the present results lend support to Davies and Parasuraman's (1982) assertion that simultaneous- and successive-type vigilance tasks provide observers with singular psychophysical challenges. From the present results, one might conclude that training programs for the improvement of monitoring skills should not be based solely on the expectation of non-specific or generalized transfer effects. Observers may learn to be vigilant, but the nature of what is learned will clearly depend upon the task category, simultaneous or successive, within which their training occurs.

REFERENCES

Becker, A.B., Warm, J.S., & Dember, W.N. (1994). Specific and nonspecific transfer effects in training for vigilance. In M. Mouloua & R. Parasuraman (Eds.), *Human performance in automated systems: Current trends.* (pp. 294-299). Hillsdale, NJ: Erlbaum.

Becker, A.B., Warm, J.S., Dember, W.N., & Hancock, P.A. (1995). Effects of jet engine noise and performance feedback on perceived workload in a monitoring task. *The International Journal of Aviation Psychology, 5,* 49-62.

Davies, D.R., & Parasuraman, R. (1982). *The psychology of vigilance.* London: Academic Press.

Holding, D.H. (1987). Concepts of training. In G. Salvendy (Ed.), *Handbook of human factors* (pp. 939-962). New York: Wiley.

See, J.E., Howe, S.R., Warm, J.S., & Dember, W.N. (1995). Meta-analysis of the sensitivity decrement in vigilance. *Psychological Bulletin, 117,* 230-249.

See, J. E., Warm, J. S., Dember, W. N., & Howe, S. R. (1997). Vigilance and signal detection theory: An empirical evaluation of five measures of response bias. *Human Factors, 39,* 14-29.

Underwood, B.J. (1966). *Experimental psychology* (2nd ed.), New York: Appleton-Century-Crofts.

Wiener, E.L. (1967). Transfer of training from one monitoring task to the other. *Ergonomics, 10,* 649-658.

Sustained Attention In Opponent-Process Color Channels

John O. Simon   Joel S. Warm   William N. Dember
University of Cincinnati

INTRODUCTION

Vigilance or sustained attention tasks require observers to detect transient signals over prolonged periods of time. A major characteristic of performance in such tasks is the event-rate effect, the finding that the accuracy of signal detections varies inversely with the rate of repetition of non-signal background events among which critical signals for detection appear (Warm & Jerison, 1984).

Posner (1978) has suggested a mechanism involving the inhibition of **psychological pathways**, or sets of internal codes and their interconnections, which can be used to account for the event-rate effect. According to Posner, repeated stimulation leads to inhibition in a pathway; accumulated inhibition interferes with the activation of central processing mechanisms, thereby degrading performance efficiency. Since the accumulation of inhibition varies directly with the rate of stimulation, performance is poorer with fast than with slow event-rates.

Galinsky, Warm, Dember, Weiler, and Scerbo (1990) attempted to attenuate the build-up of pathway inhibition by shuttling stimulation between two sensory modalities (vision and audition) carrying common information. The results of their study indicated that sensory alternation did eliminate the event-rate effect.

One goal for the present study was to determine whether a similar effect could be achieved by alternating information pathways within a single sensory channel. Color channels (red-green, yellow-blue; Kandel, Schwartz, & Jessel, 1995) were chosen for that purpose. Using the study by Galinsky et al. as a guide, and the fact that duration is a common attribute of all sensory channels, the present test of Posner's pathway inhibition model also involved temporal discriminations, but with carrier-stimulus alternation across **color** channels. It was anticipated that **cross-channel** or **heterogeneous-channel** alternation (red-blue or yellow-green) would attenuate the build-up of inhibition within carrier channels and lead to a weakened event-rate effect in comparison to stimulation that was alternated **within** a color channel or **homogeneous alternation** (red-green or yellow-blue), or maintained on the **same** color channel throughout the vigil (**no-alternation**).

METHOD

Ninety-six students, 48 men and 48 women, served as observers to fulfill a course requirement. They ranged in age from 18 years to 49 years with a mean age of 23 years. All of the students had normal color vision and normal or corrected-to-normal visual acuity. All observers participated in a 40-min vigil composed of four continuous 10-min periods of watch. In all conditions, they monitored the repetitive presentation of a 2mm x 9mm colored bar (red, green, blue, or yellow) which appeared in the center a video display terminal (VDT). Except for the bar, the VDT was uniformly black. Following Galinsky et al. (1990), neutral events were flashes of the bar lasting for 248 msec. Critical signals for detection were briefer (125 msec) flashes.

Observers in the no-alternation conditions monitored the same colored bar throughout the vigil. Each color, red, green, blue, or yellow, was assigned to equal numbers of male and female observers. In the alternating groups, bar colors were interchanged at 5 min intervals during each 10 min period of watch, consistent with the audio-visual alternation schedules employed by Galinsky et al. (1990). In the homogeneous (same channel) alternation conditions, equal numbers of observers (balanced for sex) experienced alternation in the orders red → green, green → red, blue → yellow, or yellow → blue. In the heterogeneous (cross-channel) alternation conditions, equal numbers of observers (balanced for sex) experienced alternation in the order red → blue, blue → red, green → yellow, or yellow → green.

Event-rates of 5 and 40 events/min were achieved by setting the intervals between events at 12 and 1.5 sec respectively. Ten critical signals were presented per period of watch in all conditions. The intervals between signals varied from 20 to 100 sec with a mean of 60 sec. In all conditions, five signals appeared within each half-period, ensuring that signals were distributed equally among bar colors in the homogeneous and heterogeneous alternating conditions.

Stimulus presentations were orchestrated by a Macintosh IIci computer in conjunction with a color monitor and SuperLab (v1.5) software. Observers indicated their detection of critical signals by pressing the spacebar on the computer keyboard. Responses occurring within 1.5 sec after the appearance of a critical signal were recorded automatically by the computer as correct detections. All other responses were tallied as false alarms. Observers were tested individually in an unlighted 1.96 x 1.62 x 2.18 meter sound-attenuated chamber. They were seated in front of the VDT, which was mounted on a table at eye level. Viewing distance was 1 meter. Black fabric lined the interior of the chamber. Prior to the main vigil, observers received a 10-min practice trial duplicating the task conditions they would encounter during the vigil. They were dark-adapted for 5 min prior to the practice trial.

RESULTS

Detection Scores: Global Analysis

The percentage of correct detections during each period of watch was determined for each observer in all experimental conditions. Mean percentages of correct detections under the slow (5 events/min) and fast (40 events/min) event-rates are plotted as a function of time on task in Figure 1. Data for the no-alternation, homogeneous, and heterogeneous alternation conditions are presented separately in each panel. The values for the no-alternation condition represent the overall detection scores for that condition, with no distinction made between the hues (red, green, yellow, or blue) involved. Likewise, the data for the homogeneous and heterogeneous alternation conditions represent the overall detection scores for those conditions regardless of the color pairings. In the homogeneous condition, the parings were red-green and yellow-blue, while in the heterogeneous condition the pairings were red-blue and yellow-green.

Figure 1 shows consistent event-rate differences in the no-alternation and homogeneous alternation conditions. In both cases detection scores were uniformly higher with a slow as compared to a fast event-rate, and the vigilance decrement was more pronounced with a fast event-rate. The event-rate effect was suppressed in the heterogeneous alternation condition, although a performance decrement remained evident.

These impressions were supported by means of analyses of variance performed separately on the data for the three alternation conditions, using arcsin transformations of the detection scores. In the no-alternation and homogeneous-alternation conditions, significant main effects were found for event-rate, $F(1,30) > 10.0$, $p<.01$ in each case, and periods, $F(3,90) > 8.0$, $p<.001$ in each case, and there was a significant Event-Rate x Periods interaction, $F(3,90) > 4.0$, $p <. 02$, in each case. The only significant source of variation in the heterogeneous alternation condition was that for periods of watch, $F(3,90) = 5.40$, $p<.002$.

Detection Scores: Microanalysis

While the absence of an event-rate effect in the heterogeneous or cross-channel alternation condition is consistent with expectations derived from Posner's (1978) pathway inhibition model, this result does not necessarily provide convincing support for the model. Instead, it may be an artifact of averaging scores across color-pairings rather the consequence of releasing channels from inhibition. Different performance patterns across the red-blue and yellow-green pairings in the heterogeneous condition, which may be masked by the averaging process, could also lead to the absence of an event-rate effect. To explore this possibility,

189

mean percentages of correct detections during each period of watch were determined for the red-blue and yellow-green pairings at each event-rate in the heterogeneous-alternation condition.

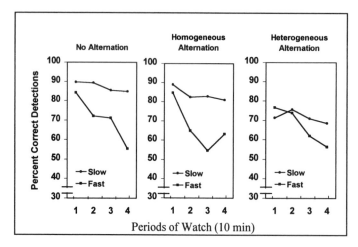

Figure 1. Percentages of signals detected at fast and slow event-rates as a function of periods of watch. Data for the three conditions are presented separately in each panel.

An analysis of variance of an arcsin transform of the data revealed a significant main effect for color-pairing, $F(1,28) = 13.64$, $p < .01$, and a significant interaction between event-rate and color-pairing, $F(1,28) = 17.71$, $p < .01$. The analysis also revealed a significant main effect for periods of watch, $F(3,84) = 7.03$, $p < .01$, as well as a significant Event-Rate x Periods interaction, $F(3,84) = 3.25$, $p < .05$.

The interaction between event-rate and color-pairing is displayed on the left in Figure 2, which shows the typical event-rate effect for the red-blue pairing, but an **inverted** effect for the yellow-green pairing; the detection scores in the latter case were higher in the fast than the slow event-rate condition. The interaction obviously stems primarily from a drop in the detection scores for the slow event-rate in the yellow-green pairing; the scores for the fast event-rate in that pairing are almost identical to those for the fast event-rate in the red-blue pairing. Here, then, is the source of the muted event-rate effect when scores were averaged for event-rates over the two color-pairing combinations in the global analysis of the heterogeneous alternation condition.

The Event-Rate x Periods interaction in the microanalysis of the heterogeneous alternation condition is shown on the right in Figure 2. Once again the vigilance decrement appears to be more pronounced in the fast than the slow event-rate condition.

False Alarm Scores. Although it is typical to examine the percentage of false alarms as well as the percentage of correct detections in vigilance studies, it is difficult to use percentage false alarm scores where event-rate is involved, because the value of the denominator in the false alarm equation is not equal in different event-rate conditions. In the present case, 390 neutral events occurred per period of watch in the fast event-rate conditions while 40 neutral events appeared per period in the slow event-rate conditions. Thus, observers in the fast conditions would need to be hyper-responsive in order to achieve the same percentages of false alarms as those in the slow event-rate conditions. For this reason, the absolute number of false alarms, rather than the percentage of false alarms, was examined in this study. Except for a main effect for periods (a decline in false alarms over time) in the heterogeneous alternation condition, $F(3,90) = 3.58$, $p < .05$, analyses of variance of the false alarm scores in each alternation condition produced no significant main effects for event-rate or periods and no significant interaction between these variables, $p > .05$ in all cases. Thus, the false alarm scores added little to the meaningfulness of the results.

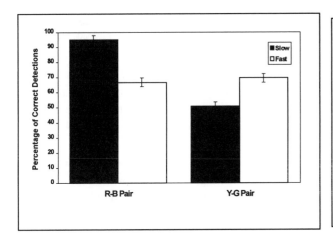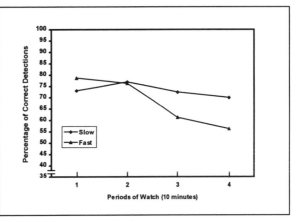

Figure 2. Percentages of signals detected at fast and slow event-rates (left) and as a function of periods of watch (right) for the two color pairings in the heterogeneous alternation condition.

DISCUSSION

Following the lead of Galinsky et al. (1990), Posner's (1978) pathway inhibition model of the event-rate effect was tested in this experiment by alternating information-carrying pathways during a watchkeeping session. In this case, the red-green, yellow-blue color channels of the visual system were selected for that purpose.

Event-rate effects on signal detection scores were obtained in the **no-alternation** condition, wherein only one portion of an opponent-process channel was employed (the red or the green component of the red-green channel or the yellow or blue component of the yellow-blue channel), and in the **homogeneous-alternation** condition, wherein both components of each opponent-process channel were used (red and green or yellow and blue). As in prior studies, a higher rate of signal detections occurred in these conditions with the use of a slow as compared to a fast event-rate, and a more pronounced vigilance decrement was observed in the context of a fast event-rate. As in the experiment by Galinsky et al., the event-rate effect appeared muted in the cross-channel or **heterogeneous-alternation** condition.

At first glance, the latter result would appear to provide a clear replication of the findings by Galinsky et al. and powerful support for Posner's (1978) pathway inhibition model. However, careful probing of the effects associated with the red-blue and yellow-green pairings in the heterogeneous-alternation condition led to a totally different conclusion. The overall event-rate effect remained evident in the red-blue pairing of that condition even though stimulation was alternated across channels. Moreover, while the detection scores for the fast event-rate in the yellow-green pairing were similar to those of the red-blue pairing, detection probability was greatly impaired for the slow event-rate within that pairing, with the result that the event-rate effect was **inverted** when yellow and green stimulation was involved. Thus, the apparent muting of the event-rate effect in the heterogeneous alternation condition proved to be an artifact of averaging across the opposite outcomes associated within the red-blue and yellow-green pairings. The impairment of slow event-rate performance in the yellow-green pairing clearly is not predictable from the pathway inhibition model. In sum, the present results neither replicate those of Galinsky et al. regarding the beneficial effects of channel alternation, findings which those authors showed to be immune from a stimulus averaging artifact, nor do they offer support for Posner's pathway inhibition model.

The inversion of the event-rate effect in the yellow-green portion of the homogeneous-alternation condition is quite unusual. Indeed, such an outcome in the presence of any type of psychophysical manipulation has not been previously reported in the literature. The reasons for that result, however, are not clear. One possible explanation may stem from the fact this study made use of what Davies and Parasuraman (1982) have described

191

as successive or absolute judgment tasks, in which observers need to consult working memory in order to separate signals from noise, with critical signals for detection being stimuli of briefer duration than usual. Such judgments are very sensitive to the perturbing effects of irrelevant color stimulation (Morgan & Alluisi, 1967). Thus, color variation may have distracted observers' temporal judgments, particularly in a slow event-rate condition, where stimulus instantiation does not occur as frequently as in a fast event-rate condition, thereby leading to poorer performance at a slow event-rate. The restriction of this effect to the yellow-green pairing may come from the proximity of these colors in the spectrum, a possibility which has been suggested by other investigators when examining differential wavelength effects on performance (Pollack, 1968).

As noted earlier, Posner's (1978) model treats the concept of psychological pathways very broadly; a key issue in testing this model is the need to identify appropriate and singular stimulus pathways which permit similar types of discriminations. The use of duration discrimination in this investigation satisfied that requirement, while also permitting a direct comparison with the work of Galinsky et al. (1990). However, the use of opponent-process color coding channels may have not met the singular pathway requirement. While the existence of such channels within the visual system is clearly documented, there are also cells within the system which respond to the intensity of light regardless of wavelength (Kandel, Schwartz, & Jessel, 1995). Moreover, the use of bar stimuli of constant width also limited stimulation to a unitary spatial frequency channel (DeValois & DeValois, 1987). Therefore, the color manipulation featured in the present study may not have restricted channel stimulation sufficiently to provide a fair test of Posner's (1978) model.

It is worth noting that the present investigation is not the only one to have been unable to support the pathway inhibition model of the event-rate effect. An earlier experiment by Lanzetta, Dember, Warm, & Berch (1987) also found that stimulus heterogeneity did not moderate this effect. In that study, observers detected occasional increments in the height of pairs of forms under conditions in which the physical structure of the forms was repeatedly varied over the course of the vigil. As in the present case, Lanzetta and his associates also suggested that lack of sufficient stimulus variation may have been the reason for their failure to support the theory. Taken together with the outcome of the Lanzetta et al. (1987) experiment, the present results lead one to wonder whether a model such as Posner's, which is so recalcitrant to operationalization and hence also to definitive falsification, can long endure as a useful heuristic for explaining the ubiquitous but elusive event-rate effect.

REFERENCES

Davies, D.R., & Parasuraman, R. (1982). *The psychology of vigilance*. London: Academic Press.

DeValois, R.L., & DeValois, K.K. (1987). *Spatial vision*. New York: Oxford University Press.

Galinsky, T.L., Warm, J.S., Dember, W.N., Weiler, E.M. & Scerbo, M.W. (1990). Sensory alteration and vigilance performance: The role of pathway inhibition. *Human Factors, 32*, 717-728.

Kandel, E.R., Schwartz, J.H., & Jessel, T.M. (1995). *Essentials of neural science and behavior*. Norwalk, CT: Appleton and Lange.

Lanzetta, T.M., Dember, W.N., Warm, J.S., & Berch, D.B. (1987). Effects of task type and stimulus heterogeneity on the event rate function in sustained attention. *Human Factors, 29*, 625-633.

Morgan, B.B., Jr., & Alluisi, E.A. (1967). Effects of discriminability and irrelevant information on absolute judgments. *Perception & Psychophysics, 2*, 54-58.

Pollack, J.D. (1968). Reaction time to different wavelengths at various luminances. *Perception & Psychophysics, 3*(1A), 17-24.

Posner, M.I. (1978). *Chronometric explorations of mind*. Hillsdale, N.J.: Lawrence Erlbaum.

Warm, J.S., & Jerison, H.J. (1984). The psychophysics of vigilance. In J.S. Warm (Ed.), *Sustained attention in human performance* (pp. 15-60). Chichester, UK: Wiley.

Olfaction and Vigilance: The Role of Hedonic Value

Keith S. Jones   Rebecca L. Ruhl   Joel S. Warm   William N. Dember
University of Cincinnati

INTRODUCTION

Olfactory stimuli can be quite salient and can play important roles in memory and cognition (Engen, 1991; Richardson & Zucco, 1989). To further examine the extent to which olfactory stimuli can influence psychological function, we initiated a series of studies on the ability of fragrances to enhance signal detection in tedious but demanding sustained attention (vigilance) tasks (Dember, Warm & Parasuraman, 1991). As performed in the laboratory, these tasks are meant to simulate core features of "real-world" tasks engaged in by quality control inspectors, radar operators and other personnel who must monitor displays for the occasional, unpredictable, occurrence of critical events or signals (Warm, 1984). Our early results were encouraging. Exposure to brief whiffs of the odor of Peppermint or Muguet enhanced signal detectability in comparison to control conditions in which observers were exposed to puffs of unscented air. In addition, accessory olfactory stimulation also served to attenuate the decrement function, the decline in the frequency of signal detections over time that characterizes vigilance performance (Dember, Warm & Parasuraman, 1996; Warm, Dember & Parasuraman, 1991). These studies suggested that exposure to fragrance may serve as an effective form of ancillary stimulation in tasks demanding close attention for prolonged periods of time.

While these initial findings were promising, they were limited by the fact that our experiments used only hedonically positive fragrances (Muguet and Peppermint). This fact led us to question the generalizability of our findings. Namely, can they be observed with other fragrances and does the hedonic value of the fragrance (positive, negative) matter? Hedonic factors have been shown to be critical in relation to olfactory processing (Richardson & Zucco, 1989). To address these issues, three new fragrances were chosen; Vanilla Bean, Clementine and Butyric Acid. The first two were pleasant scents, while the latter was chosen for it's unpleasant quality.

SUBJECTIVE RATINGS

Alertness/Relaxation and Pleasantness Ratings

Prior to initiating the main portion of the experiment, 16 students from the University of Cincinnati (eight men and eight women) rated the candidate fragrances on the dimensions of alertness/relaxation and hedonic value. Peppermint, one of the fragrances used previously, was included for comparison purposes. The alertness/relaxation scale was a 15 cm line labeled "more relaxing" at the zero point and "more alerting/stimulating" at the 15 cm point. To aid in making alertness/relaxation judgments, observers were asked to imagine that they were engaged in a tedious task. They were to note whether each fragrance, if present during the conduct of the task, would be more relaxing or more alerting/stimulating. They were to so indicate by placing an appropriate mark on the line. Marks above the midpoint of 7.5cm were considered to designate a stimulating fragrance. The hedonic scale was a 16cm line, with the zero point labeled "very unpleasant" and the 16cm point "very pleasant". Observers placed a mark on the line corresponding to how pleasant or unpleasant they found each of the four fragrances. For the hedonic scale, marks above the midpoint of 8cm were considered to designate a pleasant fragrance.

193

Each observer judged each fragrance once on each of the scales. The order in which the observers experienced the fragrances as they progressed through the rating phase was varied at random for each individual, while the sequence in which they responded to the two scales was balanced within the gender groups. Observers sampled each fragrance via a squeeze bottle containing fragrance-impregnated polyethylene pellets. Preliminary inspection of the data for both scales revealed that ratings were similar for male and female observers. Accordingly, the data were collapsed across sex prior to further analysis. Overall mean alertness/relaxation and hedonic ratings and their associated standard errors are presented in Table 1.

Table 1

Means and Standard Errors for Alertness/Relaxation and Pleasantness Ratings.

Fragrance	Alertness/Relaxation		Pleasantness	
	Mean	SE	Mean	SE
Butyric Acid	11.05	.70	1.11	.34
Clementine	4.97	.68	10.59	.79
Peppermint	9.79	.71	11.15	.54
Vanilla Bean	4.92	.76	10.82	.63

Alertness/Relaxation scale: < 7.5 relaxing; 7.5 neutral; > 7.5 stimulating.
Pleasantness scale: < 8 unpleasant; 8 neutral; > 8 pleasant.

Separate analyses of variance revealed statistically significant differences among the fragrances on both dimensions. For alertness/relaxation ratings, $F(3,45) = 22.82$, $p < .0001$; for the hedonic ratings, $F(3,45) = 77.92$, $p < .0001$. On the basis of these ratings, it is clear that Clementine and Vanilla Bean were both judged to be relaxing and pleasant, with mean ratings at least one standard error beyond the neutral point. The pattern of results for Peppermint (alerting and pleasant) replicated our earlier findings for this fragrance (Warm, Dember & Parasuraman, 1991). In contrast to the other fragrances, Butyric acid was rated as alerting and unpleasant. The mean ratings for this fragrance also fell at least one standard error beyond the neutral point.

Salience Ratings

A separate group of 12 students (six men and six women) was asked to rate the salience of the three fragrances when presented through the fragrance delivery system. In this way, observers made judgments of the presence or absence of each fragrance under conditions which duplicated those to be used in the vigilance experiment. Judgments were made on the basis of a 7-point scale ranging from "distinctly not present" (1) to "distinctly present" (7). Four observers, two men and two women, were assigned at random to one of the three fragrance conditions. Seven trials were presented with each fragrance, four of which contained the fragrance while three were vexier-versuche (blank) trials. In all cases, observers rated the blank trials as "distinctly not present". Mean salience ratings for the three conditions and their associated standard errors are presented in Table 2. It is evident in the table that all fragrances were clearly perceivable when present and that their salience ratings were similar.

VIGILANCE EXPERIMENT

Observers

Sixty-four students (32 men and 32 women) from the University of Cincinnati, served as observers in order to fulfill a course requirement. They ranged in age from 18 - 49 years with a mean of 21. All of the

194

Table 2

Means and Standard Errors for the Salience Ratings.

Fragrance	Mean	SE
Butyric Acid	6.70	.15
Clementine	6.10	.32
Vanilla Bean	6.55	.22

observers had normal or corrected-to-normal vision and passed a test for anosmia as a condition for gaining entry into the study. Observers were asked not to wear perfume/cologne on the day of the experiment and to refrain from smoking or chewing gum for 30 min prior to reporting for testing.

Experimental Design

Sixteen observers (eight men and eight women) were assigned at random to one of four fragrance conditions; Clementine, Vanilla Bean, Butyric Acid and a pure air control group. All observers participated in a 40-min vigil divided into four consecutive 10-min periods of watch.

Fragrance Delivery System

The fragrance delivery system was identical to that used in all of our prior studies (see Warm, Dember & Parasuraman, 1991). It consisted of pumps that forced air through Teflon tubing into a charcoal filter. This filtered air was then pumped into a 35 ml glass reservoir containing six polyethylene pellets which incorporated the fragrance to be used. Air from the reservoir was transmitted through additional tubing to a modified home oxygen mask worn by the observer while seated in an Industrial Acoustics sound attenuated chamber used for testing. The fragrance delivery system was located outside the chamber. Odor concentration at the mask was approximately 0.08 parts per million.

Vigilance Task

As in our earlier studies, observers monitored a video display terminal for the repetitive presentation of 1mm x 5mm vertical lines with a 1mm dot centered vertically and horizontally between them. The distance between each line and the centering dot was normally 10mm. Critical signals for detection were configurations in which both lines were 2mm farther from the centering dot than usual. Stimuli were presented at a rate of 24 events/min, with an exposure time of 150 msec. In all conditions, five critical signals were presented within each period of watch (signal probability = .02). Inter-signal intervals ranged form 20 to 240 sec with a mean of 120 sec. Stimulus presentations and response recording were orchestrated by an Apple IIe microcomputer. Observers indicated their detection of critical signals by pressing the spacebar on the computer keyboard. Responses occurring within 1.25 sec of the onset of a critical signal were recorded automatically as correct detections; all other responses were recorded as errors of commission (false alarms). During the experiment, observers received 30-second bursts of fragrance or pure air starting at 4.4 min into the vigil and then once every five minutes thereafter.

Results

Correct Detections. Mean percentages of correct detections in the four experimental conditions are plotted as a function of periods of watch in Figure 1. An analysis of variance based upon an arcsine transform of the detection scores revealed that there were significant main effects for groups, $F (3, 60) =$

2.89. *p* <.05, for periods, *F* (3, 180) =23.25, *p*<.0001, and that the Groups x Periods interaction approached significance, *F* (9, 180) = 1.80, .07>*p*>.05. Newman-Keuls tests with alpha set at .05 for each comparison indicated that performance efficiency was significantly greater in the Clementine and Vanilla Bean groups than in the air control group. While the detection scores for the Butyric Acid group were generally above those of the controls (see Figure 1), the supplementary Newman-Keuls tests indicated that they did not differ significantly from those of the control group. As can be seen in the figure, performance efficiency tended to decline over time in all groups with a tendency for maximal decline in the Butyric Acid and air control groups.

False Alarms. Errors of commission were less than 1% in all experimental conditions. Given their rarity in this study, errors of commission were not analyzed further.

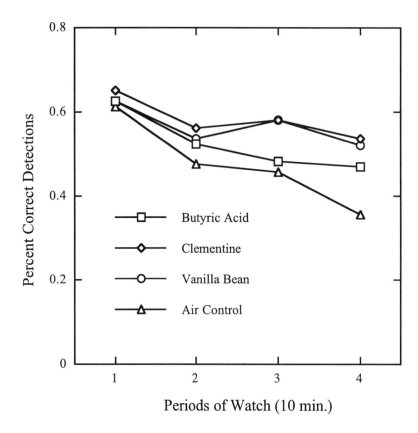

Figure 1. Percent Correct Detections for the Four Conditions as a Function of Periods of Watch.

DISCUSSION

The results of this investigation confirm and extend our previous findings demonstrating that exposure to whiffs of hedonically positive odors can enhance vigilance performance. As with the Peppermint and Muguet fragrances used in our earlier research, both the Clementine and Vanilla Bean scents were associated with elevations in the overall level of signal detection in comparison to an unscented air control group. This enhancement was not associated with a concomitant increase in errors of commission. Thus, the results show that the beneficial effects of fragrance administration are generalizable to scents other than those used in our initial investigations. However, the use of accessory olfactory stimulation to bolster vigilance performance seems to be limited to hedonically positive fragrances since Butyric Acid, a hedonically negative scent, failed to enhance performance above that of the control group. Note that Butyric

Acid was rated by observers as stimulating while the Clementine and Vanilla Bean scents were rated as relaxing. Thus, one might argue that the key dimension separating these fragrances is stimulation/relaxation and not hedonic value. This possibility seems unlikely. In our initial study, Muguet and Peppermint, both hedonically pleasant fragrances, enhanced vigilance performance in a similar manner although the former was rated as relaxing and the latter as stimulating (Warm, Dember & Parasuraman, 1991).

Given that the olfactory sense projects to the limbic system (Engen, 1991), parts of which are implicated in emotion and alertness (Carlson, 1994), one might expect that accessory olfactory stimulation enhances vigilance performance through an increase in the level of general arousal. However, as described by Dember, Warm and Parasuraman (1996), research designed to test the general arousal hypothesis has not supported it. The present results with regard to Butyric Acid reinforce that conclusion, since the supposedly stimulating fragrance failed to enhance performance efficiency. On the basis of changes in the amplitude of the N160 component of observers' evoked brain potentials when exposed to whiffs of Peppermint during the performance of a vigilance task, Dember, Warm and Parasuraman (1996) suggested that olfactory stimulation boosts vigilance performance via more efficient allocation of attention during the vigil rather than through a simple increase in general arousal. Such a possibility may help to explain the role of hedonic value in the effects of accessory olfactory stimulation on vigilance -- negative hedonics may suppress the potential benefits to attentional allocation produced by exposure to intersensory stimulation through the nose.

ACKNOWLEDGEMENTS

This research was supported by a contract with the International Flavors and Fragrances Corporation (IFF), Union Beach, NJ. Dr. Stephen Warrenberg was the contract monitor. IFF provided all fragrances.

REFERENCES

Carlson, N.R. (1994). *Physiology of behavior* (5th ed.). Boston: Allyn-Bacon.

Dember, W. N., Warm, J.S., & Parasuraman, R. (1996). Olfactory stimulation and sustained attention. In A.N. Gilbert (Ed.), *Compendium of olfactory research* (pp. 39-46). Dubuque, IA: Kendall/Hunt.

Engen, T. (1991). *Odor sensation and memory*. New York, NY: Praeger.

Richardson, J.T., & Zucco, G.M. (1989). Cognition and olfaction. *Psychological Bulletin, 105*, 352-360.

Warm, J.S. (Ed.). (1984). *Sustained attention in human performance*. Chichester, Great Britain: Wiley.

Warm, J.S., Dember, W.N., & Parasuraman, R. (1991). Effects of olfactory stimulation on performance and stress in a visual sustained attention task. *Journal of the Society of Cosmetic Chemists, 42*, 199-210.

STRESS,

WORKLOAD,

AND FATIGUE

Assessment of Task-Induced State Change: Stress, Fatigue and Workload Components

Gerald Matthews and Sian E. Campbell
University of Dundee

Paula A. Desmond
University of Minnesota

Jane Huggins, Shona Falconer and Lucy A. Joyner
University of Dundee

Use of automated systems may have adverse influences on the operator's subjective state. Automated tasks resemble vigilance tasks in that they require, typically, prolonged monitoring, but infrequent response. Such tasks tend to provoke 'stress' symptoms such as fatigue, boredom and loss of motivation (e.g. Scerbo, 1998). However, the distinctiveness of constructs such as stress, anxiety, arousal and so forth is unclear, and research in this area has been limited by the lack of a comprehensive dimensional model of subjective states. This chapter outlines the development of such a model for states of task-induced stress and fatigue, and its application in studies of manual and automated performance.

ASSESSMENT OF TASK-INDUCED STRESS STATES

The Dundee Stress State Questionnaire (DSSQ; Matthews et al., in press a) assesses affect, cognition and motivation. 767 subjects completed the questionnaire following performance of various demanding tasks, such as working memory and vigilance. Factor analysis of the item responses identified 10 correlated dimensions of subjective state. A second-order factor analysis discriminated three broad stress syndromes, labeled *task engagement*, *distress* and *worry*. Table 1 shows the definition of these secondary factors by the primary dimensions. 'Cognitive interference' refers to intrusive thoughts about the task or about personal concerns, which may disrupt performance. Studies of task-induced fatigue show that task (dis)engagement appears to overlap with boredom to a large degree, although physical fatigue symptoms are distinct from these psychological states (Matthews & Desmond, in press).

Table 1

Definition of Three Secondary Stress State Dimensions by Primary DSSQ Scales

	Task Engagement	Distress	Worry
Principal scales	Energetic arousal	Tense arousal	Self-consciousness
	Motivation	Low hedonic tone	Low self-esteem
	Concentration	Low perceived control	Cognitive interference (task-related)
			Cognitive interference (task-irrelevant)

Note. Some primary scales have additional, minor loadings on other secondary factors

In several studies the DSSQ was administered before and after performance. Different types of task provoked qualitatively different patterns of state change. Tasks overloading working memory tended to induce both task engagement and distress, whereas vigilance reduced task engagement. DSSQ dimensions relate to individual differences in performance. Table 2 summarizes data from two of the Matthews et al. (in press a) studies. In Study 1, subjects performed a working memory task requiring solution of arithmetic

199

problems while keeping ordered word strings in memory. High distress was related to both poorer recall and impaired arithmetic, whereas worry was associated with impaired arithmetic only. The second study used two visual, high event rate vigilance tasks; the 'successive' task version required integration of information across trials, but the 'simultaneous' version did not. Both versions were sensitive to individual differences in task engagement, consistent with previous studies of energetic arousal, one of the components of engagement (Matthews & Davies, 1998). Hence, the DSSQ allows the precise description of task-induced stress, and identification of the stress factors most detrimental to specific tasks.

Table 2
Performance Correlates of Stress States (Matthews et al., in press a)

	Study 1 (N = 137)		Study 2 (N = 229)	
	Arithmetic	WM	Sim. Vig.	Suc. Vig
Engagement	-09	11	17*	21**
Distress	-23**	-28**	-13	-07
Worry	-23**	-08	-11	-11

Note. WM = Working memory, Sim. Vig. = Simultaneous vigilance task. Suc. Vig. = Successive vigilance task, * $p<.05$; ** $p<.01$

Appraisal and Coping as Correlates of Stress State

According to contemporary 'transactional' theories, stress symptoms reflect appraisal and coping processes. Appraisals of task environments will be influenced most strongly by task demands, and also by more general appraisals of working conditions. Our research at Dundee has routinely used a slightly modified version of the NASA-TLX (Hart & Staveland, 1988) to assess appraisals of task demands. Overall workload was assessed as a simple arithmetic mean of the six rating scales. Table 3 shows how the NASA-TLX scales correlated with post-task state. Distress was more strongly related to overall workload than were engagement and worry. Engagement related mainly to workload patterning; i.e. high effort and high appraised demands, coupled with positive assessment of performance and low frustration.

Table 3
Correlations between Stress States and Workload Ratings (Matthews et al., in press a; N =567)

Frustration	Workload (mean)	Mental demands	Physical demands	Temporal demands	Poor performance	Effort	
Engagement	20**	44**	23**	29**	-48**	56**	-34**
Distress	59**	30**	28**	39**	40**	20**	55**
Worry	00	-10*	03	-07	-12**	-19**	20**

Note. Data aggregated across three studies, using vigilance and visual and auditory memory tasks,
    * $p<.05$; ** $p<.01$

Stress reactions also depend on the person's active attempts to cope with the stressor. Matthews et al. (in press a) showed that measures of typical coping with workplace task demands predicted stress states during workplace performance. We have also developed a questionnaire measure of three fundamental aspects of coping with task demands: *task-focus* (e.g. focusing on performing well), *emotion-focus* (e.g. worrying about performance) and *avoidance* (e.g. deciding not to take the task seriously). Matthews et al. (in press b) found that engagement related to task-focus and low avoidance, distress to emotion-focus, and worry to both emotion-focus and avoidance.

These correlational data do not support strong conclusions about causality, but it is plausible that appraisal and coping influence stress reactions. Awareness of subjective state may also feed back to influence appraisal and coping. Matthews et al. (in press a) conclude that stress states represent relational descriptions of the major adaptive challenges posed by task environments: handling overload (distress), commitment of effort to the task (engagement) and self-regulation (worry).

STRESS AND FATIGUE STATES IN SIMULATED DRIVING

Patternings of State Change Induced by Driving

Next, we discuss two studies from a series (reviewed by Matthews & Desmond, 1997) in which drivers performed in demanding, potentially stressful conditions. The studies used subsets of the DSSQ scales, so we present data for primary scales, rather than for the three secondary factors. Figure 1 shows state scale *change*, in standardized units, in two studies, with scales grouped by secondary factor (see Table 1). In the first study, drivers were exposed to a loss of control manipulation: the car skidded frequently. Loss of control provoked increased distress (e.g. tension) and worry (task-related cognitive interference), but had little effect on engagement (energy). These state changes may reflect the subject's awareness of overload (distress) and evaluation of personal goals (worry). In the second study, drivers were fatigued through performance of a high-workload secondary task for about 25 minutes. Again, subjects showed increased distress. However, with this different source of demand, task engagement deteriorated, and worry was not

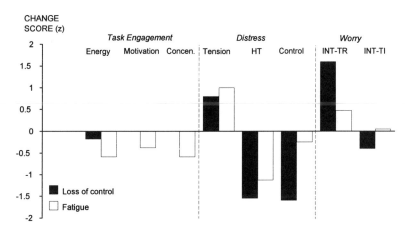

Figure 1. State changes induced by loss of control (*N*=80) and fatigue (*N*=80) manipulations. (HT = Hedonic tone, INT-TR = Task-related cognitive interference, INT-TI = Task-irrelevant cognitive interference. Motivation and concentration were not measured in the loss of control study).

much affected. The study also found that task-focused coping was considerably reduced by the fatigue manipulation. The fatigue transaction seems to be associated with overload and reduced effort commitment. Although both types of drive were 'stressful', the pattern of subjective state change was qualitatively different in each case.

The subjective state changes found in these studies parallel effects of stress on objective performance, assessed by measures such as heading error and speed of response to secondary task stimuli (see Matthews & Desmond, 1997). In the loss of control study, performance deficits of stress-vulnerable drivers may be a consequence of distraction by worries about driving (task-related cognitive interference). In the fatigue study, performance deficits appeared to be linked to loss of motivation and a reduction in active regulation of effort.

Implications for Automated Driving

Driver stress and fatigue appear to be multifaceted, and vary with the demands of the traffic environment. Hence, consequences of automation may vary with the nature of the driver-environment interaction. Automation reduces opportunities for effortful, task-focused coping, and so we might expect the most common outcome to be reduced task engagement (Matthews & Desmond, 1997). If automation reduces appraisals of workload, it should also reduce distress, a generally beneficial state change. The challenge for designers is to promote reduced distress without also lowering task engagement. Singh, Molloy and Parasuraman (1993) discuss the problem of complacency, which may reflect a combination of low distress and low engagement. Conversely, if the technology is appraised as error-prone ('mistrust'), distress and worry may well ensue. Presumably, vehicular automation technology will be highly reliable, but some individuals may be especially sensitive to any suboptimal functioning.

Desmond, Hancock and Monette (in press) have shown some of these state changes in a study which compared reactions to automated and manual driving. Automation of trajectory failed three times during the 40 minute drive. Task demands were lower than in the previous studies; the road was straight throughout, rather than comprising straights and curves. Figure 2 shows that both types of drive provoked similar state changes. In this instance, it appears that the low level of effort required for both tasks provokes decreased

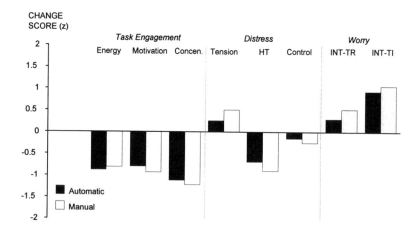

Figure 2. State changes induced in automated and manual driving (Desmond et al., in press; N = 40) (HT=Hedonic tone, INT-TR=Task-related cognitive interference, INT-TI=Task-irrelevant cognitive interference).

task engagement, and increased worry (cognitive interference). Distress increased, but less markedly than in the Figure 1 data. Presumably the automation failures prevented complacency and improvements in mood. These data suggest it is not automation per se which is critical, but task demands (as appraised by the driver).

CONCLUSION

Task-induced state changes may be described as patterned shifts in task engagement, distress and worry. State change patterning is sensitive to task and environmental demands, as illustrated here in studies of manual and automated driving. Effects of situation and person factors on state may be mediated by cognitive stress processes: operators' appraisals of task demands (workload) and their choice of coping

202

strategy. Stress processes reflect, in turn, the adaptive basis for the transaction between operator and task environment. Consequences of automation may vary considerably depending on factors such as perceived reliability and controllability of the system, provision of residual tasks for the operator, the wider occupational context for performance, and various intrapersonal factors such as personality and coping style. Hence, there is unlikely to be any simple remedy for stress-related problems associated with automation, such as boredom and complacency. Fine-grained assessment of the operator's states and cognitions is required to determine vulnerabilities to performance breakdown.

ACKNOWLEDGEMENTS

Some of this research was supported by a Medical Research Council Grant (G9510930) to the first author. The DSSQ is available to qualified psychologists from the first author.

REFERENCES

Desmond, P.A., Hancock, P.A., & Monette, J.L. (in press). Fatigue and automation-induced impairments in simulated driving performance. *Transportation Research Record.*

Hart, S.G., & Staveland, L.E. (1988). Development of a multidimensional workload rating scale: Results of empirical and theoretical research. In P. A. Hancock & N. Meshkati (Eds.), *Human mental workload.* Amsterdam: Elsevier.

Matthews, G., & Davies, D.R. (1998). Arousal and vigilance: The role of task factors. In R.R. Hoffman, M.F. Sherrick, & J.S. Warm (Eds.), *Viewing psychology as a whole: The integrative science of William N. Dember.* Washington, DC: American Psychological Association.

Matthews, G., & Desmond, P.A. (1997). Underload and performance impairment: Evidence from studies of stress and simulated driving. In D. Harris (Ed.), *Engineering psychology and cognitive ergonomics. Vol 1. Transportation systems.* Aldershot: Ashgate Publishing.

Matthews, G., Joyner, L., Gilliland, K., Campbell, S.E., Huggins, J., & Falconer, S. (in press a). Validation of a comprehensive stress state questionnaire: Towards a state 'Big Three'? In I. Mervielde, I.J. Deary, F. De Fruyt, & F. Ostendorf (Eds.), *Personality psychology in Europe* (Vol. 7). Tilburg: Tilburg University Press.

Matthews, G., Schwean, V.L., Campbell, S.E., Saklofske, D.H., & Mohamed A.A.R. (in press b). Personality, self-regulation and adaptation: A cognitive-social framework. In M. Boekarts, P.R. Pintrich & M.Zeidner (Eds.), *Handbook of self-regulation.* New York: Academic.

Matthews, G., & Desmond, P.A. (in press). Personality and multiple dimensions of task-induced fatigue: a study of simulated driving. *Personality and Individual Differences.*

Scerbo, M.W. (1998). What's so boring about vigilance? In R.R. Hoffman, M.F. Sherrick, & J.S. Warm (Eds.), *Viewing psychology as a whole: The integrative science of William N. Dember.* Washington, DC: American Psychological Association.

Singh, I.L., Molloy, R., & Parasuraman, R. (1993). Individual differences in monitoring failures of automation. *Journal of General Psychology, 120,* 357-373.

Sleep-Disordered Breathing (SDB), Daytime Sleepiness, and Complex Monitoring Task Performance

Hyon Kim   Barrett Caldwell   Jason Pionek   Terry Young
Departments of Preventive Medicine and Industrial Engineering, University of Wisconsin-Madison

INTRODUCTION

Sleep-disordered breathing (SDB) is a condition characterized by repeated breathing pauses, decreased oxyhemoglobin levels, and sleep fragmentation during nocturnal sleep. SDB is considered to be a chronic condition, and unrecognized SDB is prevalent in adult population. SDB is hypothesized to be associated with chronic behavioral morbidity including problems with decreased concentration and attention. Diminished concentration and attention due to daytime sleepiness or other consequences of SDB may impair complex monitoring task performances in day to day operation. This study explores the relationship between SDB and complex monitoring task performance in a general population. We report here on the association between unrecognized SDB and overall performance, decrement in performance, and self-reported workload in a complex monitoring computer task. In addition, data on subjective and objective measures of sleepiness were used to investigate the role of sleepiness in complex task performance.

METHODS

Subjects

From the Wisconsin Sleep Cohort Study, a longitudinal population-based study of the natural history of SDB among working adults (Young, Palta, Dempsey, Skatrud, Weber, & Badr, 1993), 107 consecutive participants (67 men and 40 women) were used in this cross-sectional study. Mean age of our study sample was 51±9. Most (99) had computer experience; 73 had computer experience greater than "moderate".

Data Collection

Complex Monitoring Task Performance : Multi-Attribute Task Battery. A revised version of the Multi-Attribute Task battery (MAT; Comstock & Arnegard, 1992) was used to assess complex monitoring task performance and decrement in performance on the simulated systems monitoring and fuel management tasks, and self-reported workload during task performance. The 25-minute MAT consisted of a low systems monitoring task load preceded and followed by both a fuel management task and higher task load on the systems monitoring task. The MAT battery signal script is a shown Table 1.

Table 1
MAT Signal Script

Elapse Time (min.)	Time	# of light changes	# of dial changes	# of pump failures
1-9	1	8	5	8
9-16		5	2	0
16-24	2	8	5	8

Workload Rating: Every 5 minutes.

204

The systems monitoring task consisted of 2 lights (green and red) and 4 moving dials. The appropriate response keys for lights and dials were labeled on the keyboard. In the normal condition, the green light remained "on" while the red light remained "off" and the dials fluctuated in a random manner around the center within one scale marker in each direction of the center. The participant was instructed to respond as quickly and accurately as possible using the corresponding keys, to 1) turn the green light on if it turned off, 2) turn the red light off if it turned on, and 3) correct the dials if they shifted above or below the normal fluctuation of one scale marker in each direction of the center. A miss was credited and the lights and dials were reset to normal if the participant failed to detect the lights and dials within 10 seconds of the change.

The fuel management task consists of 6 tanks connected by series of pumps. The participant was instructed to maintain the fuel levels in both main tanks A and B as close as possible to 2500 units. Since the fuel from tanks A and B were being lost a constant rate, the task was to transfer fuel from the lower supply tanks to the main tanks by using the pump keys on the keyboard. The activated pumps showed a green light indicating that the fuel pump was pumping fuel in the direction indicated by the arrow located next to the pump. A red light in the pumps indicated pump failures. The combined RMS error (from 2500 units sampled every 30 seconds) in the fuel levels for the main tanks were measured.

The number of misses for lights and dials and RMS error were computed for the overall 25 minute testing period and for each higher task load period (TIME period).

A self-reported workload during the task performance was assessed every 5 minutes (5 presentations) using a multi-dimensional rating scale (NASA Task Load Index; Hart & Staveland, 1988). An overall workload rating was computed by a weighted average of ratings on 6 subscales. The subscales were: Mental Demand, Physical Demand, Temporal Demand, Own Performance, Effort, and Frustration. The participants were instructed to rate their perceived exertion from "Low" to "High" on 5 of the subscales (except for Own Performance). The Own Performance subscale ranged from "Good" to "Poor". A composite rating for the overall workload rating and subscales was computed by averaging the 5 workload rating scale presentations. The overall rating and subscales from the workload rating scale presented after 5 minutes (Time 1) and 20 minutes (Time 2) were computed.

A 3-minute practice session was given prior to the 25-minute testing session.

Sleepiness: Multiple Sleep Latency Test (MSLT) and Stanford Sleepiness Scale (SSS). The MSLT (Carskadon, Dement, Mitler, Roth., Westbrook, & Keenan, 1986) was used to objectively measure sleepiness. The participant was given a 20 minute opportunity to fall asleep in a dark room while having EEG, EOG, and EMG measured. Sleep latency (in minutes) to onset of EEG sleep was scored.

The SSS (Hoddes, Zarcone, Smythe, Phillips, & Dement, 1973) was used to subjectively measure sleepiness. The SSS is scored on a Likert rating scale ranging from one, indicating "alert, wide awake," to seven, "almost asleep."

For our study, average of the 0900 hr and 1100 hr sleep latencies were used in our analysis. Average of the 0900 hr and 1100 hr SSS was also used.

Sleep Disordered Breathing: Polysomnograph Recording. The nocturnal polysomnogram consisted of continuous polygraphic recording of EOG, EEG, tracheal sounds (microphone), nasal airflow (end-tidal carbon dioxide concentration), oral airflow (thermocouple), thoracic and abdominal respiratory effort (inductance plethysmography), and oxyhemoglobin saturation (finger-pulse oximeter).

Each 30 second epoch of the recordings was scored by a team of technicians and sleep specialists for sleep, respiration, oxyhemoglobin changes, and movement. Sleep stage was scored by the criteria of Rechtschaffen and Kales (1968). Respiration was evaluated for apnea (cessation of airflow for 10 seconds or more) and hypopnea (reduction in respiratory effort accompanied by a 4% drop in oxyhemoglobin saturation). Apnea-hypopnea index (AHI) was defined as the number of apneas plus hypopneas per hour of total sleep time.   The median AHI for our study sample was 3.07 (range: 0-66). For our analysis, SDB was categorized into two groups (AHICAT): AHI<=15 (n=88) and AHI>15 (n=19).

Study Protocol. The overnight polysomnography was conducted on an average of 45 days prior to the MAT testing. On the day of the MAT testing, sleepiness, as assessed by MSLT and SSS, was determined at 0900 hr and 1100 hr. The MAT testing was administered at 1000 hr. A self-reported hours of restful sleep and quality of sleep the participant had the night before were determined prior to MAT testing.

Statistical Analysis. Multiple regression analysis was used to estimate the relationship between AHICAT and overall MAT battery performance measures (misses for lights and dials, and RMS errors) and workload rating scales while controlling for age and gender. Multiple regression analysis was also used to investigate the relationship between sleepiness variables, including self-reported hours of restful sleep and quality of sleep, and overall MAT battery performance measures and workload rating scales controlling for AHICAT, age, and gender. Repeated measures ANOVA, with performance and workload scales scores before and after the low systems monitoring task load as the repeated measure (TIME), was used to assess performance and workload decrement over time and to estimate the TIME by AHICAT interaction effect. Age and gender were included in the models. P-values <0.05 were considered to indicate statistical significance.

RESULTS

A repeated-measure ANOVA revealed a statistically significant TIME main effect (F=11.43, p=0.001) and a AHICAT X TIME interaction (F=8.13, p=0.0053) for number of misses for dials. There were no significant TIME and AHICAT X TIME effects for number of misses for lights, RMS errors, and workload rating scales.

The reported hours of restful sleep the night prior to testing, independent of AHICAT, gender, and age, was associated with several averaged workload rating scales. The higher reported hours of restful sleep was significantly associated with lower scores on the overall demand scale (coefficient of overall scale = -3.0, p=0.0053), mental demand (coefficient of mental = -3.3, p=0.014), physical demand (coefficient of physical = -3.2, p=0.023), temporal demand (coefficient of temporal = -3.0, p=0.02), effort (coefficient of overall scale = -2.8, p=0.049), and frustration (coefficient of overall scale = -3.7, p=0.01).

There were no significant relationships between AHICAT and overall MAT battery performance measures and workload rating scales. Also, there were no significant relationships between the sleepiness variables and overall MAT battery performance measures and workload rating scales.

DISCUSSION

The main finding of this study supports the hypothesis that undiagnosed sleep-disordered breathing is associated with a low task load induced decrement in signal detection during a complex monitoring task. Our findings, however, did not show a relationship between SDB and overall task performance. The results suggest SDB may impair the ability to detect subtle changes rather than impair the ability to perform globally in a complex monitoring task. In addition, the decrement in performance associated with SDB can be seen in a complex task that is relatively short in duration.

Our findings do not suggest that perceived and physiological sleepiness compromise overall performance and performance over time in a complex monitoring task. These results suggest that other consequences of SDB such a hypoxemia or sleep fragmentation may play a role in the decrement in signal detection.

We found negative associations between reported number of hours of restful sleep the night prior to the complex task performance with perceived overall, mental, physical, and temporal demand, effort, and frustration experienced during the complex monitoring task, independent of SDB, age, and gender. The results suggests that a reduction in the self-reported hours of restful sleep can compromise perceived mood and physical demand during a complex task that requires continuous monitoring.

In summary, we found that unrecognized, untreated sleep-disordered breathing indicated by an apnea-hypopnea index greater than 15 in the general population is significantly related to decrement in signal detection in complex monitoring tasks.

REFERENCES

Carskadon, M. A., Dement, W.C., & Mitler, M. M., Roth T., Westbrook P. R., & Keenan S. (1986). Guidelines for the multiple sleep latency test (MSLT): A standard measure of sleepiness. *Sleep, 9*, 519-524.

Comstock, J.R., & Arnegard, R.J. (1992). *The Multi-Attribute Task Battery for human operator workload and strategic behavior research* (Technical Memorandum 104174). Hampton, VA: NASA Langley Research Center.

Hart, S. G., & Staveland, L. E. (1988). Development of NASA-TLX (Task Load Index): Results of empirical and theoretical research. In P. A. Hancock & N. Meshkati (Eds.), *Human Mental Workload* (pp. 139-183). Amsterdam: North-Holland.

Hoddes, E., Zarcone, V., Smythe, H., Phillips, R., Dement, W. C. (1973). Quantification of sleepiness: a new approach. *Psychophysiology, 10*, 431- 436.

Rechtschaffen, A., & Kales, A. A. (1968). A manual of standardized terminology, techniques and scoring system for sleep stages of human subjects. U.S. Government Printing Office, Washington, D.C. NIH Publication No. 204.

Young, T., Palta, J., Dempsey, J. B., Skatrud, S., Weber, S., & Badr S. (1993). The occurrence of sleep-disordered breathing among middle-aged adults. *New England Journal of Medicine, 328*, 1230-1235.

Operator Alertness And Human-Machine System Performance During Supervisory Control Tasks

Steven A. Murray
SPA WAR Systems Center
San Diego, CA

Barrett S. Caldwell
University of Wisconsin
Madison, WI

INTRODUCTION

Increased reliance on automation in the design of human-machine systems is based largely on the assumption that reducing task demands will reduce the probability of human overload and lead to more reliable system performance. Automated systems, however, shift the operator's job from continuous control to intermittent or supervisory control, inducing boredom and monotony; personal involvement with the state of the system is also reduced. The supervisory control environment can therefore diminish the operator's ability to respond to abnormal conditions following extended periods of work underload, creating a new path for system failure (e.g., Ryan, Hill, Overline, & Kaplan, 1994).

Krulewitz, Warm, and Wohl (1975), using a variable-demand vigilance task, found that subjects transitioning from a slow to a fast event rate performed more poorly in the immediate post-transition period than control subjects who had performed the task at the fast rate throughout the test session. This result indicated that a shift between two workload levels may impose performance costs over and above those of either workload level alone; i.e., resources are required for the adapting process itself.

Our research focused on the performance of workstation operators under conditions of protracted work underload, so the general issue of supervisory control and the specific results of the Krulewitz, et al. (1975) study were of interest to us. The literature of vigilance performance (e.g., Davies & Parasuraman, 1982) has shown that human capabilities vary in both tonic and phasic fashion, over a range of time scales, and that these fluctuations can become more prominent in settings with little external stimulation. It was reasonable to expect that such operator fluctuations would influence the degree of the efficiency of the adapting process found by Krulewitz, et al., and others. We were specifically interested in evaluating performance changes to high workload events as a function of the operator alertness states that *preceded* those events. If the availability of cognitive resources was a function of operator alertness during work underload, then task performance should be directly related to measures of operator alertness state during these underload periods. In particular, higher rates of performance errors would be related to lower levels of pre-task alertness, as operators failed to attend to task components that they could normally handle.

AN EXPERIMENT IN WORKLOAD TRANSITIONS

An air defense scenario was chosen for the experiment, typical of the kind performed in the Navy and elsewhere, and included both a primary task and a secondary task. The primary task was to process aircraft inbound to two ship icons, shown at fixed positions on a conventional color monitor. Aircraft symbols originated from random points around the borders of the screen and proceeded inward toward the ships. The subject's job was to classify each aircraft symbol as a friend or an enemy, to destroy enemy aircraft before they advanced too close to the ships, and to avoid shooting down any friendly aircraft. Subjects classified aircraft by selecting a symbol with a trackball, matching its displayed name to a list of friends and enemies, and designating it with a pair of pushbuttons. The secondary task required the subject to acknowledge an auditory tone in a headset, by pressing a button. It was reasoned that any increased demand for processing by the primary task would result in diminished performance on the secondary task, as limited resources were re-allocated. Because the hypothesis predicted that cognitive resources would vary as a function of pre-task alertness, the secondary task - referred to as the *Fail to Acknowledge* measure - represented the major index of high workload task performance for the experiment.

Two measures of operator alertness were employed during work underload periods: the same secondary task used as the workload performance measure and heart rate. The secondary task - referred to as the *No Response* measure - represented a behavioral index of vigilance (i.e., signal detection performance) that could relate the current results with the larger body of vigilance research. Because this measure was identical to the *Fail to Acknowledge* task, however, a correction for subject response bias was necessary to compare these performances between high and low workload conditions. The *No Response* data were therefore also expressed for this experiment using the *A-prime* statistic, as an operational derivative of *d-prime* (e.g., Wickens, 1984). *Heart Rate* was selected as the most directly interpretable psychophysiological measure available, in beats-per-minute. In general, low levels of workload tend to lower arousal levels and reduce heart rate (e.g., Braby, Harris, & Muir, 1993).

During work underload periods, an average of 1 - 2 aircraft were shown on the display at any one time, with periods of no activity at all. High workload events lasted 6 - 8 minutes and involved an average of 7 - 9 aircraft on the display at once. Intervals between high workload events were controlled randomly by the simulation software, and three events per hour were administered. Each subject was given three training periods and then completed four two-hour task sessions over a two-day interval (i.e., eight hours of data for each subject).

RESULTS

Twelve subjects completed the planned protocol. No significant session effects were noted in the data, so all four sessions were pooled for the analysis. Each high workload period was parsed into one-minute segments, effectively converting the performance measures to performance rates (i.e., events per minute). These rate scores were then converted to efficiency measures by dividing the respective rate by the average number of aircraft on the display during that minute. These procedures were necessary to standardize for intra- and inter-trial fluctuations in workloads. Finally, data for each alertness and performance measure were normalized and points more than ± 3 standard deviations of each mean were excluded from the analysis as outliers.

Cross-correlations were performed between each of the operator alertness measures - *No Response, A-prime,* and *Heart Rate* - and the performance measure *Fail to Acknowledge.* All correlations were significant at $p < .01$ (i.e. a correlation coefficient greater than ± 2.58 times the standard error), as shown in Table 1.

Table 1
Cross-correlation Results

	No Response	*A-prime*	*Heart Rate*
Fail to Acknowledge	0.581	-0.392	-0.242

As a final step, surface plots were generated for these alertness-performance comparisons by sorting the data set and using least squares data-fitting methods, to graphically convey the underlying patterns in the results. In each of the Figures (1 - 3, below), decreasing alertness is shown to the right of the X-axis, the sequence of one-minute periods for the high workload condition is shown along the Y-axis, and poorer task performance (i.e., increasing failure to respond to the auditory tones) is shown moving up the Z-axis.

Increasing error rates for the *Fail to Acknowledge* performance measure were consistently related to increasing error rates of the pre-task *No Response* measure, as shown in Figure 1. A "band" structure, i.e., a grouping of consistent scores, can be clearly seen across all periods of the high workload condition. In addition, an unusually high error rate for the *Fail to Acknowledge* task can also be seen during the ramp down period, for those trials that began at the lowest levels of pre-task alertness.

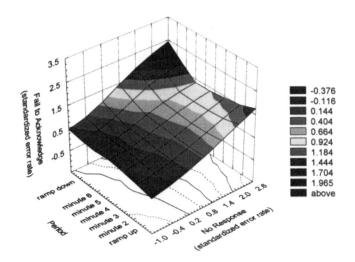

Figure 1. *Fail to Acknowledge* Performance as a Function of Pre-task *No Response* Rate.

Reduced signal sensitivity, as indexed by lower values of *A-prime* performance, was also consistently related to higher error rates for the *Fail to Acknowledge* measure (Figure 2), although the band pattern was not quite as clear as that for the *No Response* measure. This figure, however, serves to demonstrate that this relationship represents an effect other than operator response bias.

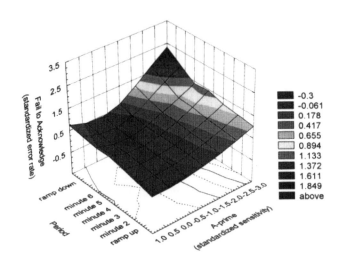

Figure 2. *Fail to Acknowledge* Performance as a Function of Pre-Task *A-prime* Performance.

Finally, *Heart Rate* also showed a clearly interpretable relationship with *Fail to Acknowledge* performance, as depicted in Figure 3. This result was very similar to the patterns found for the previous measures of operator alertness, and also shows a significant increase in error rate late in the high workload events for those trials initiated at very low levels of the *Heart Rate* alertness measure.

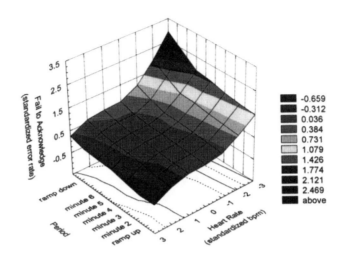

Figure 3. *Fail to Acknowledge* Performance as a Function of Pre-task *Heart Rate*.

DISCUSSION

In summary, clear performance patterns were found for the statistically significant cross-correlations between the high workload performance measure, *Fail to Acknowledge,* and two presumed measures of pre-task alertness (i.e., as the *A-prime* measure was derived from the *No Response* data). These results conformed to expectations from the literature concerning alertness fluctuations during periods of work underload, as a wide range of values was obtained for each measure. Furthermore, the different measures showed consistently significant and surprisingly similar patterns of results concerning task performance during protracted (i.e., 6 - 8 minute) time periods; the impact of pre-task alertness was neither momentary or transitory.

The use of a secondary task as the principal performance measure was useful, in that both vigilance performance and cognitive resource availability could be tested with a single metric, reducing the number of measurement intrusions in the scenario. These data, however, only demonstrate a pattern of changes in general operator capabilities that can be related to alertness, which is not the same as demonstrating specific, operationally meaningful changes in task performance. Further work remains to be done to examine other components of primary task performance, such as the number of enemy aircraft allowed to fly to the ships, the number of friendly aircraft shot down, etc., to better understand the implications of this reduction in general resources.

SUMMARY

Both the scientific literature and industrial safety reports have highlighted a growing hazard of increasingly automated systems in that long periods of work underload, typical of automated job settings, tend to induce states of low alertness in system operators (e.g., Pope & Bogart, 1992). When workload levels increase, as they inevitably do during times of anomalous system conditions, such states can interact with task performance in perilous ways.

The results of this experiment provided a positive demonstration of the influence of pre-task alertness on subsequent task performance and, furthermore, showed that this influence could extend over periods of at least several minutes. In other words, the "initial conditions" of operator state can propagate through a task, over time. This evidence was particularly strong regarding secondary task performance, demonstrating that prolonged narrowing of attention - indicative of reduced cognitive resources - can occur as a function of lower pre-task alertness levels.

REFERENCES

Braby, C.D., Harris, D. & Muir, H.C. (1993). A psychophysiological approach to the assessment of work underload. *Ergonomics, 36*(9), 1035-1042.

Davies, D.R. & Parasuraman, R. (1982). *The psychology of vigilance.* London: Academic Press.

Krulewitz, J.E., Warm, J.S., & Wohl, T.H. (1975). Effects of shifts in the rate of repetitive stimulation on sustained attention. *Perception and psychophysics, 18,* 245-249.

Pope, A.T. & Bogart, E.H. (1992). Identification of hazardous awareness states in monitoring environments. In *SAE 1992 transactions: Journal of aerospace, section 1- volume 101,* (pp. 449-457). Warrandale, PA: Society of Automotive Engineers, Inc.

Ryan, T.G., Hill, S.G., Overline, T.K., & Kaplan, B.L. (1994). Work underload and workload transition as factors in advanced transportation systems. *Workshop proceedings* (EGG-HFSA-11483). Idaho Falls: Idaho National Engineering Laboratory.

Wickens, C.D. (1984). *Engineering psychology and human performance.* Ohio: Charles E. Merrill.

Dynamics of Supervisory Control Task Performance: SCAMPI Project Summary

Barrett S. Caldwell
University of Wisconsin-Madison
Madison, Wl

INTRODUCTION

The Supervisory Control Alertness Monitoring and Performance Indicators (SCAMPI) project was begun at the University of Wisconsin-Madison (UW) in early 1995. SCAMPI was developed in the context of a new UW research center bringing researchers from engineering, social, life, and medical sciences together to cooperate on improving human-system interactions. The UW Center for Human Performance in Complex Systems (CHPCS) was identified by both government and corporate participants as a unique opportunity for pursuing novel directions in technology developments and implementations to reduce human error and accidents in complex technological systems. One project specified by the Advanced Research Projects Agency (ARPA: now DARPA) was a need to create "adaptive automation" technologies which were capable of predicting and counteracting deficits in operator alertness while performing computer-based supervisory control tasks. Equipment and performance monitoring in adaptive automation is believed to be able to allow fewer human operators to interact more effectively with complex system processes.

I was originally involved in SCAMPI development efforts on the basis of three relevant links to research conducted in my human factors group performance laboratory. The first link was based on my interests in information flow and methods to improve information presentation in human-machine interfaces (HMI) for cooperative task performance. The second link involved my development of quantitative approaches to analyzing human responses to changing cognitive task demands. These approaches are based on feedback control engineering systems tools using second-order differential equations of system responses to changing input functions. The third link incorporated research being conducted by a group at the Naval Research and Development (NRaD) Center in San Diego (including my graduate student Steven Murray), looking at human supervisory control of multiple autonomous robots, and real-time neural network algorithms for identifying operator state during task performance. By spring 1995, the SCAMPI project was tasked to demonstrate feasibility in three technology development areas:

• Ambulatory, lightweight technologies to collect physiological data (including EEG and EKG) to be transmitted in "noisy" radio frequency environments;

• Analysis tools to identify and predict short-term (minute scale) deficits in operator alertness and associated human performance decrements:

• Strategies to incorporate negative feedback signals to improve the quality and reliability of human supervisory control HMI and adaptive automation.

Technologies to support wearable computing capabilities and ambulatory physiology data collection have already been in existence for several years, and continue to improve in performance and comfort (Bass, 1998). Therefore, SCAMPI was not required to conduct specific technology development efforts in physiological data collection hardware. Similarly, neural science and physiological psychology researchers have focused on collecting data from specific brain and other sites that are thought to relate to human cognitive performance tasks (Makeig, Elliott, & Posal, 1993; Wilson, Fullenkamp, & Davis, 1994). Thus, my tasks as leader of the SCAMPI project were to simply select candidate technologies capable of nonintrusively collecting data from 2-8 physiological sites for real-time signal processing. The unique contributions of SCAMPI, which were imagined to take place in a 12-18 month time span, were the identification of appropriate signals for processing and relating to brief decrements in operator alertness and degraded task performance. Although SCAMPI did eventually succeed to some degree in achieving its intended goals, the results of the project have uncovered several new, exciting, and still unresolved issues for cognitive ergonomics, human factors engineering, and system development research and practice.

WORKLOAD TRANSITIONS AND ULTRADIAN CYCLES

Initial funding of SCAMPI began with an ARPA contract from March through September, 1995. By summer 1995, several candidate hardware options for collection and signal processing of EEG EMG, and EKG data had been acknowledged. However, preliminary SCAMPI activity had determined that traditional vigilance paradigm research designs had two critical shortcomings in the context of adaptive automation systems. The signal detection paradigm at the heart of vigilance research assumes discrete presentations of distinct (and usually low probability) events which are averaged over relatively long time periods. Many human supervisory control tasks, however, consist of quantitative levels of task demands requiring varying levels of operator involvement. In fact, the very concept of supervisory control tasks being modeled using the vigilance paradigm has come into question (Roth, Mumaw, Vicente, & Burns, 1997). Examining the interaction between operator state and continuously varying task requirements also points out the importance of workload transitions, or the human operator's speed and effectiveness in changing performance styles to new levels of task demands (Ryan, Hill, Overlin, & Kaplan, 1994). Continuous task systems, such as air traffic or process engineering control, requires determination of operator alertness in ultradian (much less than 24 hours) task performance cycles much shorter than vigilance paradigm signal detection sampling rates.

The SCAMPI performance tasks were selected and modified to specifically capture the dynamics of tasks with "waves" of task demands lasting between 5 and 15 minutes. The NASA Multi-Attribute Task Battery (MATB), and air traffic control tasks (including the Target / Threat Identification Display, TTID), are reasonable analogues for the variety of supervisory control tasks conducted onboard ships, in control rooms, and in other complex systems settings. A third performance task, TRACON, was used for other SCAMPI technology development work (see Caldwell, Derjani Bayeh, Schein, & Hodge-Diaz, 1997), and was not an experimental task protocol. Both MATB and TTID have been used in other cognitive ergonomics and human performance studies. A major contribution of SCAMPI was to introduce these tasks into research studies being conducted by neural science and epidemiological researchers. Unfortunately, as is discussed below, these new research relationships were not without significant problems.

Not surprisingly, the use of variable and multiple cognitive processing tasks leads to very different styles of statistical analysis for physiology / task performance relationships than traditional signal detection and choice reaction tasks. With much smaller task performance units, fine-grained analyses can be conducted by increasing the length of the task, or adding more subjects. SCAMPI tried to do both of these, using only a few physiological channels (2-3 scalp sites for EEG, as well as EKG) and doing more comparisons against performance over time and across subjects. The MATB task was used in conjunction with an ongoing UW study examining sleep disordered breathing and apneas in an epidemiological study of state workers (Kim, Young, Caldwell, & Pionek, 1998). Out of a proposed 130 workers, SCAMPI has collected performance and physiology data for over 100 persons with varying levels of sleep apneas. Each of the participants performed the MATB for a 25-minute task which included two major workload transition periods of automation start and stop. An 8-minute period of high automation (fuel tank levels controlled by the computer) was inserted between two 8-minute periods of lower levels of task automation.

The TTID task collected data at NRaD from 12 participants for four two-hour task segments each (spread over 2 days). Although the sample of TTID task performance is smaller in terms of participants, the total amount of physiology data collection (approximately 5800 person-minutes) is in fact over twice as great as the MATB sample (to date, approximately 2750 person-minutes). The advantage of these studies is that the unit of task performance analysis can be focused down to segments as small as 3-8 minutes. In contrast to traditional vigilance decrement studies, failures to detect relevant signals were able to be averaged over shorter task segments, due to the higher event rates of these tasks. Quantitative dependent variables such as response time, which can be uniquely calculated for each event, permit further study of more task segments per individual participant. Both MATB and TTID studies have specified continuous data collection of both physiological and task performance data in order to support these analyses.

PROJECT MANAGEMENT LESSONS LEARNED

Unfortunately, my SCAMPI experiences have highlighted many of the barriers facing projects attempting distributed research involving multiple disciplinary groups. The first, and perhaps most important, lesson for young investigators with new project management responsibilities is that of maintaining a focus on the overall goals of the research in the face of logistical problems. The original vision of SCAMPI was a three-year project which would yield definitive and complete signal processing and correlational time series analyses of EEG, EKG, and task performance data. In order to achieve this goal, SCAMPI was organized in terms of overlapping and semi-dependent tasks, so that delays in one study would not severely hamper overall project success. This coordinated approach requires coherent sources of funding, and common analysis methods across studies. The SCAMPI project could not maintain either of these features. Neural science or epidemiological researchers traditionally do not utilize physiological and performance data in the same manner as human factors researchers. Therefore, common assumptions of data collection, preprocessing, or summary analysis can easily be misinterpreted, and underlying guiding principles questioned. The workload transitions study approach suggests repeated measures as well as between subjects research designs, where the emphasis is on consistent relationships across individuals at a fine grain of analysis. Neural science paradigms emphasize more individual patterns of physiological behavior, often with only qualitative comparisons; epidemiology focuses on quantification of aggregated behavior at a more coarse level of generalization.

Technology development and implementation feasibility work schedules operate on fundamentally different time scales from more basic research efforts. SCAMPI began with an assumption that major data collection and analysis milestones would be achieved in six-month cycles. These project schedules were important for ensuring that the eventual SCAMPI results would be able to be compared and interpreted in ways that could inform adaptive automation technologies. There are questions about the deep structure of EEG activity, or the nature of the feedback loops between visual sensory input and resulting cognitive processing, which are not addressed in SCAMPI. The fact that such analyses are possible, however, can be a major distraction to project advancement.

The advantages of collecting large volumes of data for analysis also have significant negative correlates. Each study collected over 3 GB of physiology data, which is of itself a significant issue. Within the neural science research community, no single standard or reference authority of data reduction and analysis exists. Therefore, for each possible analysis which can be done, researchers can (and in these studies have) try to dismiss entire sets of results because of a belief that a certain type of artifact rejection should have been utilized. When this project began, desktop computers were only beginning to be able to have the computing power to process long periods of highly sampled data. Even now, relatively few software programs are able to handle concatenation or processing of files in the realm of tens of megabytes while maintaining the requirements for real-time processing. In the MATB study, these problems combined with the demands to perform feasibility studies on a limited budget and tight timeline resulted in a catastrophic computer hard drive and motherboard crash. Approximately five months were required to identify, rectify, and overcome the hardware and software problems associated with novel EEG / EKG data collection and storage techniques. It is not clear how much actual data were lost due to this delay. Because the original motherboard and hard drive were replaced with far superior equipment (200 MHz Pentium vs. 100 MHz 486-DX4; over a sixfold increase in hard drive storage), analysis capabilities since the crash have been greatly improved. Progress is approximately 9-12 months behind expectations, but analysis of MATB workload transition performance segments still seems feasible.

PATTERNS OF POSITIVE RESULTS

Despite delays and difficulties, several positive SCAMPI results were found. Both studies were able to uncover relationships between physiological measures and components of task performance in 3-8 minute cycles. Results from both studies are presented in greater detail in other conference presentations in this volume (Kim, Young, Caldwell, & Pionek, 1998; Murray & Caldwell, 1998). It is important to note that no single physiological measure was uniformly linked to performance, and that the relationships between physiology and performance were limited to specific subtasks, rather than overall task success. In other words, the patterns of results demonstrated predictability for assessing facets of complex tasks over short periods of performance degradation.

An unexpected, and yet beneficial, set of results from the studies was the preliminary identification of physiological variables which could be normalized across study participants, rather than having to be recalculated for each individual. This finding has significant value for adaptive automation development, as it suggests that neural learning networks to optimize responses to physiological changes may be able to start with a higher level of initial accuracy based on population norms. The strongest evidence for these results comes from the NRaD TTID study, although additional results from UW MATB analyses are still pending.

NRaD TTID Study

Participant physiology data and secondary task performance during the single minute before 58-minute segments of high task loading in the TTID study were analyzed for their predictive ability. As mentioned above, no single physiological site was able to significantly predict all relevant types of performance variability. Pre-task heart rate measures were able to significantly predict failures to acknowledge new aircraft during high task load periods. Other components of performance degradation, such as "cycle time" (time required to initially process new aircraft) or errors in querying aircraft, were linked to EEG measures of pre-task alertness (measured at Pz). An additional important finding of this study was the ability to detect the effects of pre-task state on operators' ability to respond to workload transitions. Heart rate and heart rate variability measures were able to significantly predict a range of performance degradation types, although no one measure was superior overall. The patterns of significant predictors of workload transition performance degradation were similar, but not identical, to significant predictors for the high task load periods in general.

UW MATB Study

Both subjective and objective data from subjects in the UW MATB study were able to identify performance patterns across segments of the 25-minute task. The number of sleep apneas and hypoxias experienced by the subject the night before the task was able to classify participants who continued their level of task performance after a period of low workload (increased automation) from those whose performance degraded after the automation period ended. However, this result was only true for one component of the MATB task: detection of out of range conditions for continuous scale display ("dials") readings. As in the TTID study, physiological measures were linked to performance in relatively short (8-minute) periods of task performance. Subjective measures of overall workload, mental and temporal demands (via NASA TLX ratings) in the last 5-minute task segment were also related to hours of sleep reported by the participant. These measures of perceived workload were not accompanied by reductions of overall task performance, suggesting that the TLX ratings were able to capture the reduced cognitive overhead experienced by fatiguing individuals with diminishing ability to perform complex, integrated tasks. Ongoing

216

analyses are continuing to examine the relationships between sleep apneas, real-time physiology measures, and participant responses to workload transitions.

CONCLUSION

Increased computing power makes detailed physiological data collection possible, supporting capabilities for adaptive automation in complex HMI tasks. Therefore, the primary limitations of linking real-time physiological measures to operator performance capabilities are the complexities of the physiology-performance relationships themselves. Links between physiology and task activity are identifiable, but there are subtle multifactor relationships that depend heavily on the type of task performance and the dynamics of task and workload transitions. With sufficient measurement and event resolutions, SCAMPI studies were able to use as little as one minute of physiological data to predict significant variance in future task performance from 1 to 10 minutes later. Important concerns resulting from SCAMPI projects include the pre-eminence of workload transitions and dynamic cognitive resource allocations, rather than vigilance decrements, as issues for further study. Future system development must emphasize relating vectors of physiology data to vectors of component task performance, based on the goals and success criteria of the tasks themselves.

REFERENCES

Bass, T. (1998). "Dress Code." *Wired, 6* (4, April), pp. 162ff.

Caldwell, B. S., Derjani Bayeh, A., Schein, A., & Hodge-Diaz, T. (1997). Developing supervisory control interface tools for task and workload management and operator response dynamics. In M. Mouloua & J. M. Koonce (Eds.), *Human-automation interaction: Research and practice* (pp. 137-142). Mahwah, NJ: Lawrence Erlbaum Associates, Inc.

Makeig, S., Elliott, F. S., & Posal, M. (1993). First demonstration of an alertness monitoring/management system (Technical Report 93-36). San Diego: Naval Health ResearchCenter.

Kim, H., Young, T. Caldwell, B., & Pionek, J. (1998). Sleep-Disordered Breathing (SDB), daytime sleepiness, and complex monitoring task performance. In M.Scerbo, & M. Mouloua (Eds.), *Automation technology and human performance: Current research and trends.* This volume.

Murray, S. A., & Caldwell, B. S. (1998). Operator alertness and human-machine system performance during supervisory control tasks. In M.Scerbo, & M. Mouloua (Eds.), *Automation technology and human performance: Current research and trends.* This volume.

Roth, E. M., Mumaw, R. J., Vicente, K. J., & Burns, C. M. (1997). Operator monitoring during normal operations: Vigilance or problem-solving? *Proceedings of the Human Factors and Ergonomics Society 41st Annual Meeting (Albuquerque),* 158-162. Santa Monica, CA: Human Factors and Ergonomics Society.

Ryan, T. G., Hill, S. G., Overlin, T. K., & Kaplan, B. L. (1994). Work underload and workload transition as factors in advanced transportation systems (Technical Report EGG-HFSA-11483). Idaho Falls, ID: Idaho National Engineering Laboratory / EG&G Idaho.

Wilson, G. F., Fullenkamp, P., & Davis, I. (1994). Evoked potential, cardiac, blink, and respiration measures of pilot workload in air-to-ground missions. *Aviation, Space, and Environmental Medicine, 65,* 100-105.

The Influence of Ascending and Descending Levels of Workload on Performance

Brooke Schaab
U.S. Coast Guard Research and Development Center

INTRODUCTION

The influence of varying levels of workload on performance has been a prevalent topic in human factors and information processing research. The seminal law of Yerkes-Dodson proposed that both high and low levels of perceived stress degrade performance, with optimal performance occurring at some intermediate level. The findings reported in this study were obtained when a task was counterbalanced between increasing and decreasing levels of workload experienced during a simulated air traffic control task. It was found that performance differed depending upon whether participants experienced increasing or decreasing levels of workload.

TASK DESCRIPTION

Ninety-six undergraduates simulated radar "spotters" in a surveillance aircraft. Their task was to monitor a computer display and identify aircraft in the area as a "friend" or an "enemy." A representation of the task is presented in Figure 1. The center of the computer screen displayed a small red circle representing an aircraft carrier. Black squares, representing aircraft, traveled toward the carrier at a speed of .5 mm/sec. Participants used a mouse to click on these "aircraft." This produced a display box on the screen with four pieces of information: type of aircraft (Jet or Prop); altitude; speed; and identification number. The order of the information remained constant. The participant's task was to use either one, two, three, or four pieces of the information displayed to make a decision on whether the aircraft was a "friend" or an "enemy." For each participant the number of pieces of information used to make a decision (one, two, three, or four) remained constant throughout the task. A copy of the criterion used to make the decision and identify the aircraft was permanently displayed. The number "1" key was pressed if the criteria were met for identifying an "enemy." If the criteria were not met the number "2" key was pressed. Each respondent was exposed to 4, 8, 12, and 16 aircraft that were displayed on the screen for 3 min each, and the number of aircraft were presented in both ascending and descending order for every participant. The order, ascending or descending, was counterbalanced. After 12 min the participants experienced all 4 levels of the number of planes displayed (4, 8, 12, and 16 or 16, 12, 8, and 4). They took a short break and completed the NASA-TLX. The last 12 min of the task repeated the first but the number of aircraft on the screen was presented in reverse order. The TLX was readministered at the conclusion of the last 12 min segment.

Increases in workload were produced through the addition of the number of planes that required monitoring (within subjects). Information load was manipulated by increasing the number of criteria used for identifying the aircraft as "friend" or "enemy" (between subjects).

RESULTS

Significant order-by-number of planes-by-half interactions were found for the number of planes acquired ($F(3,240)=92.31, p<.01$); the number of planes acquired and identified correctly ($F(3,240)=64.48, p<.01$); and reaction time ($F(3,240)=93.03, p<.01$). Overall, performance improved when the number of planes displayed increased in ascending order (4, 8, 12, then 16 planes displayed), with more planes acquired with each increase, more identified correctly, a larger percent correct of those acquired, and a decrease in reaction time. Performance showed a similar pattern in the first and second half, although overall

218

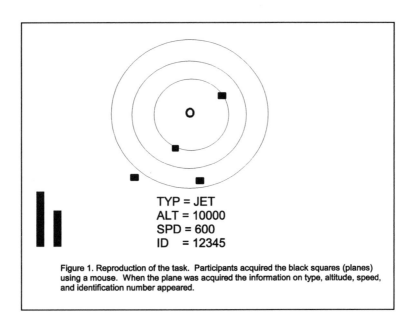

Figure 1. Reproduction of the task. Participants acquired the black squares (planes) using a mouse. When the plane was acquired the information on type, altitude, speed, and identification number appeared.

performance was better in the second half. Performance differed depending on whether the descending order (16-12-8-4) of the planes occurred during the first or second half of the task. If it was encountered during the first half, the number of planes acquired and the number acquired correctly were the lowest when 16 planes were displayed. There was a slight increase with 12 planes displayed and then a gradual decline was seen with 8 and 4 planes. In contrast, when the descending order occurred during the second half, the 16 planes displayed condition resulted in more planes accessed and accessed correctly than in any other condition. Performance than showed a gradual decline with 12, 8, and 4 planes displayed. Regarding reaction time, when the ascending order was presented before the descending order, reaction time decreased with the ascending order and increased with the descending order. If the descending order was presented first, reaction time showed a decrease for 16, 12, and 8 planes, followed by a flattening out of performance.

The poor performance when the task began with the highest number of planes on the screen may have contributed to this 3-way interaction. This does not totally account for the results. It appears that when a task progresses from easy to difficult performance improves. By contrast when a task starts as hard and becomes easier participants initially put more effort into the task but they begin to reduce their effort as the task becomes easier. Fowler (1980) found a similar pattern of performance with air traffic controllers. Errors tended to be more prevalent during periods of low workload that immediately followed a high workload period.

CONCLUSIONS

This research suggests that changes in presentation of the same task from a high to a low level of difficulty or from a low to high level of difficulty can change performance results. Additional study of this issue is necessary before drawing any conclusions.

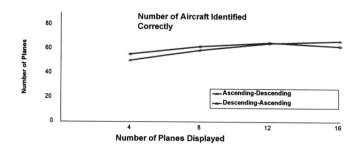

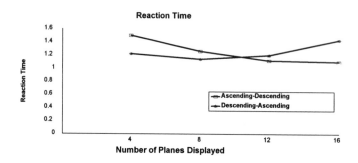

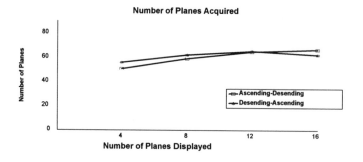

Figure 2. Significant order-by-number of planes-by-half interactions.

REFERENCES

Fowler, F. D. (1980). Air traffic control: A pilot's view. *Human Factors, 22*, 645-654.

DESIGN

AND

INTERFACE ISSUES

Multisensory Feedback in Support of Human-Automation Coordination In Highly Dynamic Event-Driven Worlds

Nadine B. Sarter
Institute of Aviation - Aviation Research Laboratory
University of Illinois at Urbana-Champaign

INTRODUCTION

The evolution of modern technology from reactive tools to powerful and independent agents has created problems which are related to breakdowns in human-automation communication and coordination (Eldredge et al., 1991; Rudisill, 1991; Sarter & Woods, 1994; 1997; Wiener, 1989). Several factors are known to contribute to these breakdowns. Human operators sometimes have an inaccurate and/or incomplete model of the functional structure of the automation which can make it difficult or even impossible for them to predict, monitor, and interpret system status and behavior (Carroll & Olson, 1988, Sarter & Woods, 1994; Wiener, 1989). At the same time, systems sometimes fail to complement operators' expectation- or knowledge-driven information search by providing them with external attentional guidance, especially in the case of events and activities that were not (explicitly) commanded by the user (Sarter, 1995; Sarter, Woods, & Billings, 1997; Wiener, 1989). To address the latter problem, the communicative skills of modern technology need to be improved. They need to be enabled to play a more active role in sharing information with their human counterparts concerning their status, behavior, intentions, and limitations in a timely manner. One promising candidate for achieving this goal is the (re)introduction and context-sensitive use of multisensory feedback. In particular, peripheral visual and tactile feedback appear underutilized but powerful means for capturing and guiding operators' attention and for supporting the parallel processing of considerable amounts of relevant information in highly complex dynamic domains. Benefits and potential costs associated with these feedback mechanisms are currently examined in the context of simulation studies of pilot interaction with modern flight deck technology.

THE PROBLEM

Modern automation technology involves increasingly high levels of autonomy, authority, complexity, and coupling. Systems can initiate actions on their own without input from their operator. They can also take actions that go beyond or even counteract those explicitly commanded by the system user. In some cases, systems can communicate with each other and engage in cooperative activities without operator involvement (see current plans for future air traffic management operations).

Operational experience and research in the aviation domain (e.g., Eldredge et al., 1991; Rudisill, 1991; Sarter & Woods, 1994; 1995; Wiener, 1989) as well as recent incidents and accidents involving advanced technology aircraft (e.g., Dornheim, 1995; Sparaco, 1994) have shown that these system properties and capabilities can create difficulties if they are not combined with effective communication and coordination skills on the part of the automation. For example, pilots are known to sometimes lose track of the status and behavior of their "strong but silent" automated counterparts. The result is automation surprises and mode errors (Sarter & Woods, 1995; 1997; Sarter, Woods, & Billings, 1997).

One recent accident illustrates this potential for mode confusion on highly automated aircraft. In this case, the crew is believed to have selected a particular lateral navigation mode to comply with an ATC clearance. Due to system coupling, their selection had not only the desired but also an unintended effect – a transition in vertical modes, resulting in an excessively high rate of descent. This mode transition and its effect on aircraft behavior went unnoticed -- the pilots did not expect and monitor for it, and the system did not capture their attention or ask for their consent to both transitions. The airplane descended far below the intended flight path and ultimately crashed into a mountainside.

This example (as well as many other cases of breakdowns in human-automation interaction) involves a lack of effective system feedback and a failure of human and machine to explicitly communicate and agree upon their intentions and coordinate their actions. While some kind of feedback on system status and activities is available in most cases, it is presented in a form that does not match human information processing requirements, abilities, and strategies. Current indications of uncommanded mode transitions illustrate this problem quite well. (Changes in) The active mode configuration is currently indicated by so-called flight mode annunciations – cryptic alphanumeric indications (e.g., VNAV SPD or HDG SEL) on the Primary Flight Display. Most systems indicate mode transitions by means of a box that appears around the flight mode annunciation that has just changed. This box disappears after a short time, leaving only the indication of the new active mode. Unless the pilot happens to look at the display at the time of transition, it can be difficult for him/her to notice that something has changed in the system set-up. There are no changes in global features of the display that would support immediate recognition of a transition through pre-attentive reference (Treisman, 1985). Also, the box that is supposed to indicate a transition is not a very salient cue. It does not differ significantly from its immediate surroundings (for example, in most cases, hue, brightness, and saturation of the letters forming the flight mode annunciation (FMA) and of the box around the FMA are very similar), and the surround in which the box appears is not sufficiently homogenous to allow the signal to stand out (Duncan & Humphreys, 1989). Furthermore, the detection of this cue requires a specific head orientation and visual focus which is impossible for pilots to maintain for an extended period of time given competing visual demands. To better support operators in maintaining mode awareness, automated systems need to be enabled to capture and guide operators' attention in a timely and efficient manner.

MULTISENSORY FEEDBACK IN SUPPORT OF MODE AWARENESS

One way to address the problem of breakdowns in human-automation coordination is to exploit the various different sensory channels available to humans by providing tactile, auditory, and peripheral visual feedback. Recent trends in feedback design appear to move in the opposite direction, thus aggravating already existing difficulties. Some sources of tactile and peripheral visual feedback that were available on earlier aircraft (e.g., moving throttles, trim wheels) have been replaced by yet more visual indications of the underlying processes -- a trend that has been shown to create difficulties for pilots in certain circumstances (Folkerts & Jorna, 1994, Last & Alder, 1991). Pilots complain that they need to focus on and read everything instead of being able to pick up information at-a-glance and/or in parallel with other ongoing activities.

To address these complaints and actual difficulties, we need to better understand the benefits, disadvantages, and compatibility of the various sensory modalities that are available to human operators and combine them to create an integrated suite of feedback mechanisms. To date, most research and development has focused on visual and auditory feedback. Two other modalities that are (increasingly) underutilized on the flight deck and in most other domains have received far less attention -- peripheral visual and tactile information presentation. Numerous findings from basic research suggest that tactile and peripheral visual cues can be processed in parallel with other sources of information and represent powerful means of capturing attention (Yantis & Hillstrom, 1994; Theeuwes, 1994; Zlotnik, 1988). In fact, it has been suggested that 90% of all visual stimulation is obtained in the periphery (Malcolm, 1984), helping us perceive the destination of our next saccade and thus serving attentional (re)orientation and control (McConkie, 1983). Tactile feedback is underutilized in most operational settings even though "the skin is rarely ever busy" and although tactile feedback is omni-directional – an important benefit in information-rich environments where operators often have to divide their visual attention across numerous displays. The under-utilization of tactile feedback may be explained, in part, by the fact that only recently, the development of small tactors has made tactile feedback a feasible option for many operational domains.

To date, most empirical evidence on the effectiveness of multisensory feedback has been collected in the context of rather Spartan laboratory settings. The few attempts to introduce peripheral visual displays to the aviation domain were limited to the task of vehicle guidance and control. Examples of such displays are instrument landing aids (e.g., PVD) or peripheral attitude indicators (e.g., Malcolm horizon). To determine the ability of this form of feedback to support other tasks and functions such as attention capture, studies are currently under way in our laboratory to determine whether peripheral visual and vibro-tactile feedback are sufficiently salient and can be reliably detected amidst competing stimuli in an in formation-rich highly dynamic environment such as the modern flight deck. Potential costs and limitations (e.g., attentional narrowing in highly demanding situations) associated with these forms of feedback are being examined. Finally, we are exploring pilots' ability to not only detect *that* a transition occurred but to also identify its *nature* based on variations in the nature of feedback cues (e.g., location, duration).

CONCLUSION

The evolution of modern automation technology from reactive tools to highly independent and powerful agents brings with it the need for more effective human-machine communication and coordination. However, most advanced systems do not possess the communicative skills that are required to know when and how to share information with their human operators. To date, most efforts to address the resulting difficulties with human-automation coordination focus on the modification or addition of yet more visual displays or on new approaches to operator training to support the formation of mental models and thus system monitoring. Operational experience shows that difficulties continue to exist, and visions for future operations (e.g., highly flexible air traffic operations involving an increased number of human and machine agents) suggest that they may become even more frequent and severe. To prevent this from happening, new means of supporting communication and cooperation need to be explored. One possible approach -- the (re)introduction and context-sensitive orchestration of multisensory feedback -- was discussed in this paper as a promising candidate for moving towards more human-centered feedback design (Billings, 1997) and the creation of truly collaborative human-machine teams where both human and machine partners play an active role in keeping each other informed about events and changes in their intentions and actions.

ACKNOWLEDGMENT

The preparation of this manuscript was supported, in part, by a research grant from the Federal Aviation Administration (96-G-043; Technical Monitors: Dr. Eleana Edens and Dr. Tom McCloy).

REFERENCES

Billings, C.E. (1997). *Aviation Automation: The Search For A Human-Centered Approach*. Hillsdale, N.J.: Lawrence Erlbaum Associates.

Carroll, J.M. & Olson, J.R. (1988). Mental Models in Human-Computer Interaction. In M. Helander (Ed.), *Handbook of human-computer interaction* (pp. 45-65). Elsevier Science Publishers.

Dornheim, M.A. (1995). Dramatic Incidents Highlight Mode Problems in Cockpits. *Aviation Week and Space Technology, January 30*, 6-8.

Duncan, J. & Humphreys, G.W. (1989). Visual search and stimulus similarity. *Psychological Review, 96*, 433-458.

Eldredge, D., Mangold, S., & Dodd, R.S. (1991). *A review and discussion of flight management system incidents reported to the Aviation Safety Reporting System*. (DOT/FAA/RD-92/2). Washington, D.C.: Federal Aviation Administration.

Folkerts, H.H. & Jorna, P.G.A.M. (1994). *Pilot Performance in Automated Cockpits: A Comparison of*

Moving and Non-Moving Thrust Levers. NLR Technical Report 94005 U. Amsterdam, The Netherlands: National Aerospace Laboratory.

Last, S. & Alder, M. (1991). *British Airways Airbus A-320 Pilots' Autothrust Survey.* Paper presented at Aerospace Technology Conference and Exposition, Long Beach, CA, September.

Malcolm, R. (1984). Pilot disorientation and the use of a peripheral vision display. *Aviation, Space, and Environmental Medicine, 55(3),* 231-238.

McConkie, G. (1983). Eye movements and perception during reading. In K. Rayner (Ed.), *Eye movements in reading: Perceptual and language processes.* New York: Academic Press.

Rudisill, M. (1991). *Line pilots' attitudes about and experience with flight deck automation: Results of an international survey and proposed guidelines.* In R.S. Jensen (Ed.), Proceedings of the 8th International Symposium on Aviation Psychology. Columbus, OH.

Sarter, N.B. (1995). *'Knowing When To Look Where': Attention Allocation on Advanced Automated Flight Decks.* In R.S. Jensen (Ed.), Proceedings of the 8th International Symposium on Aviation Psychology. Columbus, OH.

Sarter, N.B. & Woods, D.D. (1994). Pilot interaction with cockpit automation II: An experimental study of pilots' model and awareness of the flight management system (FMS). *International Journal of Aviation Psychology, 4(1),* 1-28.

Sarter, N.B. & Woods, D.D. (1995). "How in the world did we ever get into that mode?" Mode error and awareness in supervisory control. *Human Factors, 37(1),* 5-19.

Sarter, N.B. & Woods, D.D. (1997). Teamplay with a powerful and independent agent: A corpus of operational experiences and automation surprises on the Airbus A-320. *Human Factors, 39(4),* 553-569.

Sarter, N.B., Woods, D.D., & Billings, C.E. (1997). Automation surprises. In G. Salvendy (Ed.), *Handbook of human factors and ergonomics* (2nd edition). New York, NY: Wiley.

Sparaco, J. (1994). Human Factors Cited in French A320 Crash. *Aviation Week and Space Technology, (1/3/94),* 30.

Theeuwes, J. (1994). Stimulus-driven capture and attentional set: Selective search for color and visual abrupt onsets. *Journal of Experimental Psychology: Human Perception and Performance, 20(4),* 799-806.

Treisman, A. (1985). Pre-attentive processing in vision. *Computer Vision, Graphics, and Image Processing, 31,* 156-177.

Wiener, E.L. (1989). *Human factors of advanced technology ("glass cockpit") transport aircraft.* (NASA Contractor Report No. 177528). Moffett Field, CA: NASA-Ames Research Center.

Yantis, S. & Hillstrom, A.P. (1994). Stimulus-driven attentional capture: Evidence from equiluminant visual objects. *Journal of Experimental Psychology, 20(1),* 95-107.

Zlotnik, M.A. (1988). *Applying electro-tactile display technology to fighter aircraft - Flying with feeling again.* In Proceedings of the IEEE 1988 National Aerospace and Electronics Conference NAECON 1988 (pp. 191-197). New York: IEEE Aerospace and Electronics Systems Society.

The Development of an Automated Monitoring System Interface Associated with Aircraft Condition

John E. Deaton and Floyd Glenn
CHI Systems, Inc.

INTRODUCTION

The work to be described here concerns initial efforts to develop aircrew interfaces for mechanical fault information that will be provided from advanced sensors and processors. The ultimate goal of this program is to build an aircrew interface that optimally aids the aircrew in managing in-flight mechanical emergencies. The approach adopted by the current researchers to develop the aircrew interface is to sequence through the following paradigm:

- Identify information and control needs of aircrew for optimal fault management,
- Identify all relevant information that can be supplied by the best available sensors and signal processors, and then
- Build the intelligent interface system that forges the gap between the information and control needs and the information that can be supplied by sensors and signal processors by applying appropriate information processing technology.

The key to a successful implementation of mechanical diagnostic capabilities within the aircraft crew-station environment is effectively integrating and presenting the information derived from these sensors in a manner that is timely, useful, and readily interpretable by the aircrew. To accomplish this task, a better understanding of the demands required for aircrew to perform within the aircraft environment is a necessity. In other words, the aircrew interface design problem consists of determining what information to present to the aircrew and how to present it.

In order to determine what information should be presented to the aircrew, it is necessary to establish both what information the aircrew needs and wants and also what information can feasiblely be generated to satisfy their needs/desires. Since essentially none of the currently generated advanced sensor outputs are suitable for aircrew presentation due to the complexity of interpretation required, this process is expected to warrant iterative refinement. As a starting point, it is important to identify the general types of information that the aircrew needs, then try to characterize the kinds of information that can be generated in the identified categories. Next, one must assess the utility to the aircrew of those specific kinds of information, etc. It is assumed that all information that might be presented to the aircrew would have to be generated via some kind of aiding algorithm which would use advanced sensor data as its primary input. Such algorithms could vary from extremely simple ones like current chip detector lights which just indicate that some anomaly has been detected, to very sophisticated algorithms which would tell the aircrew precisely what to do to respond optimally to the detected problem.

Advanced mechanical diagnostic technologies are emerging from the Navy scientific and technology community that soon will allow both advanced ground-based diagnostics and mobile data access and onboard real-time processing of data to accurately determine the health of aircraft mechanical systems (summary reviews of this new technology are available in Stevens, Hall, & Smith, 1996; Parry, 1996; Marsh, 1996, and Nickerson, 1994). Through the use of a combination of sensors, software, and displays, it is now possible to track component wear and fatigue trends, monitor for conditions that indicate impending failure, and alert the aircrew. Furthermore, this tracking and monitoring capability will allow accelerated wear and fatigue trends to be identified, so that the aircraft can be flown within its design parameters.

This new diagnostic technology, sometimes called HUMS (Health and Usage Monitoring Systems), has been used by both UK and Norwegian operators during helicopter ferrying services to North Sea oil platforms. Through implementation of this capability in North Sea operations, improvements in flight safety have been realized (Chamberlain, 1994). HUMS systems have allowed detailed views of rotor system track and balance, power train, drive train and aircraft structural conditions.

The U.S. Navy has recently conducted an Air Vehicle Diagnostic Systems (AVDS) program (Chamberlain, 1994; Nickerson & Chamberlain, 1995) to investigate the potential benefits of HUMS technology to improve safety and maintenance in aging Naval rotorcraft fleets (particularly H-53, H-46, and SH-60) and to facilitate the Navy's transition to a paradigm of "condition-based maintenance" (CBM). CBM is intended to replace the current usage-based maintenance policy whereby major maintenance actions and component replacements are scheduled according to recorded aircraft component usage (typically flight hours, operating hours, takeoffs, landings, etc.) as compared to expected component usage life. Since usage-based maintenance policies are necessarily quite conservative, a conversion to a CBM policy is expected to produce significant savings in maintenance costs along with an improvement in aircraft safety. The AVDS program was designed to produce new technologies that detect flight critical failure modes onboard in real-time. Such technologies must also target high maintenance cost drivers, reduce vibration and increase dynamic component and avionics life.

In order to develop an empirical perspective on aircrew information requirements, two studies were conducted with CH-46 aircrews at the Naval Air Station (NAS) North Island using a motion-base simulator in order to: (1) determine procedures currently used by aircrew in mitigating in-flight mechanical systems emergencies, (2) determine the decision processes employed by aircrew in resolving these emergencies, (3) gather aircrew evaluations of new mechanical fault diagnostic technologies, and (4) determine how information available from emerging mechanical diagnostic systems could best be displayed to aircrew to predict and mitigate these emergencies. The goal of the first investigation was to conduct systematic interviews with flight crews regarding their information requirements for using new diagnostic systems to predict and mitigate in-flight mechanical system emergencies (see Deaton, Glenn, Federman, Nickerson, Byington, Malone, Stout, Oser, & Tyler, 1997(a), and Deaton et al., 1997(b) for a summary of that study). A second study, initiated during the writing of this paper, also utilized aircrew from NAS North Island. The goal of this second study was: (1) to develop a clear conception of the kind of information that we believe we can eventually offer to rotorcraft aircrews via a real-time, embedded intelligent agent integrated with HUMS on one side and an aircrew interface on the other, (2) to demonstrate that this potential information is of significant value and interest for the target aircrews, and (3) to demonstrate that we can effectively support presentation of and interaction with this information in an acceptable aircrew interface.

STUDY METHODOLOGY

Participants of both studies consisted of eight CH-46 crews from NAS North Island. Each crew consisted of an aircraft commander, a copilot, and a crew chief. Both studies required aircrews to fly a pre-determined mission scenario in the CH-46 simulator. The mission scenario was representative of operational missions currently flown by the aircrew. During the scenarios, major mechanical system failures were introduced, with primary and secondary indications (e.g., engine lube pump failure, imminent transmission failure, an engine sprag clutch failure, an impending fuel pump failure, etc. and several false alarms). The first study did not include an actual HUMS system. In this case, participants were asked how mechanical diagnostic information could potentially aid them in avoiding or mitigating these failures. Data were gathered from both interviews as well as questionnaires. The second study, however, included information provided to the aircrew via an interface not currently available in the cockpit (this interface was called HINTS for HUMS Interface Systems). Questionnaire and debrief data will assess aircrew attitudes and value assessments to distinguish between the relative values of different kinds of information. The second study also was designed to determine what we can about aircrew reactions to the new HINTS interface.

The following section summarizes what we have learned thus far in the program. Much of what is discussed was abstracted from the first North Island study. Additional information will be made available in the very near future as data are analyzed from the second North Island study.

SUMMARY OF RESULTS

- Aircrew are most concerned about the reliability of such new diagnostic systems; if confidence in the system is less than they determine to be satisfactory for decision-making, they might resort to mitigating mechanical faults much as is currently done and utilize such technology only as a secondary indicator
- Crewmembers expect this new system to make the most potential contributions to the area of fault diagnosis; it was regarded as a system that would aid in reducing the time that is currently spent in diagnosing mechanical failures; it has the potential for reducing, if not eliminating, misdiagnoses, troubleshooting, and internal communications
- HUMS has been regarded to be of greater value to the less experienced crewmember, the crewmember that might not have had to deal with a specific mechanical problem in the past
- There was controversy associated with the amount of information that should be provided by such a system; some pilots would like this system to provide them with as much detailed information as possible, others only wanted to be informed of the specific fault in more general terms
- The prime impact that this technology would have on the prediction of mechanical emergencies was identified as the advantages that early detection would have on the continuation or abortion of a mission; if this system provides the crew with an indication of the amount of time remaining before critical failure, the crew will be able to reach mission-related decisions sooner and with less risk to personnel and equipment
- HUMS was perceived to have added value for its positive effect on workload, stress, and fatigue for all crewmembers
- Aircrew identified very specific action recommendations that they would expect of such a system; providing crews with action recommendations have the advantages of reducing problem-related communications, decreasing workload, and reducing the lag time between action determination and action initiation
- Aircrew were relatively prolific in their thoughts and suggestions regarding interface design issues; as a result, a preliminary interface design has been developed and will be used in the second North Island study (see next section)

PRELIMINARY INTERFACE DESIGN

The pilot station is a 10.4-inch, TFT Color, LCD monitor with touch panel that is connected to and driven by a Pentium 133 laptop computer. By touching the screen where a button is displayed, the pilots can select the various screens that provide fault analysis, secondary indications and action recommendation information as it becomes available to the HINTS interface. The touch screen will be located on the glare shield in front of the magnetic compass. This display will be used in place of the existing Master Caution Advisory (MCA) panel, which will be hidden from aircrew view. The crew chief station will be a Pentium 133 laptop computer, containing the same analysis, secondary indications and action recommendations screens as the pilot station. This station is completely independent of the pilot station, and allows the crew chief to view different information from that displayed on the pilot's screen. A touch panel is not included. The crew chief uses a mouse attached to the computer for selections.

When the pilot and crew chief stations are turned on a modified MCA panel will be the initial screen on the display. This MCA display has yellow indicator lights which maintain the same functions of the current MCA panel with regards to alerting the aircrew of a problem. Some of the indicator lights have been moved to allow for added indicators. Blue indicator lights deal with HUMS detected problems in the #1 and #2 engines, drive train and forward and aft rotor heads. In addition to the HUMS indicator lights, we added a white Instrument Check button. As an initial test for the system, the user may select any of the indicators to obtain a current system status. For example, the user may select #2 Engine. Text information will be displayed stating that conventional and HUMS sensors detect no abnormalities in the engine.

When a fault is detected the associated indicator light will flash red. The user will select the indicator to obtain the information selection screen. The purpose of this screen is to simply provide the user with a choice of data they wish to view with regards to the fault. The text block shows a historical record of the fault as it progresses. Below the text block are the buttons that will allow the user to either return to the MCA screen, or obtain additional data related to the abnormality. These data include HINTS analysis information, fault secondary indications and HINTS generated action recommendations. The user can return to the MCA screen at any time, and if there are new alerts on the MCA, this button will flash red. The analysis, secondary and recommendations screens cannot be accessed until data are available. As the information becomes available, the associated buttons will flash red. Once information becomes available to a particular screen, the user can return to that screen at any time.

When the analysis button flashes, the user may select the button to display the HINTS analysis screen (see Figure 1). At the top of the screen, the WCA indicator title will be shown to allow the user to track which abnormality, assuming multiple alerts, is being investigated. The text block will display information that can include HUMS information, conventional sensor indications and what could be the source of the problem. The details button will allow the user to obtain more qualitative information on the information displayed in the text block. As this information becomes available, the details button will flash red. The user will select the button and enter the screen that will provide the information either in graphical or text format. The MCA, secondary and recommendation buttons will flash red as new information becomes available.

When the secondary indications button is selected, the HINTS interface will display a screen similar to the analysis screen. When the fault is first detected this screen will provide the user with secondary information that will allow for the development of a confidence level. This information will include, but not be limited to, expected secondary indications and impact on overall aircraft operation. As the secondary indications are detected, the screen will automatically update to display the data. Details will provide more qualitative information on the secondary (e.g., oil pressure 20 psi below normal operating range, rate of pressure decrease is 1 psi/minute, etc.) as it becomes available. The MCA, analysis and recommendation buttons will flash red when new information becomes available.

The recommendation screen (see Figure 2) is a single text block that will provide action recommendations that will assist the aircrew in dealing with the problem. These recommendations can include actions to extend aircraft flight (e.g., single engine flight envelop data) and applicable NATOPS

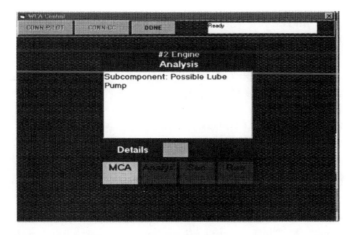

Figure 1. HINTS Analysis Screen.

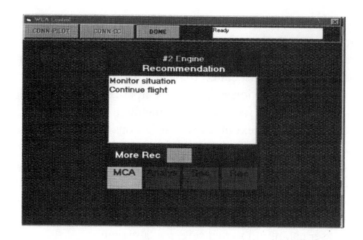

Figure 2. HINTS Recommendations Screen.

procedures. As more information is obtained on the fault, this information will automatically update to deal with the situation as it is understood at that point. Users have the option of going directly to the action recommendations screen if they desire and bypass all other information. The MORE button will allow the user to progress to the next page to view further action recommendations. MCA, analysis and secondary buttons will flash as new information becomes available.

CONCLUSIONS

The investigations reported here represent an important first step towards identifying the informational requirements of aircrew using advanced sensor technology to determine the status of the mechanical systems on their aircraft. These informational requirements have provided the basis for the initial development of an effective interface utilizing advanced sensor technology that will hopefully improve aircrew safety and enhance mission effectiveness.

REFERENCES

Chamberlain, M. (1994). U.S. Navy pursues air vehicle diagnostics research. *Vertiflite*, 40(1), 16-21.

Deaton, J.E., Glenn, F., Federman, P., Nickerson, G.W., Byington, C., Malone, R., Stout, R., Oser, R., & Tyler, R. (1997a). Mechanical fault management in Navy helicopters. *Proceedings of the 41st Annual Meeting of the Human Factors and Ergonomics Society*. Albuquerque, NM: Human Factors and Ergonomics Society.

Deaton, J.E., Glenn, F., Federman, P., Nickerson, G.W., Byington, C., Malone, R., Stout,R., Oser, R., & Tyler, R. (1997b). Aircrew response procedures to in-flight mechanical emergencies. *Proceedings of the 41st Annual Meeting of the Human Factors and Ergonomics Society*. Albuquerque, NM: Human Factors and Ergonomics Society.

Marsh, G. (February 1996). The future of HUMS. *Avionics Magazine*, 22-27.

Nickerson, G. (1994). *Conditioned-based maintenance: A system perspective*. Presented at the 1994 ASME/STLE International Tribology Conference and Exhibition, Lahaina, HI.

Nickerson, G., & Chamberlain, M. (1995). *On-board diagnostics: More than a red light on a panel*. Presented at the Future Transportation Technology Conference, Irvine, CA.

Parry, D. (February 1996). Evaluating HUMS. *Avionics Magazine*, 28-32.

Stevens, P., Hall, D., & Smith. E. (1996). *A multidisciplinary research approach to rotorcraft health and usage monitoring*. Presented at the American Helicopter Society 52nd Annual Forum.

Performance Effects of Display Incongruity in a Digital and Analog Clock Reading Task

J. Raymond Comstock, Jr.
NASA Langley Research Center

Peter L. Derks
College of William and Mary

INTRODUCTION

In an era of increasing automation, it is important to design displays and input devices that minimize human error. In this context, information concerning the human response to the detection of incongruous information is important. Such incongruous information can be operationalized as unexpected (perhaps erroneous) information on which a decision by the human or operation by an automated system is based. In the aviation environment, decision-making when faced with inadequate, incomplete, or incongruous information may occur in a failure scenario.

An additional challenge facing the human operator in automated environments is maintaining alertness or vigilance (Comstock, Harris, & Pope, 1988). The vigilance issue is of particular concern as a factor that may interact with performance when faced with inadequate, incomplete, or incongruous information. From the literature on eye-scan behavior we know that the time spent looking at a particular display or indicator is a function of the type of information one is trying to discern from the display (Harris, Glover, & Spady, 1986). For example, quick glances are all it takes for confirming that an indicator is in a normal position or range, whereas a continuous look of several seconds may be required for confirmation that a complex control input is having the desired effect. Important to consider is that while an extended look takes place, visual input from other sources may be missed. Much like an extended look, the interpretation of incongruous information may require extra time.

The present experiment was designed to explore the performance consequences of a decision making task when incongruous information was presented. For this experiment a display incongruity was created on a subset of trials of a clock reading laboratory task. Display incongruity was made possible through presentation of "impossible" times (e.g. 1:65 or 11:90). Subjects made "same" "different" decisions and keyboard responses to pairings of Analog-Analog (AA), Digital-Digital (DD), and Analog-Digital (AD), display combinations. For trials during which display incongruities were not presented, based on prior research (Miller & Penningroth, 1997) comparing digital and analog clock displays, it would be expected that the Digital-Digital condition would result in the shortest response times and the Analog-Analog and Analog-Digital conditions would have longer response times. The performance consequence expected on trials with incongruous times would be very long response times.

METHOD

Subjects

Twenty university students participated in the experiment. The median age of the subjects was 21.5 years, and there were 10 male and 10 female subjects. All subjects reported normal or corrected to normal vision. Subjects were paid for their participation in the experiment.

Apparatus and Stimuli

The time comparison task consisted of a left and right time display presented simultaneously on the CRT screen. The time displays could be both digital, both analog, or a combination of the two. When digital, the displayed times were digits one cm in height showing the hour and minutes (e.g. 9:30), and were based on a twelve-hour clock. The distance between the center of the left and right time display areas was 10.5 cm. The font was a modified bold Arial with serifs on the numerals "1" and "7". When analog, a round clock-face was displayed with a 7.7-cm diameter round dial with hands showing the hour and minutes. The long hand was 4.4 cm (3.6 cm from center) and the short hand was 2.5 cm. These hands were tapered with a maximum width of approximately 0.25-cm near the center of the clock face. Twelve large tick marks were shown at each hour position and smaller tick marks were shown at the minute increments. Numerals were not displayed on the analog clock face. The digital time display numerals were white. The analog clock face and tick marks were yellow and the clock hands were white. The background for the entire screen was dark gray. The time displays were presented on a 35.56 cm (14 inch) diagonal Magnavox VGA computer monitor at a screen resolution of 640 X 350 pixels. Viewing distance (subject to monitor) was about 46 cm. The software generating the stimuli and recording response times and accuracy was running on a Compaq 386-20 desktop computer.

Procedure

Once seated in front of the computer monitor, subjects were shown examples of the analog and digital time displays. Then instructions for the "Time Display Comparisons" task were presented both on screen and verbally by an experimenter. The instructions stated that "The task you are about to begin will present two Time Displays near the center of the screen." "You should press the **S** key if the times shown are the Same, or the **D** key if the times shown are Different." There were a total of 96 trials with an inter-trial interval of about three seconds. Prior to each new time display comparison a countdown digit at the center of the screen signaled two seconds and one second before the next presentation. This was included to direct attention to the center of the screen which would otherwise have been blank between presentations.

All displayed times, for both digital and analog presentations, were in even five minute increments (e.g. 2:10 or 1:35, never 1:32). Display incongruity was made possible through presentation of "impossible" times (e.g. 1:65 or 11:90). Display incongruities were present on 16 of the 96 trials in the Digital-Digital, Analog-Digital, and Digital-Analog conditions. Subjects were not briefed on the possibility of seeing "impossible" times because responses to incongruous or unexpected information was one of the foci of the study.

RESULTS

The initial analyses addressed differences between the analog and digital display conditions when no incongruity trials were present. As illustrated by the mean response times (RT) shown in Figure 1, the Digital-Digital display condition RT is significantly shorter than for the Analog-Analog display condition ($F(1,18)=13.49, p<.01$). In addition, the cases in which the displayed times were different in each of these display conditions resulted in significantly longer RTs ($F(1,18)=10.87, p<.01$). As can be seen in Figure 1, these longer times for detecting a "different" condition did not carry over to the Analog-Digital condition where still longer RTs were found. Since there were no differences found or expected between Analog-Digital and Digital-Analog conditions (swapping of positions left and right) the results from both of these conditions are reported as the Analog-Digital condition. As would be expected, the Analog-Digital condition did reflect significantly longer RTs than the Analog-Analog condition ($F(1,18)=33.91, p<.01$). Analyses by subject gender showed no differences in RT in any display condition.

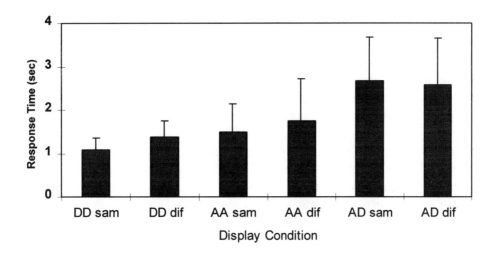

Figure 1. Response Time and St. Dev. by display condition (sam=same, dif=different).

The RTs for the first five Analog-Digital incongruity trials are shown in Figure 2. As would be expected, these RTs showed a much longer (slower) response time (a mean of over 4 seconds) to the initial incongruity exposure (Trial 11).

These incongruity trials differ significantly ($F(4,85)=15.08$, $p<.01$), and Duncan post hoc tests ($p<.05$) showed a significant difference between trials 11 and 20 and between these two and subsequent incongruity trials. As can be seen in Figure 2, subsequent incongruity trial response times gradually began to look like standard trials as these trials were no longer viewed as incongruous to the subjects.

Incongruity trials were also presented in the Digital-Digital condition. These trials did not result in the very long RTs associated with the Analog-Digital condition. Perhaps this is because the subject could do a simple pattern match as opposed to the mental transformation required when comparing the analog and digital displays. There were no Analog-Analog incongruity trials in this experiment because of the way the incongruity trials were generated.

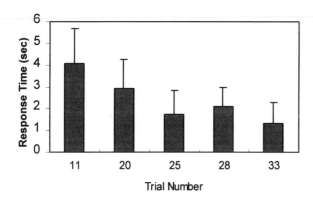

Figure 2. Response Time and St. Dev. On Analog-Digital Incongruity Trials.

DISCUSSION

The present experiment showed that the shortest response times were found for "same / different" comparisons made for Digital-Digital time displays for young (median age 21.5) subjects. Significantly longer RTs were found for Analog-Analog comparisons, and still longer times for Analog-Digital comparisons. The longer RTs for the Analog-Digital case are most likely explained by additional mental operations involved in conversion of one of the two times before the "same / different" judgement can be

233

made. In the case of the Digital-Digital or Analog-Analog comparisons a rapid pattern match would be all that was needed. The additional time required of the analog display condition may reflect the added complexity of the clock-face and hands relative to the simple digital numbers.

Trials containing an "incongruity" consisting of an "impossible" time on the digital side of the Analog-Digital pairing resulted in markedly longer RTs. As subjects saw subsequent presentations of these anomalous cases the RTs gradually became shorter. It is interesting to note the length of the initial incongruity RTs. The means exceeded four seconds. In this experiment there were no other tasks to perform therefore long decision times did not have a penalty with regard to other tasks. It would be interesting to explore incongruities in a multitask environment where long RTs on one task would negatively impact other tasks.

The present experiment, although using a different type of discrimination task, confirms the findings of Miller and Penningroth (1997), and showed that in general digital display comparisons were done more quickly by the young subjects used in both studies. It would be interesting to conduct the same analog and digital display condition experiments with older subjects whose early clock reading experience was all with analog clocks.

In the present experiment all incongruity trials consisted of an "impossible" time on one of the digital time representations on selected trials. It is possible to create an analog incongruity in future experiments by reducing or eliminating the length difference between the long and short hands on the clock face. In cases in which clock hand length discrimination is made more difficult, longer response times and increased error rate would be expected.

The importance of the findings of this experiment are threefold: (1) display incongruity can be created in a laboratory task environment, (2) much more time is required for subjects to process an unexpected incongruity, and (3) there is a time cost for comparisons made in which mental conversions are required, as in the Analog-Digital and Digital-Analog pairings in the present experiment. While all information was presented visually in this study, the latter finding concerning mental conversion or transformation of information has wider implications. One such case would be detection of errors when pilots receive air-traffic-control information aurally but this information is entered and checked visually.

REFERENCES

Comstock, J. R., Jr., Harris, R. L., Sr., & Pope, A. T. (1988). Physiological assessment of task underload. *Second Annual Workshop on Space Operations Automation and Robotics (SOAR '88),* NASA CP-3019. Washington, D.C.: National Aeronautics and Space Administration, 221-226.

Harris, R. L., Sr., Glover, B. J., & Spady, A. A., Jr. (1986). *Analytical techniques of pilot scanning behavior and their application.* NASA Technical Paper 2525, NASA Langley Research Center, Hampton, Virginia.

Miller, R. J. & Penningroth, S. (1997). The effects of response format and other variables on comparisons of digital and dial displays. *Human Factors, 39*(3), 417-424.

Beyond Supervisory Control:
Human Performance in the Age of Autonomous Machines

Elizabeth D. Murphy and Kent L. Norman
University of Maryland

INTRODUCTION

For over 30 years, the National Aeronautics and Space Administration (NASA) has been placing unmanned spacecraft in low-Earth orbit for various scientific purposes. Over the years, the role of human operators in these systems has evolved from semi-manual to supervisory control (see, e.g., Moray, 1986; Sheridan, 1976, 1997). Although supervisory control distances the operator from the moment-to-moment operations, it still requires that operators be present to respond to alerts from the system. Until very recently, ground monitoring-and-control of these missions has required the support of human operators on a 24-hour-a-day basis. As advances in technology make it possible to build spacecraft that can operate even more autonomously than their predecessors, a continuous human presence in the control room is no longer needed.

With advances in the technology of ground operations, the required monitoring and control can be handled to a large degree by automated systems in a "lights-out" (or "hands-off") unattended context. In the full lights-out concept, operations would be run autonomously, around the clock, seven days a week. In a partial lights-out environment, there might be one daytime shift staffed by human operators/analysts. At the end of that shift, control-room personnel turn the lights off and go home. When a problem develops that is outside the capabilities of either on-board or ground systems, however, a human expert (or team of experts) is called upon, via a paging system, to perform fault diagnosis and resolution.

Little or no attention has been paid to the cognitive issues associated with the full lights-out mode of operation. The primary issue addressed in the present research is that of human performance, i.e., how to bring the human into the full situational context quickly after a period of prolonged detachment from a particular mission. Given that full context can be provided, a plausible view of the future holds that human operators will be neither monitors nor supervisors of autonomous systems, but consultants who are called in on an exception basis to solve problems beyond the scope of the automation. The present research is designed to investigate implications of that hypothesis.

AUTONOMOUS OPERATIONS

In the unmanned spacecraft domain, high levels of autonomy are already being introduced, both onboard and on the ground (e.g., Abedini, Moriarta, Biroscak, Losik, & Malina, 1995; Aked & Pylyser, 1996). These advanced software capabilities take engineers and analysts beyond supervisory control into a new paradigm, where they are still part of the system but are needed only when the "intelligent" automation calls for human help. They do not need to be physically present at any particular support facility, and they are not tasked with monitoring mission operations.

Autonomous operations depend on the design, development, and continuing operation of software processes both onboard the spacecraft and embedded in the ground operations system. Onboard capabilities may include automatic safeing of the spacecraft in response to unknown conditions and rule-based fault detection and resolution, such as those long in place on the Voyager 1 spacecraft (Sawyer, 1998). Alternatively, monitoring and fault detection might also be performed within the ground operations system by "surrogate controllers," that is, automated processes that perform functions formerly performed by human ground-control operators (Truszkowski, 1996). These processes are also known as "software agents" (e.g., Bradshaw, 1997). Agents developed at NASA-Goddard are distinguished from other automated processes

235

by the following characteristics (Truszkowski, 1996):

Autonomy – the agent performs its functions in the background, without requiring human interaction for routine functioning

Migration – the agent can shift some of its task load across distributed nodes

Cloning/Spawning – the agent can make a copy of itself to support parallel processing

Collaboration – the agent interacts with other agents and with users

Learning – the agent builds up its knowledge base from case-by-case experience and from user feedback, showing improved performance over time

In the Lights-Out Ground Control Operations System (LOGOS), currently under development at NASA's Goddard Space Flight Center, software agents acting as surrogate controllers perform various, interrelated functions, such as accessing the schedule of scientific observations and spacecraft communication events; entering archival information into the database; preparing scripts for spacecraft command-and-control activities; as well as fault isolation and resolution. The human analyst or a team of human analysts is paged by the paging agent when LOGOS is unable to resolve an unforeseen anomaly detected by the Fault Isolation and Resolution Expert (FIRE) agent. Although ground-control personnel may check on spacecraft status at any time from a remote location, weeks or months may elapse between anomalies that require human expertise and intervention.

OVERVIEW OF THE LIGHTS-OUT ANALYST'S ROLE

The LOGOS user interface is designed to support both browsing of system status and active resolution of anomalies. In *browse mode*, the analyst or mission manager reviews the operations of spacecraft and ground systems. The analyst may be interested in recent interactions among agents, especially if a fault condition or anomaly has been handled by the team of agents. In such a case, the analyst is likely to be interested in the agent system's justification for its decisions and for any actions taken on spacecraft subsystems. The analyst may be interested in the current status of subsystems in the context of a specified number of orbits or a specified time period. Historical data and trends may be of interest in helping the analyst develop a sense of how various subsystems have been performing. Reports may be consulted on any human interventions that have occurred, including identifying information, reason for the intervention, and outcome of the intervention. The analyst's purpose in browsing is to gain assurance that mission operations are proceeding as they should be, both on the spacecraft and on the ground, according to past and current command-and-control schedules.

Active mode is invoked when the analyst has been paged to respond to some condition that cannot be handled by on-board or ground automation. (It is important to remember that the anomaly may be on the ground or, if on board, may be caused by some fault(s) in the ground system.) The first phase of activity is focused on assessing the situation (i.e., gathering information about the condition of on-board and ground subsystems to confirm the existence of a fault and, then, to identify/isolate the fault). The analyst builds his or her situation awareness through a process of examining the current status of on-board and/or ground subsystems. The analyst may also want to refer to historical data and other reference material. The analyst determines whether the fault/anomaly reported actually exists, whether there is some other fault/anomaly, whether the system has issued a false alarm, or whether there is no anomaly. In the second phase, the analyst proceeds to determine a course of action if an anomaly has actually occurred. The candidate course of action is tested through the spacecraft simulator and implemented if simulation is successful. The analyst completes and submits a report on the anomaly and actions taken to correct it.

COGNITIVE ISSUES IN AUTONOMOUS OPERATIONS

When the human analyst is paged, presumably after a prolonged period of non-involvement with a particular orbiting mission, what can be expected of human performance? As described in the overview, the analyst's tasks will involve confirming the anomaly reported by the agent community, problem solving to resolve the anomaly, and formulation of a command set to correct the anomaly. Going boldly beyond supervisory control, into the realm of agent-based, autonomous machines, requires consideration of the implications of this new paradigm:

- *Situation Awareness*: If the human expert has not been monitoring the operation of the automation, what information will (s)he need and how long will it take to develop an adequate understanding of the situation? Is it possible to develop adequate understanding without continuous monitoring?
- *Potential for Overreliance on Automation*: How will being called upon "out of the blue" to deal with an emergency affect the system engineer's ability to resolve problems? To what extent will the system engineer be likely to accept a software agent's diagnosis and recommended action, if any, without verifying the nature of the problem?
- *Information Display Techniques*: How should information be displayed? How effective are various graphical techniques? Is textual/columnar data adequate, or is performance enhanced by graphical displays?
- *Influence of Agents*: How will the presence and recommendations of so-called "intelligent agents" affect human performance?

These issues are under investigation in a field study at NASA's Goddard Space Flight Center and in an experimental setting, the Laboratory for Automation Psychology (LAP) at the University of Maryland.

CHANGING PARADIGM, CHANGING METAPHOR

In the supervisory-control paradigm, it made sense to talk about the operator's being "out of the loop." Terminology from manual control (i.e., closed loop, open loop) provided a natural metaphor. The operator was said to be "out of the loop" if he or she was monitoring a process but not manually controlling it. Now, however, autonomous operations are taking the operator/analyst completely out of the loop for extended periods of time. Does it make sense to continue to use the phrase "out of the loop" to describe this situation? Continuing use of this framing metaphor may constrain our thinking and lead us off on unproductive or counterproductive design paths.

A better metaphor or analogy for the paged analyst may be to the physician who is an expert diagnostician but who does not see a particular patient for six months or more, if the patient is healthy. This analogy, the medical model, requires a shift in thinking that parallels the shift from supervisory control operations to autonomous operations. Physicians do not routinely monitor patient status, but work in the paged mode as needed, consulting with the patient's history and running various tests to ascertain the status of subsystems. Assuming that an adequate context can be provided via visual and, perhaps, audio display technology, it should be possible for the analyst to become situationally aware in a short time. The analyst in a lights-out environment will essentially serve as an expert consultant/troubleshooter, a role different from those of the process operator or the supervisory controller. The required level of expertise will not come cheaply, but will be used rarely.

USABILITY STUDIES

To provide situational context to the paged analyst, NASA-Goddard has developed a number of visualization techniques within the Visual Analysis Graphical Environment (VisAGE). Available visualizations include 2-dimensional (2D) and 3-dimensional (3D) representations generated by test data. For example, the set of visualizations includes both 2D and 3D alphanumeric representations of the data, 2D and 3D histograms, 2D and 3D strip charts, and a 2D pie chart.

During planned usability studies, operational personnel will interact with the VisAGE tool and respond to items on a questionnaire. Participants will be asked to describe the information they consult in isolating and resolving anomalies in their own operational environments and to answer factual questions based on the visualizations displayed to them. These questions will test the effectiveness of the various visualizations in conveying information to an operator/analyst in a simulated lights-out context. To assess their levels of trust in automation, participants will be pre- and post-tested on their attitudes toward automation. A similar set of evaluations will be conducted on the VisAGE tool with undergraduate students at the University of Maryland. Measures of all participants' performance will include the accuracy of extracted information and response time.

EXPERIMENTAL RESEARCH

A research environment under development in the LAP mimics some of the LOGOS capabilities. Known as MOCHA (Mars Observer Calls Home Again), it assumes the existence of an agent community that is capable of monitoring spacecraft operations and of paging a human expert when necessary. In experimental trials to be conducted in the coming months, three independent variables will be manipulated: 1) the level of automated aid provided to the participant, who plays the role of the off-site analyst; 2) the selector of the visual representation(s) viewed by the participant; and 3) agent reliability.

Automated aid will have three levels: 1) one 2D representation of the data; 2) one 3D representation of the data; and 3) a 2D and a 3D representation of the data. The selector will be the participant (self) or a simulated software agent (automated). Agent reliability will vary between 50% and 75% diagnostic accuracy. Each participant will be assigned at random to an experimental condition. The task will be to confirm/disconfirm the existence of the agent-reported anomaly and to assign a level of confidence in each confirm/disconfirm decision. Dependent variables will include accuracy in confirming or disconfirming the presence of an anomaly, response time, and level of confidence in the decision to confirm or disconfirm.

It is expected that the two-representation condition will support better performance than will either of the one-representation conditions, but only if the 2D and 3D displays provide different information. It is expected that results for the dependent measures will correlate significantly with trust in automation and spatial visualization ability (SVA), both of which will be assessed before the experimental trials. SVA will be assessed by a standard test administered on line, and trust will be assessed by questionnaire. It is expected that software agents will facilitate performance when participants report high trust in the automation.

SUMMARY

Autonomous capabilities in complex, automated systems take the human operator/analyst beyond supervisory control into a new role, that of off-site, expert troubleshooter. This new paradigm requires a shift in framing metaphors, from out-of-the-loop process control to the medical model. Consideration of cognitive issues will help to ensure adequate design support for the paged analyst. Research underway at NASA-Goddard and the University of Maryland includes usability studies and experiments designed to investigate human performance issues in the context of autonomous spacecraft operations. Results of this research will provide feedback to software developers on the effectiveness of data- visualization techniques and initial findings on the effects of agent-based software on human performance.

ACKNOWLEDGMENTS

This work is supported by a grant from the NASA-Goddard Space Flight Center to the University of Maryland (NAG5-3425). Walt Truszkowski is the technical monitor.

REFERENCES

Abedini, A., Moriarta, J., Biroscak, D., Losik, L., & Malina, R. F. (1995). *A low-cost, autonomous ground station operations concept and network design for EUVE and other earth-orbiting satellites.* Berkeley, CA: Center for EUV Astrophysics, Technology Innovation Series (Publication Number 667).

Aked, R., & Pylyser, E. (1996). An operational concept for a small autonomous satellite. *Fourth International Symposium on Space Mission Operations & Ground Data Systems "SpaceOps 96"* (SP-394, Vol. 2, pp. 905-912). Noordwijk, The Netherlands: ESA Publications.

Bradshaw, J. M. (1997). An introduction to software agents. In J. M. Bradshaw (Ed.), *Software agents* (pp. 3-46). Cambridge, MA: AAAI Press/MIT Press.

Moray, N.(1986). Monitoring behavior and supervisory control In K. R. Boff, L. Kaufman, and J. P. Thomas (Eds.), *Handbook of perception and human performance* (Vol. II: Cognitive processes and performance, pp. 40-1 - 40-51). New York: Wiley.

Sawyer, K. (1998). Voyager 1 craft speeds toward edge of sun's influence. *The Washington Post*, March 9, p. A3.

Sheridan, T. B. (1976). Preview of models of the human monitor/supervisor. In T. B. Sheridan & G. Johannsen (Eds.), *Monitoring behavior and supervisory control* (pp. 175-180). New York: Plenum.

Sheridan, T. B. (1997). Supervisory control. In G. Salvendy (Ed.), *Handbook of human factors and ergonomics* (2nd ed., pp. 1295-1327). New York: Wiley.

Truszkowski, W. (1996). "Lights out" operations: Human/computer interfaces/interactions (HCI). 1996 *Technology workshop – Autonomous "lights out" operations workshop: Operational challenges and promising technologies* (Presentation viewgraphs, pp. 355-367). Greenbelt, MD: Mission Operations and Data Systems Directorate, NASA-Goddard Space Flight Center.

Pilot Fault Management Performance: An Evaluation of Assistance

Paul C. Schutte and Anna C. Trujillo
NASA Langley Research Center

INTRODUCTION

Human error has been attributed to over 70% of all aircraft accidents (Boeing, 1996). These errors are often mistakes in reasoning, what Reason calls Rule-based and Knowledge-based errors (Reason, 1990). Most of these mistakes are believed to be due to poor situation awareness (Endsley, 1995). Often the correct recognition of the situation leads almost directly to the correct response (Klein et al., 1993). When flight crews follow procedures that correspond to the specific system malfunction alert provided by an alerting system (e.g., Boeing's EICAS), they are very likely to do the right thing. The number of accidents averted due to this improved rule-based response has not been documented, but the decrease in the accident rate after implementation of improved alerts and their associated procedures is not likely to be coincidental. Thus, airframe and avionics manufacturers have gone to great lengths to account for possible failure modes and to develop corresponding checklists (i.e., procedures).

However, not all failures can be anticipated. One such unexpected event occurred in 1989 when United Airlines Flight 232 lost all hydraulics due to a tail engine failure in the DC-10 (NTSB, 1990). The crew had to determine the most appropriate way to deal with the situation since no procedures existed for this failure. While the crew's performance was heroic, some evidence suggests that they were not aware of the total ineffectiveness of the wheel and column and, most importantly, of the excess drag on one side of the aircraft. The aircraft was using asymmetric thrust to maintain straight and level flight. On landing, the crew reduced the throttles together. This allowed the asymmetric drag to dominate the dynamics of the aircraft which caused it to cartwheel. Had the crew maintained asymmetric thrust, the aircraft might have been able to land safely.

In 1984, a group of researchers at NASA Langley Research Center set out to develop a decision aid that would assist flight crews in dealing with inflight system failures on the aircraft. The concept developed, Faultfinder, provided explicit information for skill-based (i.e., monitoring) and knowledge-based (i.e., model-based) reasoning in order to augment the current rule-based (i.e., procedures) reasoning. Evaluations of the concept proved that it was successful in correctly detecting and diagnosing failures (Schutte, 1989; Shontz et al., 1993). But could the flight crew use this information in an unanticipated and untrained-for scenario? Could this information provide operational value (e.g., savings in equipment, fuel, or time)? The answers to these questions are pivotal in the decision to invest time and money in developing such concepts for commercial use. This paper describes an evaluation of the Faultfinder fault management concept in a full mission, full workload simulation.

FAULTFINDER

The Faultfinder fault management concept was developed to enhance the flight crew's understanding of novel failures (Schutte & Abbott, 1986). Faultfinder addresses two of the four fault management tasks (Rogers et al., 1997) and was instantiated in two computer prototypes: fault monitoring, which determines if an abnormality has occurred; and fault diagnosis, which determines why an abnormality has occurred.

Fault Monitoring

The fault management process is usually triggered by the occurrence of symptoms or events. These symptoms indicate in some degree of detail when the parameter is abnormal rather than normal. To detect abnormal parameters, the fault monitor compares the actual system state (as represented by the sensor readings) to the expected system state (computed from a quantitative model that simulates the normal functioning of the physical system), and defines a symptom whenever there is a predetermined significant difference between them, such as "Fuel-Flow is high." The monitor can also relay event data such as "EGT has exceeded its operational limit". A complete description of the monitor can be found in Schutte (1989).

Fault Diagnosis

The goal of the diagnostic process is to provide information on the origin of a failure and on the effect of that failure on other systems. The effect of a failure can be due to a functional propagation of the failure (e.g., the engine flames out, causing the engine-driven hydraulic pump to stop running) or due to a physical propagation of a failure (e.g., the engine explodes damaging the hydraulic line.) The model-based reasoning process can identify the origin or responsible component of the failure (e.g., the compressor, fuel line). This is accomplished through the qualitative simulation of either functional or physical propagation of the fault from the responsible component. A complete description of the diagnostic process can be found in Abbott (1991).

METHOD

Subjects

Sixteen airline-line pilots from four major US airlines served as test subjects. All subjects had experience in "glass cockpit" aircraft and overwater operations. Two pilots (one for each display condition) were not ETOPS (Extended Twin-engine Operations) qualified but they were qualified for multi-engine (greater than 2) operation over water. The average age of the subjects was 47 years and the average number of years of airline experience was 21. The total flight hours for subjects averaged 14175 hours while the total hours for being pilot-in-command of the aircraft averaged 6412 hours. The subjects participated as Captain/Pilot-Not-Flying on an ETOPS oceanic flight. A confederate first officer well versed in the operation of the simulator acted as Pilot-Flying. He obeyed the Captain; however, he offered no help in decision making.

Apparatus

The Advanced Civil Transport Simulator (ACTS) at the NASA Langley Research Center was used in this experiment. The ACTS is a twin-engine transport class flight-deck simulator. It has a desk-top layout with touch sensitive display system schematics and menus on the center three of five main CRT displays. The Flight Management System (FMS) is similarly implemented with touch sensitive screens on flat panel displays located on the desk-top in front of each pilot. Most of the look and feel of the flight deck displays are similar to a Boeing 747-400. The simulator has performance characteristics similar to a Boeing 757 but it is an all electric aircraft (i.e., no hydraulics) with side-stick controllers instead of wheel and column.

Scenarios

A multiple-fault scenario was used in this experiment. The first fault in the scenario involved the left engine, and the second fault involved a failure in the right fuel system which affected the right engine. The ETOPS scenario began over-water, enroute between Los Angeles and Honolulu. An over-water route was chosen because it would force the crew the manage the fault as opposed to selecting the nearest airport and landing as soon as possible. Standard ETOPS rules applied, such as landing the aircraft at the nearest ETOPS alternate airport with the loss of any primary system (such as an engine). The nearest alternate for the route flown depended on where the aircraft was in relation to the time mid-point between origin and destination.

This scenario began approximately 80 nm east of the DIALO waypoint, which was the time mid-point for the route. The alternate airport west of DIALO was Barbers Point on the main island of Hawaii. The aircraft was headed westbound at an altitude of 37,000 ft. and Mach number of 0.79. The weather throughout the scenario was clear. The autothrottle and autopilot were engaged with both lateral navigation and vertical navigation provided by the FMS.

Approximately 19 minutes into the scenario (58 nm west of DIALO), the left engine on the aircraft flamed out. The aircraft could not maintain altitude on just one engine, and the subject had to descend to under 30,000 ft. in order to restart the engine as the checklist for engine flameout directed. If the subject did not decide to restart the engine by a certain point in the experiment, the experimenter would ask him to state his intentions and then direct him to attempt to relight the engine, balance the fuel tanks, and return to altitude. This procedure insured that the aircraft would be at 37,000 ft. and MACH 0.79, with balanced fuel tanks and with the crossfeed valve closed (tank-to-engine-fuel-feed) prior to the next failure.

At 203 nm west of the DIALO waypoint, a large fuel leak developed in the right engine. There was no checklist for a fuel leak. The leak was located in a fuel line connection downstream of the electronic fuel controller yet upstream of the engine fuel flow sensor. As a result, the fuel controller attempts to pump more fuel into the engine, maintaining constant thrust for about 1 minute. Then the leak overwhelms the controller's capability and thrust starts to decay. The reduced thrust causes the autothrottle system to advance both throttles to maintain airspeed. At this point the size of the fuel leak (plus engine fuel consumption) on the right engine roughly equals the fuel consumption on the left engine. Thus, the tank levels are equal to each other even though thrust and fuel flow are much different between each engine. The only way to stop fuel from leaking from the aircraft is to close the fuel valve leading to the right engine, thus shutting the right engine down.

Display Conditions

There were two display conditions for this experiment. One was a baseline condition where the information elements of a modern "glass cockpit" aircraft were presented. The second condition had the same information with the addition of information elements provided by the Faultfinder fault management concept. These elements were integrated into the baseline display so as to address only information content rather than format. This was seen as a bias against the Faultfinder condition since the displays were not optimized for this information and thus presented more clutter.

Indications–Baseline. When the left engine on the aircraft flamed out, all of the engine parameters reduced to near zero (windmill). This was followed by a left engine oil pressure low caution, an auto throttle disconnect due to the engine flame out caution and both left electrical generators off-line advisory messages. Both cautions were accompanied by the caution aural alert. When the fuel leak developed there were no indications in the flight deck for approximately one minute. After that, both throttles started to advance. Right engine parameters, fuel flow, and oil pressure decreased. The indications remained at about 80% of normal for the conditions and throttle settings. The aircraft could not maintain altitude and airspeed even at full throttle although the decrease was subtle. Fuel quantity was decreasing at the same rate from both tanks but if the right engine was not shut down, the flight management system would alert the crew when the calculated fuel (based on fuel flow) and the measured fuel based on the tank quantity differed by at least 2000 lbs. of fuel (a FUEL DISAGREE alert message). The crew could then elect to use the more conservative estimate based on the tank quantity or stay with the calculated estimate.

Indications–Faultfinder. All indications in the Faultfinder condition were identical to the baseline with the following additions. During the flame out, all of the left engine tapes, with the exception of oil temperature and oil quantity, turned yellow since they were all below the tolerance bars established for the Faultfinder monitoring system. For the flameout, the diagnostic system determined that the failure was in the fuel controller. During the first minute of the fuel leak, the only indications were a monitoring message that fuel usage was abnormally high. Within that minute the diagnostic module of Faultfinder determined that the problem was in the fuel line (responsible component), and that it had propagated to the fuel usage sensor which is located in the fuel controller. After one minute, both throttles started to advance. Right engine parameters, fuel flow, and oil pressure all decreased. These decreases were below the monitoring tolerance level so the parameter displays turned yellow. These out of tolerance parameters were also indicated on the Caution and Warning display as advisory messages. The fuel display showed the fuel line between the fuel shutoff valve and the engine as the responsible component (indicated by a red outline) and that the engine was affected (indicated by a yellow outline). The engine, throttle, and FMS behavior was the same as in the baseline condition.

Data Recording

All aircraft performance parameters and system parameters were continuously recorded. Pilot inputs into the simulator were also recorded. Audio and video tapes were made and transcribed.

Experiment Design

The experiment was a single factor, between-subject design. The display condition was varied between subjects (8 subjects baseline, 8 subjects Faultfinder). The scenario contained two types of failures, one with an appropriate checklist and one without.

RESULTS

The results were categorized into performance (correct or not), time to perform (if correct), and fuel at destination. The performance criteria were correct diagnosis (fuel leak), correct system response (shut down right engine), and correct mission response (divert to alternate). Table 1 shows the performance and time results per condition.

Table 1
Performance Scores

Condition	Correct Diagnosis		Correct System Response		Correct Mission Response	
	number	avg. time	number	avg. time	number	avg. time
Baseline	5	56 min.	3	33 min.	7	40 min.
Faultfinder	5	21 min.	6	27 min.	5	32 min.

Also calculated were the fuel-at-destination scores which depended on the choice of destination (Honolulu or Barbers Point) and whether the subject shut down the engine (a confederate air traffic controller kept them on track so deviations from declared route were minimal). The average fuel at destination was 5039 lbs. for the baseline case and 6321 lbs. for the Faultfinder case.

DISCUSSION

The results showed that providing additional information regarding system status did make a difference in operational performance. The most important action for the subject to take was to arrest the fuel leak; only three subjects in the baseline condition did so as compared to six in the Faultfinder condition. The importance of arresting the fuel leak is evidenced by the higher fuel scores at destination even though more subjects choose to divert in the baseline case than in the Faultfinder case. It was somewhat surprising that there was no difference in the number of correct diagnoses between conditions but the times to a diagnosis indicate that it was not *whether* the subjects determined the cause but *when* they did it. As predicted, when the subjects knew what was going on, they were better able to make the correct system response.

One interesting result not shown above was the response of two subjects in the Faultfinder condition who correctly diagnosed the leak and correctly shut down the engine but did not divert to the nearest alternate as mandated by ETOPS rules (both subjects were ETOPS rated). In the baseline case, all three subjects who shut down the engine chose to divert. This gives rise to the possibility that the additional systems information may affect mission performance. This finding agrees with Rogers' et al. (1996) discussion of the propagation of information between different operational levels. Because this response was unexpected, it was not addressed in the standard experiment debriefing and therefore the reason that the pilots did not divert may never be known. However, one could speculate that the subjects knew that the problem was a leak, that the leak was secured, and that the engine was still good (although handicapped). Therefore, the subsystem was operational, just not operating. This, coupled with the fact that the alternate was only 100 miles closer and the passengers were expecting to go to Honolulu, may have led the subjects to press to destination. This was not necessarily a wise decision given that the right engine had failed earlier in the flight but it is understandable.

Only one subject (Faultfinder condition) felt that the engine failure and fuel loss were related. This subject elected not to shut the engine down because he felt that there was some contaminant in the fuel. Most, but not all, of the information provided by Faultfinder could have supported this hypothesis. When the FUEL DISAGREE message appeared, the subject persevered in his "contaminant" hypothesis even though it did not explain the fuel loss.

In summary, it appears that providing this additional fault management information has operational value even when the displays are not optimized for the information. As Tenney et al. (1997) pointed out in their analysis of the data, the additional information may have assisted the crew in performing more knowledge-based processing rather than simply skill- or rule-based processing. To be sure, there is more research to be done regarding issues such as false alarms and missed alerts, and over-reliance on such a system, however this research and others like it (Trujillo, 1997) demonstrate significant potential.

ACKNOWLEDGMENTS

The authors wish to acknowledge the support of John Barry, Dr. William Rogers, Dr. Yvette Tenney, Capt. Skeet Gifford (ret.), Capt. Dave Simmon (ret.) and Myron Sothcott without whom this experiment would not have been possible.

REFERENCES

Abbott, K. H. (1991). *Robust fault diagnosis of physical systems in operation* (NASA-TM-102767). Hampton, VA: NASA Langley Research Center.

Boeing Commercial Airplane Group. (1996). *Statistical Summary of Commercial Jet Aircraft Accidents, World Wide Operations, 1959-1995.*

Endsley, M.R. (1995). Toward a theory of situation awareness in dynamic systems. *Human Factors, 37(1),* 32-64.

Klein, G. A., Orasanu, J., Calderwood, R., & Zsambok, C. E. (1993). *Decision making in action: Models and methods.* Norwood, NJ: Ablex Publishing Co.

National Transportation Safety Board. (1990). *Aircraft accident report: United Airlines Flight 232 McDonnell Douglas DC-10-10 Sioux Gateway Airport, Sioux City, Iowa, July 19, 1989* (NTSB-AAR-90-06). Washington, DC: National Transportation Safety Board.

Reason, J. (1990). *Human error.* New York: Cambridge University Press.

Rogers, W. H., Schutte, P.C., & Latorella, K.A. (1996). Fault management in aviation systems. In R. Parasuraman & M. Mouloua (Eds.), *Automation and human performance: Theory and applications* (pp. 281-317). Mahwah, NJ: Erlbaum Associates.

Schutte, P. C. & Abbott, K. H. (1986). An Artificial Intelligence Approach to Onboard Fault Monitoring and Diagnosis for Aircraft Applications. In *Proceedings of the AIAA Guidance and Control Conference*, Williamsburg, VA, August 18-20, 1986.

Schutte, P. C. (1989). Real time fault monitoring for aircraft applications using quantitative simulation and expert systems. In *Proceedings of AIAA Computers in Aerospace VII Conference*, Monterey, CA.

Shontz, W. D., Records, R. M., & Choi, J. J. (1993). *Spurious symptom reduction in fault monitoring* (NASA-CR-191453). Hampton, VA: NASA Langley Research Center.

Tenney, Y. J., Pew, R.W., Rogers, W.H. & Salter, W.J. (1997). *Evaluation of Faultfinder, a Decision Aid for Fault Diagnosis in the Glass Cockpit*. Report No. 8191, Contract No. NAS-20653, Hampton, VA: NASA Langley Research Center.

Trujillo, A.C. (1997). Pilot Performance With Predictive System Status Information, *Proceedings in IEEE International Conference on Systems, Man, and Cybernetics*, IEEE Piscataway, NJ.

Blending Theory and Practice:
Design Approaches for Uninhabited Air Vehicles

Thomas C. Hughes and Cynthia D. Martin
Veridian, Veda Operations

Cynthia Dominguez and Michael Patzek
Air Force Research Laboratory/Human Effectiveness Directorate

John M. Flach
Wright State University

INTRODUCTION

For several years we've been working on integrating domain information into the design process. There has been increased emphasis on "up-front" analysis to gain a more complete understanding of the operational domain so subsequent design decisions can be justified based on their impact on system performance within a given mission context. Additionally, pressures to automate new systems to reduce manpower costs and increase system reliability have resulted in design of systems that are extremely difficult for operators to understand and control. Such conditions are often the result of insufficient insight into operational domains and inadequate consideration of the relative contributions to successful operation that human operators and intelligent automation techniques might provide. With the emergence of Uninhabited Air Vehicles (UAV) we run the risk of falling into the same trap. Some suggest that the default philosophy for UAVs is to automate everything. This leaves the operator to monitor the system and intervene as mission situations demand. Such approaches are fundamentally flawed because they fail to recognize and capitalize upon the unique contribution that the human operator provides, particularly in highly complex and dynamic work domains such as those anticipated with UAVs. The present paper describes our approach toward a design process that attempts to incorporate detailed domain information into the design that will hopefully lead to an integrated, highly coupled collaborative system interface for UAV system operators.

One Air Force mission in which UAVs are seen as particularly appropriate is the Suppression of Enemy Air Defenses (SEAD). The objective of the SEAD mission is to neutralize, destroy, or temporarily degrade enemy surface based air defenses by destructive and/or disruptive means. The use of a UAV as opposed to a traditional manned system is a particularly attractive alternative due to the extremely high attrition rates typically associated with the SEAD mission. The application of Uninhabited Combat Air Vehicles (UCAV) in a SEAD mission could effectively employ sensors and weapons in high threat areas where vulnerability of manned aircraft represents an unacceptable risk.

THE UCAV PROBLEM

Piloted systems are extremely complex, dynamic and full of risk. UAVs add to this complexity with the challenge of remote operations. In addition, a UCAV must be viewed as a *collaborative* system in that its application will involve coordination among multiple intelligent agents. These agents will include both human operators and automatic control systems. A UCAV system is an example of what Sheridan (1997) would call a multiple task telerobotics control system. The remoteness of the operator relative to the battle and the autonomous capability of the UCAV are principle challenges relative to effective collaboration, and thus unique challenges for design. The UCAV is also an example of an *adaptive* system because it needs to function in a nondeterministic, dynamically changing environment. Combat is not a static encounter, but a dynamic interaction between two competing agents. The environment that exists when a mission is planned will likely be vastly different from the environment at the time the mission is executed. Plans must be updated, revised, and/or discarded in response to a dynamically changing context.

The need to be adaptive to dynamically changing situations is another design challenge. Traditional techniques toward crew systems analysis (i.e., mission decomposition, functional flow, task analysis, etc.) represent "linear" or deterministic approaches toward assessment of the mission, focusing on rigid sequencing of crew activities given various mission situations. A weakness with this approach is its failure to capture the adaptive nature of human behavior when interacting with complex systems (i.e., changing strategies and a range of approaches for accomplishing a task or mission). Such approaches also tend to oversimplify such human behavior, placing assumptions on the analysis that eliminate many of the qualities of operator behavior upon which system performance ultimately depends. The design of interfaces to complex systems within such domains must not only provide an effective representation for predictable events, but must also support operators when required to adapt to these unexpected events. How does one design for the unexpected? This requires a perspective that considers the widest range of possible events. To deal with unexpected events, cognitive systems engineering tends to focus on constraints or boundary conditions, rather than on events or activities.

COGNITIVE SYSTEMS ENGINEERING

Rasmussen's *cognitive systems engineering* framework is uniquely suited to the collaborative, adaptive requirements of the SEAD mission. A unique attribute of the cognitive systems engineering approach is an attempt to integrate across the component analyses to achieve a global understanding of a distributed control system. Top down relations along the diagonal in Figure 1 help to uncover the underlying intentionality of the work domain, why things are done or the reasons why things are significant. Bottom up relations along the diagonal help to uncover the causal structure within the work domain. This causal structure provides a production like description of how things can be accomplished within the work domain. These causal relationships can be critical in the design of automated systems for the well behaved (rule-based or routine) aspects of the work that can free up human resources to be focused on supervising the "messier" aspects of the work.

Within the cognitive systems engineering approach the top-down and bottom-up relations are complementary perspectives that contribute to a complete understanding of the work space. A primary goal of the cognitive system engineering approach is to make the links along this diagonal explicit within the system design. Unlike classical and modern approaches to control systems that begin with restrictive assumptions (e.g., linearity, a quadratic cost functional, etc.) cognitive systems engineering focuses on the assumptions themselves. Classical and optimal control make assumptions that turn messy problems into well defined problems that allow quantitative solutions. But cognitive systems engineering is all about the "messiness," because that is why operators are included in most control systems --- to deal with the messiness of unanticipated variability (Flach et al., in press).

The approach adopted for our analysis of the SEAD domain incorporated many of the traditional activities associated with crew systems development. These activities were modified to reflect the cognitive systems engineering perspective. The Crew-Centered System Design Process (CSDP) provided a baseline process upon which to development a detailed analysis plan. Developed through the Air Force's Crew-Centered Design Technology program, the CSDP is a systematic, thorough and inspectable process for crew system development. We have integrated activities associated with this process within the Rasmussen's CSE framework as presented in Figure 2, and discussed in greater detail in the following sections. In addition to traditional analytical techniques, such as mission decomposition and functional flow analysis, cognitive task analysis techniques such as concept mapping, critical incident analysis, and critical decision methods were also applied to gain direct insight into the cognitive characteristics of system and operator behaviors (Martin, Hughes, & Williams, 1998).

247

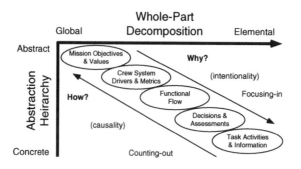

Figure 1. An illustration of how different aspects of a cognitive task analysis fit within the overall analysis space defined by the abstraction and decomposition dimensions. Adapted fromRasmussen, Pejtersen & Goodstein, (1994).

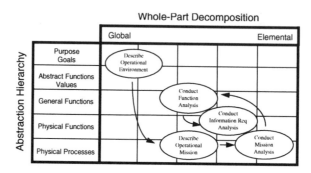

Figure 2. CSDP Activities mapped to Rasmussen's cognitive systems engineering framework.

Domain Analysis

Describe Operational Environment. In performing the domain analysis we conducted a broad review of the SEAD mission, its goals and purpose. We looked first at the operational initiatives and how the SEAD role is currently executed. Our analysis focused first on a review of available literature to gain insight into the SEAD objectives, specific weapon systems associated with SEAD and specific tactics for accomplishing SEAD objectives. Sources for this information included Air Force Manual 1-1, AF Doctrine, system technical manuals that provide detailed descriptions of systems and their performance capabilities, Dash-3 Tactics Manuals that describe the application of these systems within the context of operational missions, and other source materials available in the general literature dealing with the SEAD mission. A review of these materials provided a basic understanding of the SEAD mission, insight into some of the critical operational issues associated with SEAD, and specific performance characteristics of selected Air Forces assets applied within the SEAD mission.

Armed with this broad understanding of the SEAD mission, we contacted available subject matter experts (SME) and conducted preliminary interviews to augment information available in the literature. Given that UCAV systems currently do not exist, we drew upon the expertise of individuals with SEAD experience in piloted systems such as F-4G Wild Weasel, and the F-16 HARM Targeting System (HTS). The interviews focused on their experiences in performing SEAD missions, insights into the operational environment and issues associated with performing SEAD functions. As a result, we learned a great deal about the purpose and goals of SEAD. First, unlike traditional combat missions, the measure of success within the SEAD domain is not necessarily "kills per sortie," but rather the degree to which safe passage was afforded to friendly forces being supported. Current systems accomplish this by denying enemy acquisition and guidance radar the ability track friendly aircraft. Advanced anti-radiation missiles (ARM) use enemy radar signals as beacons for weapon guidance. If a threat system emits in an attempt to track a friendly aircraft, the ARM will home in on the source of the radar emission. Knowing this, air defense system operators will shut down to avoid being tracked. Therefore, whether the ARM actually hits the surface threat, it has successfully accomplished its mission by denying the enemy system's ability to acquire and track friendly aircraft. In the electromagnetic battlefield, denial of information is key mission survivability. Therefore, the *goal of SEAD is to deny and disrupt enemy access to the electromagnetic information upon which their air defense systems are based.* The development of systems to support the SEAD mission must maintain this goal above all others to ensure system success.

Activity Analysis

Describe Operational Mission. The domain analysis was conducted to increase our understanding of the behavior-shaping constraints that define the SEAD domain, independent of any specific situation or work context. By contrast, the goal of activity analysis was to investigate work behavior within the context of specific SEAD mission contexts and situations. By analyzing specific sets of work situations and problems to be resolved, the analysts gains detailed insight into the functional relationships between elements present in the domain space.

Based on insight gained during the preliminary domain analysis, system analysts, with the assistance of an in-house operations expert, we described the operational mission by generating a detailed mission scenario. Using the Extended Air Defense Simulation (EADSIM), a mission level, computer-based simulation model, a mission profile was created. The mission profile provided a graphical representation of the proposed mission that outlined threat and target laydowns, identified mission players (F-16 Block 50 aircraft equipped with the High-speed Anti-Radiation Missile (HARM) Targeting System (HTS), Rivet Joint, and F-15 Strike aircraft) and notional aircraft routings. The development of the mission profile provided the context, in both space and time, against which specific mission situations could be investigated in more detail. Accompanying the mission profile was a mission narrative that provided a notional narration of the mission profile in a "There I was . . ." form.

Mission/Function Analysis. A functional decomposition of the SEAD mission was conducted to establish the hierarchical relationship between high-level mission phases and segments and their supporting functions, sub-functions, and tasks. The mission event/function sequence was developed from the integration of the information obtained from SMEs using 1) concept mapping techniques and 2) critical incident analysis methods to elicit domain knowledge. Based on *the function analysis* of SEAD mission functions a series of critical mission attributes were identified (See Table 1). While the focus of the mission decomposition was "what" is done, the emphasis during the SME interview session was placed on "why" specific actions are taken and how specific events or cues drive decision making behavior.

Information Requirements Analysis. To further address information requirements associated with mission situations, SME were probed using a critical decision method (Klein, Calderwood & MacGregor, 1989) to elicit specific insight into information supporting decisions and how decisions are influenced by this information. Table 2 includes many of the probes used to elicit information from our mission experts. The probes are broken down into two categories, one focusing on situational assessment and information associated with these assessments, and the other with specific decisions and characteristics of those decisions.

Throughout our sessions, it became apparent that mission planning represents a critical component to ensure mission success. Our mission experts indicated that the mission plan represents a point of departure. The plan, as developed, will be executed until the situation evolves into a condition to which the original plan is ill-equipped to handle. It is for this reason that situation assessment is so critical. The

Table 1
Critical SEAD mission attributes

Critical Mission Attributes	Description
Stimulating the Environment.	• Accurate location of threat systems are dependent upon threat emissions, operator must decide how to stimulate system to emit
Fluid/Reactive Targeting	• Operator must adapt to change in the threat environment. Expectation based on intelligence brief must be reevaluated as situations emerge
SAM Launch Reaction.	• A success SEAD tactic is to force threats to engage SEAD assets while disengaging strike assets, operators must respond to such engagements
IR/AAA Threats.	• A primary threat to SEAD assets are passive threat systems that do not emit EM energy. SEAD assets must be able to avoid, detect, and evade such threats
System Malfunctions.	• System operator must remain aware of system health and effect of system malfunctions on capability and mission success.
Decoy Site/Anti-HARM Tactic.	• In an attempt to distract and confuse the SEAD operator, decoy emitters are often deployed. SEAD operators must assess emissions to determine their validity.
Airborne Re-Planning.	• Based on an evolving battlefield, the SEAD operator must be able to conduct in-flight mission replanning in response to unknown or unanticipated events.

assessment is conducted to determine the extent to which the mission situation *maps* to the situation anticipated during mission planning. If mapping to original plan is poor, then the crew must modify the plan to respond to the dynamics of the mission situation. Furthermore, the process of mission planning is critical to the ability of crews to effectively adapt to situation dynamics. One SME reported that the primary purpose of mission planning is to avoid surprise. Survival is dependent upon "staying ahead of the game" and not being caught off guard.

Effective mission planning is the means by which crews prepare for the unanticipated, develop contingencies, and contemplate situational possibilities and alternative responses. The characteristics of system behavior toward which cognitive systems engineering is focused during system design, are addressed operationally during mission planning. For this reason, mission planning represents an extremely fertile environment for gaining insight into the dynamics of the operational realm.

Table 2
Probes used during critical decision method session with SEAD subject matter experts.

Category	Probes
Situations Assessment	• What is being assessed? • Why is it being assessed (e.g., decision, monitor, coordinate, etc.)? • How is assessment conducted (e.g., information, inferences, assumptions, expertise, etc.)? • What is the product of the assessment? • What actions result from the assessment (e.g., decisions, reactions, etc.)?
Decision Making	• What is the decision to be made? • What are potential options? • What are relevant clues to guide the decision maker? • How does the decision maker assess the quality of the decision? • What are the consequences of a sub-optimum decision? • Are opportunities available to modify or correct decisions?

CONCLUSIONS

We report here on work in progress towards applying cognitive system engineering to the design of UAVs. Our focus has been to understand the SEAD mission. Lack of existing UCAVs performing the SEAD mission calls for a blending together of knowledge elicited from pilots performing this mission with other aircraft, knowledge elicited from UAV operators performing other missions, in depth analysis of material related to the SEAD mission, and knowledge from other experts with insight into other domain areas that will also serve to shape the application UCAVs within the SEAD mission (i.e., technologists, IADS analysts, threat system operators, etc.). There are no simple recipe-based solutions to the design of these systems, just as there are no simple steps that an operator will follow to ensure SEAD mission success with UCAVs. Rather, we are iteratively attempting to identify and analyze the boundary conditions that will define these systems. Application of these findings to spiral processes of modeling, simulation, and evaluation within realistic SEAD scenarios will be necessary.

REFERENCES

Flach, J., Eggleston, R., Kuperman, G. & Dominguez, C. (In press). *SEAD and the UCAV: A Preliminary Cognitive Systems Analysis*. AFRL Technical Report. Wright-Patterson AFB, OH.

Klein, G.A., Calderwood, R., & MacGregor, D. (1989). Critical decision method for eliciting knowledge. *IEEE Transactions on Systems, Man and Cybernetics, Vol 19*(3), 462-472.

Martin, C.D., Hughes, T.C., & Williams, R. (1998). *Lethal Uninhabited Air Vehicle (UAV) Preliminary Information Requirements Report*. Veridian Report No. 63482-97U. Wright-Patterson AFB, OH.

Sheridan, T.B. (1997). Speculations on future relations between humans and automation. In R. Parasuraman & M. Mouloua (Eds.), *Automation and human performance: Theory and applications* (pp. 449 - 460). Mahwah, NJ: Erlbaum.

Rasmussen, J., Pejtersen, A., & Goldstein, L. (1994). *Cognitive systems engineering*. New York: John Wiley & Sons, Inc.

TECHNOLOGY

AND

AGING

Monitoring Automation Failures:
Effects of Age on Performance and Subjective Workload

Dennis A. Vincenzi and Mustapha Mouloua
University of Central Florida

INTRODUCTION

Automation can be defined as the execution by a machine agent (usually a computer) of a function that was previously carried out by a human (Parasuraman & Riley, 1997). The development and application of highly reliable automated systems in today's world has changed the role of the human from an active system operator to one of a passive system monitor, a role for which humans are not well suited (Parasuraman, 1997). The introduction of automation technology in aviation systems has resulted in many benefits such as increased safety and greater fuel efficiency. Automation technology is advancing at a phenomenal rate and shows no indications of slowing across a variety of human-machine systems. While some types of failures and errors have been eliminated by automation, new types of failures and errors have been enabled (Billings, 1997). Increased automation has resulted in several behavioral problems such as automation-induced complacency, increased workload and loss of situation awareness as evidenced by both ASRS and NTSB reports as well as numerous experimental studies

Aging

With the onset and improvement of medical technologies, the average life span of the individual has increased dramatically. The fastest growing segments in the population today are individuals over the age of 65. Currently there are 23.5 million people in the United States over the age of 65. This group has almost doubled in size since 1900. The "85 and over" group has increased in size by 17 times since 1900. This "85 and over" group is the segment of the elderly population that is responsible for the rapid growth in size of the elderly population in the United States (Small, 1987). It is quite common for an individual to reach the age of 65 in the U.S., but to live past the age of 85 was not so common until the recent advent of improved medical science and technologies.

It is estimated that by the year 2025, approximately 20% of the U.S. population will be 65 years of age or older (Harbin, 1991). Certain areas relevant to aging, mostly physiological in nature, have been researched quite thoroughly. One aspect of aging that is not well documented or understood is how people will react, adapt and develop with respect to cognitive ability in this new high tech automated world that is taking shape around us. Along with the development and advancement of life prolonging medical technologies, advanced electronics and automation, individuals who will be living longer must develop new skills and adapt to these new technologies that are thrust upon them.

The "Age 60" Rule

The "Age 60" rule was placed into effect almost 40 years ago. The FAA has endorsed and supported the "Age 60" rule since its inception in 1958 even though it has no justification to do so. The "Age 60" rule does not permit pilots over the age of 60 to be in control of any commercial aircraft. This means that upon turning 60 years of age, pilots employed in commercial aviation can no longer hold the position of pilot or copilot. It is a "Catch 22" in that pilots must retire at the age of 60 regardless of physical health or past performance. When the FAA is asked questions concerning empirical evidence obtained on pilots over age

60 that would support these actions, this evidence is not available because all commercial aviation pilots must retire at the age of 60. There is presently no empirical data available to support this rule; therefore this FAA mandate is still in dispute.

Previous research using a single task and dual task paradigm has indicated that aging had no effect on performance (Vincenzi, Muldoon & Mouloua, 1997). This study was designed to examine whether performance differences related to age would be obtained under high task load (multi-task) conditions. The present study also sought to evaluate perceived workload levels experienced by both young and old adults. It was hypothesized that the older group would exhibit poorer performance than the younger group. It was also hypothesized that the subjective workload experienced by the older group would be greater than the subjective workload experienced by the younger group

METHOD

Participants

A total of 24 participants, 12 between the age of 18 and 25 (mean age = 20.25), and 12 over the age of 65 (mean age = 65.42). The 12 young adults were recruited from the undergraduate population at the University of Central Florida. The 12 old adults were recruited from LIFE (Learning Institute Foe Elders) at the University of Central Florida. All participants had normal (20/20) vision or corrected to normal vision and did not have prior experience with the flight simulation task.

Materials

The experiment used a revised version of the Multi-Attribute Task Battery (MAT) developed by Comstock and Arnegard (1992). The MAT is a multi-task flight-simulation package of the component tasks of compensatory tracking, system monitoring, resource management (fuel management), communications, and scheduling. The modified version that was used included only the compensatory tracking, system monitoring and resource management. The monitoring task consisted of a simulated automated system that kept track of temperature and pressure for two simulated engines. The compensatory tracking task involved maintaining a floating cross hair on a central target. The task was designed such that if the cross hairs were placed in the center and left alone, they would drift and require the participant to constantly adjust and manipulate the position of the cross hairs in order to maintain a central position. The resource management consisted of a fuel management task where the participant had to manipulate and maintain fuel levels in two sets of fuel tanks, one set for each simulated engine. The fuel levels in the tanks had to be maintained at specific levels, and therefore required continual attention in order to achieve this task. The MAT was run using a 486DX 66 MHz IBM compatible computer, 14 inch VGA color monitor, and an analog joystick. The three tasks were displayed in separate windows on the monitor.

Design

Upon arrival to the lab, all participants were required to complete a biographical questionnaire consisting of demographics, medical health, medications used, etc. Each participant was subjected to a 10-minute manual training session and three 30-minute automated sessions. The 10-minute training sessions consisted of complete manual control of tracking, resource management and system monitoring, whereas the three 30-minute sessions consisted of manual tracking, manual resource management, and automated system monitoring. The participants were required to achieve a minimum detection rate of 50% on the monitoring task before they were allowed to proceed to the actual three 30-minute automated sessions, thereby insuring a minimum proficiency standard among all participants. The experiment lasted approximately two hours (10-minute manual training session, three 30-minute simulation sessions, and debriefing).

RESULTS

System Monitoring Performance

The mean malfunction detection rates obtained for the elderly and young adults in the manual (training) condition were 69.67% and 75.5%, respectively. These means do not differ significantly when analyzed with an ANOVA indicating that our efforts to insure equivalent knowledge and task performance between both young and old participants was successful. The mean malfunction rate in the manual condition (72.58%) was significantly higher than the mean malfunction detection rate in the automated condition (48.58%), $F(1,44) = 16.77$, $p < .001$, indicating that overall, individuals perform better on the task of monitoring when the system is under manual control as opposed to when the system is under automated control and the operator is simply a passive monitor of the automated system. These findings agree with previous research done in this area.

The mean malfunction detection rates obtained for the manual session, 1st automated session, 2nd , automated session and 3rd automated session were 72.58, 51.17, 45.92 and 48.54, respectively. The main effect of session was significant, $F(3,66) = 11.85$, $p < .05$, indicating that as the time under automated control increased, detection of system malfunctions changed significantly. As time under automated control increased, detection of automation failures decreased overall.

The mean malfunction detection rates obtained for the main effect of age group was 63.1 and 46.0 for the young and old groups respectively. These means differ significantly, $F(1,22) = 4.42$, $p < .05$, indicating that the younger age group performed significantly better overall than the older age group with regard to correctly detecting system malfunctions.

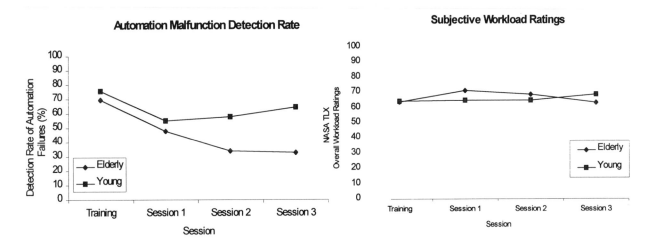

Figure 1. Mean Detection Rate of Automation Failures as a Function of Age Group and Session.

Figure 2. Mean subjective workload ratings Over sessions.

The mean malfunction detection rates obtained for the elderly group in the manual session, 1st automated session, 2nd automated session and 3rd automated session were 69.67, 47.58, 33.83 and 32.92 respectively. The mean malfunction detection rate obtained for the young group in the manual session, 1st automated session, 2nd automated session and 3rd automated session were 75.50, 54.75, 58.00 and 64.17

respectively. A significant interaction between session and age group was obtained, $F(3,66) = 3.15, p < .05$, indicating that as the time under automated control increased, the behavior of the young and old age groups grew increasingly different. From looking at a graph of the means in Figure 1, we can see that the elderly age group shows an immediate drop in detection rate of automation failures from training to session 1. The elderly group then exhibits another drop in the detection rate in session 2 and appears to level off in session 3. This drop is indicative of the classic vigilance decrement exhibited by participants in previous research. The young age group exhibits a decrease in detection rate from training to session 1, a slight rise in detection rate in session 2 and another rise in detection rate in session 3.

Subjective Workload

Even though the performance for the young and old age groups differed significantly, no significant differences in subjective workload were obtained in this experiment (Figure 2). Subjective workload for both groups across sessions remained relatively unchanged.

CONCLUSIONS

The first hypothesis, that older individuals would perform poorer than younger individuals, was supported. The detection rate of both age groups grew increasingly different over time. Detection rate of automation failures was higher for the younger group than for the older group. Detection rate increased significantly for the younger group as a function of time on task, whereas the older group showed a decline over time. Performance cost of automation-induced complacency is more pronounced in the older group than the younger. The second hypothesis that the subjective workload experienced by the older group would be significantly greater than the subjective workload experienced by the younger group was not supported. Subjective workload does not vary as a function of age. Both the older and younger group experienced comparable high levels of subjective workload. Performance under automation control was significantly lower than under manual control. This confirms previous work on automation-induced monitoring inefficiency (Parasuraman, Molloy, & Singh, 1993). Even though the detection rate of the younger group was significantly higher than for the older group, monitoring of automated systems remains poor regardless of age group. The older group exhibited the expected performance under high workload conditions but the younger group began to improve as a function of time on task indicating that more cognitive resources were available in the younger group or the younger group was able to allocate cognitive resources more efficiently.

Reducing the Performance Gap

One proposed method of maintaining high levels of monitoring performance lies in proper implementation of adaptive task allocation or adaptive automation. Adaptive automation refers to a system capable of dynamic, workload-triggered reallocations of task responsibility between human and machine (Hilburn, Jorna, Byrne, & Parasuraman, 1997). The traditional approach to allocation of functions in a human-machine system was originally determined by a dichotomous relationship, which gave full control of a task either to the human, or to the machine. In today's technological society it is debatable whether some of the functions that need to be performed are better executed by a human or by a machine (Fitts, 1951). According to proponents of adaptive systems, the benefits of automation can be maximized and the costs minimized if tasks are allocated to automated subsystems or to the human operator in an adaptive, flexible manner rather than in an all-or-non fashion (Rouse, 1988). An ideal situation would be one where the operator could switch the control of a task from manual to automated when workload conditions are high. Once the operator's workload was reduced, the operator could then continue to perform the task in manual mode, thereby maintaining familiarity with the system and preserving the operator's cognitive ability and baseline skill level.

Directions for Future Research

One direction for future research would be to solicit experienced young and old pilots and subject them to similar high workload, multi-task situations. A serious potential confound with most research performed in this area is that the subject pool includes a mixture of both pilots and non-pilots. Restricting the subject pool to pilots only would eliminate this potential confound and allow meaningful examination of interactions between age, workload and expertise. A potential application of this research would be to determine if the FAA's "Age 60" rule has any merit. The FAA has enforced the "Age 60" rule which effectively prevents a pilot from flying a commercial aircraft as of age 60. If older and younger pilots were to exhibit similar performance on monitoring of automation failures, this could potentially refute the validity of the "Age 60' rule. As mentioned previously, since the fastest growing segment of the population is the elderly population and proliferation of automation and technology shows no indication of slowing, further research in this area is needed.

REFERENCES

Billings, C.E. (1997). *Aviation Automation: The search for the human-centered approach.* Mahwah, NJ, Erlbaum.

Comstock, J.R. & Arnegard, R.J. (1992). *Multi-attribute task battery.* (NASA Technical Memorandum 104174). Hampton, VA. NASA Langley Research Center.

Fitts, P.M. (1951). Engineering psychology in equipment design. In S.S. Stevens (Ed.), *Handbook of experimental psychology* (pp. 50-56). Hillsdale, NJ: Erlbaum.

Harbin, T.J. (1991). Environmental toxicology and the aging visual system. In D. Armstrong, M.F. Marmor & J.M. Ordy (Eds.), *The effects of aging and environment on vision* (pp. 219-224). New York: Plenum Press.

Hilburn, B., Jorna, P.G., Byrne, E.A., & Parasuraman, R. (1997). The effect of adaptive air traffic control (ATC) decision aiding on controller mental workload. In M. Mouloua & J. Koonce (Eds.), *Human-automation interaction: Research and practice* (pp. 42-47). Mahwah, NJ: Erlbaum.

Parasuraman, R. (1997). Human use and abuse of automation. In M. Mouloua & J. Koonce (Eds.) *Human-automation interaction: Research and practice.* Mahwah, NJ, Erlbaum.

Parasuraman, R., Molloy, R., and Singh, I.L. (1993). Performance consequences of automation-induced "complacency." *International Journal of Aviation Psychology, 3,* 1-23.

Parasuraman, R., & Riley, V. (1997). Humans and automation: Use, misuse, disuse, abuse. *Human Factors, 39*(2), 230-253.

Rouse, W.B. (1988). Adaptive aiding for human/computer control. *Human Factors, 30,* 431-438.

Small, A. (1987). Design for older people. In G. Salevendy (Ed.), *Handbook of human factors,* (pp. 495-504). New York: John Wiley and Sons.

Vincenzi, D.A., Muldoon, R.V., & Mouloua, M. (1997). Effects of age and workload on monitoring of automation failures in multi-task environments. In M. Mouloua & J. Koonce (Eds.), *Human-automation interaction: Research and practice* (pp. 118-125). Mahwah, NJ: Erlbaum.

Age Differences in the Effects of Irrelevant Location Information

David F. Pick
Purdue University Calumet

Robert W. Proctor
Purdue University

INTRODUCTION

Understanding the conditions in which the elderly are able to ignore irrelevant information is of both theoretical and practical importance (e.g., Plude & Hoyer, 1985). The theoretical value of such knowledge is to clarify basic processes of selective attention and how these processes change with age. The practical value is to improve the way in which information is presented to elderly persons.

Most studies of the ability of the elderly to ignore irrelevant information report a deficit. Rabbitt (1965) used a task in which subjects sorted cards into two stacks according to whether the letter A or B was present. The location of the target letter varied unpredictably from one card to the next, and the number of irrelevant letters on each card in the pack being sorted varied from zero to eight. The older adults were slower than the younger adults, with this difference increasing as the number of irrelevant letters increased. Comalli, Wapner, and Werner (1962) used the Stroop color-naming task to evaluate the interference effects produced by irrelevant information as a function of age. They used three cards each containing 100 items. The first card had the words red, green, and blue, which were to be read as quickly as possible. The second card had the three colors in patches that approximated the size of a word, and these colors were to be named as quickly as possible. The third card was a traditional Stroop card with the three-color words printed in non-corresponding colors of ink, with subjects required to name the ink colors. An Age x Card interaction was obtained, with the older subjects performing significantly worse than younger subjects only on the Stroop card. Other studies using the version of the Stroop color naming task in which RTs to individual stimuli are recorded have confirmed that older subjects show particularly large Stroop effects (Dulaney & Rogers, 1994; Spieler, Balota, & Faust, 1996).

Using an entirely different methodology, McDowd and Filion (1992) evaluated the effects of an irrelevant tone on the GSR response of young and old subjects. They presented tones at various times during an interesting radio program. Two types of instruction were given: Ignore the tones and concentrate on the program, or count the tones and ignore the program as much as possible. The GSR response of the younger group of subjects habituated to the tone in six presentations under the instructions to ignore it, while they went through all 20 tones before the GSR response disappeared when the tones were to be counted. The older subjects were still responding with large GSRs in both instructional conditions at the end of the 20-tone sequence. Like the results of Rabbitt (1965) and Comalli et al. (1962), these results were taken as an indication that aging results in a reduced ability to ignore irrelevant information.

An exception to the general finding that irrelevant information is more distracting for older than for younger subjects was reported by Simon and Pouraghabagher (1978). Subjects made speeded left or right choice responses to an X or 0 presented visually at a centered location. Simultaneous with the letter, a tone whose location was irrelevant to the correct response was presented to the left or right ear. Both young and old subjects showed a typical correspondence effect for the irrelevant location information (often referred to as the Simon effect): Reaction time (RT) was faster when the location of the tone corresponded with that of the response than when it did not. However, the magnitude of the correspondence effect did not vary as a function of age (young, $M = 26$ ms; old, $M = 23$ ms), even though the older subjects' mean RT was more than 100 ms slower than that of the younger subjects.

It is possible that the processing of irrelevant location information is not affected by age. However, the procedure used by Simon and Pouraghabagher (1978) differed from that used in most other investigations in that the irrelevant information was presented in a different sensory modality than the relevant information. The experiment reported here examined the effect of irrelevant location information on performance using a variation of the Simon and Pouraghabagher (1978) procedure in which the irrelevant location information was presented in the same modality as the relevant information. Two conditions were examined, one in which the stimuli were presented visually and another in which they were presented auditorily.

METHOD

Subjects

The subjects were 16 undergraduate students and 16 older volunteers. All subjects reported having normal or corrected to normal vision and hearing. Subjects in the younger group (age range 18 - 20, mean age 19.1) were enrolled in introductory psychology courses at Purdue University and participated to satisfy in part a course requirement. Subjects in the older group (age range 55 - 85, mean age 65.6) were recruited from the Retired and Senior Volunteer Program and from Purdue employees and their spouses.

Apparatus and Stimuli

IBM-compatible PCs were used to generate and present all stimuli and record all responses. Auditory stimuli were 880 Hz and 220 Hz tones presented to the left or right ear through headphones at 90 dB (A). Visual stimuli were red and green solid circles of 6.4 mm presented 6-cm left or right of center on a VGA monitor. Viewing distance was approximately 55 cm. Responses were made by pressing the 'z' or '/' key of a standard keyboard with the appropriate left or right index finger.

Procedure

All subjects performed two blocks of 16 practice trials and 152 test trials. The blocks differed only in the stimuli used and the sensory modality used to present error feedback. Half of the subjects were randomly assigned to perform on a block of trials using auditory stimuli first, followed by a block of trials using visual stimuli; the other half of the subjects performed the opposite sequence of blocks. On auditory trials, the word 'ERROR' (presented center screen) was used to inform subjects when an error was made or when a response was too slow. On visual blocks of trials, the PC bell was sounded when an error was made or a response was too slow.

Half of the subjects pressed the left key in response to a high-pitch tone and the right key in response to a low-pitch tone in the auditory task, and half performed with the opposite mapping. Half of the subjects pressed the left key in response to a red circle and the right key in response to a green circle in the visual task, and half used the opposite mapping. All subjects were instructed to respond as fast as possible without making too many errors.

Each trial began with the presentation of a stimulus to the left or right. The stimulus stayed on until a response was made or 1.5 s had elapsed. Feedback lasted 500 ms on trials where an incorrect response had been made. A new trial began 2 s later. Four types of trials were possible, involving two stimulus locations (left or right) and two response locations (left or right). Trial type was randomly selected without replacement for each four-trial sequence.

RESULTS

Mean correct RTs and percentages of error (PEs) were obtained for each subject as a function of stimulus-response correspondence (corresponding, non-corresponding) and modality (auditory, visual). For the younger adults, mean RTs for corresponding and non-corresponding trials were 379 and 424 ms for the auditory modality and 488 and 495 ms for the visual modality. For the older adults, mean RTs for corresponding and non-corresponding trials were 471 and 556 ins for the auditory modality and 547 and 580 ms for the visual modality.

The RT data showed a main effect of age, $F(1,30) = 11.63$, $p = .002$, $MSE = 23,268$, with RTs being slower for the older group $(M = 539$ ms) than for the younger group $(M = 447$ ins). Significant main effects were also obtained for correspondence, $F(1,30) = 132.3$, $p < .001$, $MSE = 443$, and modality, $F(1,30) = 31.03$, $p < .001$, $MSE = 5,035$. Responses were faster when stimulus and response location corresponded $(M = 471$ ms) than when they did not $(M = 514$ ins), and responses were faster to auditory stimuli $(M = 457$ ms) than to visual stimuli $(M = 528$ ms).

The Age x Correspondence, $F(1,30) = 19.2$, $p < .001$, $MSE = 443$, and Modality x Correspondence, $F(1,30) = 33.8$, $p < .001$, $MSE = 458$, interactions were also significant. The correspondence effect was larger for the older subjects (59 ms) than for the younger subjects (26 ms) and for auditory stimuli (65 ms) than for visual stimuli (20 ms). A similar difference in correspondence effects between auditory and visual versions of the task was obtained by Proctor and Pick (in press). The two-way interaction of Modality x Age approached significance, $F(1,30) = 2.57$, $p = .120$, $MSE = 5,035$, indicating that the difference in RTs between age groups tended to be larger in response to auditory (112 ins) than to visual (72 ms) stimuli. There was a tendency for the auditory correspondence effect to increase relatively more than the visual correspondence effect for the older adults in comparison to the younger adults, but the three-way interaction of Modality x Age x Correspondence was non-significant, $F < 1.0$.

For the younger adults, mean PEs for corresponding and non-corresponding trials were 0.8% and 7.1% for the auditory modality and 2.7% and 2.4% for the visual modality. For the older adults, mean RTs for corresponding and non-corresponding trials were 0.8% and 6.0% for the auditory modality and 1.6% and 3.9% for the visual modality. No main effect of age was present in the PE data, $F < 1.0$. This outcome is somewhat surprising in that much of the RT literature on aging would predict a speed-accuracy trade off where increased speed is accompanied by higher error rates for younger subjects. Main effects were obtained for modality, $F(1,30) = 7.85$, $p = 0.009$, $MSE = 4.14$, and correspondence, $F(1,30) = 64.96$, $p < 0.001$, $MSE = 5.53$. More errors were made to auditory stimuli $(M = 3.7\%)$ than to visual stimuli $(M = 2.7\%)$ and when stimulus and response location did not correspond $(M = 4.9\%)$ than when they did $(M = 1.5\%)$. Significant Modality x Correspondence, $F(1,30) = 55.24$, $p < 0.001$, $MSE = 3.2$, and Modality x Correspondence x Age, $F(1,30) = 8.46$, $p = 0.007$, $MSE = 3.2$, interactions were also obtained. The correspondence effect in the PE data was much larger for the auditory modality than for the visual modality, with this difference more pronounced for the younger subjects than for the older ones.

DISCUSSION

In agreement with most other studies of irrelevant information and aging, we found the effect of irrelevant location information to be larger for the older adults than for the younger adults. This result was evident for both visual and auditory stimulus presentations. Our findings imply that irrelevant location does not differ from other types of irrelevant information in showing age-related effects. Our finding is in contrast to that of Simon and Pouraghabagher (1978), which showed irrelevant location to produce correspondence effects of similar magnitude for younger and older adults. The major difference between Simon and Pouraghabagher's method and ours was that in their study the irrelevant location information was conveyed in a different sensory modality than the relevant stimulus information, whereas in our study it was not. If it turns out to be the case that elderly adults show no deficiency in ignoring irrelevant information presented in a different modality from that to which they must attend for the relevant information, this would suggest that the deficit observed when the irrelevant information is in the same modality as the relevant information does not reflect a general deterioration of the ability to selectively attend. An applied implication would be that it is particularly important for older adults to have information to which they should attend be presented in a different sensory modality than that in which distracting stimuli are occurring.

REFERENCES

Comalli, P.E., Jr., Wapner, S., & Werner, H. (1962). Interference effects of Stroop color-word test in childhood, adulthood, and aging. *Journal of Genetic Psychology, 100,* 47-53.

Dulaney, C. L., & Rogers, W. A. (1994). Mechanisms underlying reduction in Stroop interference with practice for young and old adults. *Journal of Experimental Psychology: Learning Memory, and Cognition, 20,*470-484.

McDowd, J.M., & Filion, D.L. (1992). Aging, selective attention, and inhibitory process: A psychophysiological approach. *Psychology and Aging, 1,* 65-7 1.

Plude, D.J., & Hoyer, W.J. (1985). Attention and performance: Identifying and locating age deficits. In N. Charness (Ed.), *Aging and human performance* (pp. 47-99). New York: Wiley.

Proctor, R.W., & Pick, D.F. (in press). Lateralized warning tones produce typical irrelevant-location effects on choice reactions. *Psychonomic Bulletin & Review.*

Rabbitt, P.R. (1965). An age-decrement in the ability to ignore irrelevant information. *Journal of Gerontology, 20,* 233-238.

Simon, J. R., & Pouraghabagher, A. R. (1978). The effect of aging on the stages of processing in a choice reaction time task. *Journal of Gerontology, 33,* 553-56 1.

Spieler, D. H., Balota, D. A., & Faust, M. E. (1996). Stroop performance in healthy younger and older adults and in individuals with dementia of the Alzheimer's type. *Journal of Experimental Psychology: Human Perception and Performance, 22,* 461-479.

Performance Differences Between Older and Younger Adults for a Virtual Environment Locomotion Task

**Daniel P. McDonald, Dennis A. Vincenzi, Robert V. Muldoon,
Robert R. Tyler, Amy S. Barlow, and Janan Al-Awar Smither**
University of Central Florida

INTRODUCTION

The early simulators could best be described as "eye - hand" coordination devices used primarily to teach and reinforce rule based responses. Since then, the goal has been to create more "real world-like" simulations. We are moving toward being able to create cost-effective virtual environments (VE), which can be used for everything from military training applications to industrial simulation. VE gives us the capability for conducting complete mission rehearsals, enabling the use of a wide variety of instructional strategies. On the other side of the spectrum, the entertainment industry is rushing ahead to give consumers what they want, a more engaging environment within which to reap enjoyment. It is likely that low-cost VE will be commonplace in the near future. Anticipated widespread use of VE deems it necessary to begin addressing the limitations and capabilities of both today's and tomorrow's technologies for the target populations.

The U.S. Army and others been involved in ongoing VE related research addressing psychophysical and fidelity issues (Hays & Singer, 1989; Levison & Pew, 1993; Singer, Ehrlich, Cinq-Mars, & Papin, 1995; Lampton, Gildea, McDonald & Kolasinski, 1996; Rinalducci, 1996), spatial knowledge acquisition and transfer (Witmer, Bailey & Knerr, 1996; Singer, Allen, McDonald & Gildea, 1997), team training and situation awareness (Ehrlich, Knerr, Lampton & McDonald, 1997), and simulator sickness (Kennedy, Lane, Lilienthal, Berbaum & Hettinger, 1992). However, most research evaluating VE has been conducted using young military personnel or university students.

While VE training and performance research marches on, studies addressing performance of populations such as older adults is lagging. This in spite of the elderly being the largest growing segment of the population in the United States. Currently there are 23.5 million individuals over the age of 65. This group has nearly doubled in size since the early 1900's. The "85 and over" group has increased in size nearly 17 times since the early 1900's. This "85 and over" group is the segment of the elderly population that is responsible for the rapid growth in size of the elderly population in the United States. With the advent of recent advances in medical technologies, it is not unusual for an individual to reach the age of 85 in the U.S. today. It is estimated that by the year 2025, approximately 20% of the U.S. population will be 65 years of age or older (Harbin, 1991). People are living much longer, and retiring at an older age than before. Moreover, with the uncertain future of the social security system, older persons may be forced to continue working past the age of 65. If VE technology becomes a commonplace medium for training and a number of other life-related functions more older adults will find themselves being required to use it. However, little is known about how age-related differences may interact with VE characteristics.

The effectiveness of VE for training or other applications has been attributed largely to the level of fidelity of the VE system. Regian, Shebilske, and Monk (1993) maintain that training transfer requires both preservation of the visual spatial characteristics of real world and that the interface preserves the link between motor actions and the effects in VE. User capabilities and limitations are important factors to consider in the interface. User characteristics can have a bearing the system's overall effectiveness. For example, a person who has a greater than average difficulty resolving images under lower levels of luminance may not be able to resolve images in a low luminance HMDs, even though the physical fidelity of that device may be suitable for most other persons for performing the task. According to Bjorn, Kaczmarek, and Lotens (1995), it is important to have a thorough understanding of the capabilities of the human sensory system and to use this knowledge in the design of virtual worlds and the deriving technical specifications for VE equipment. If VE is to be a useful tool for the older population we must consider the changing sensory, cognitive and motor functioning capabilities, as well as experiential influences affecting performance.

VEPAB

The United States Army Research Institute has developed a set of perceptual and psychomotor tasks designed to assess human performance and the effects of immersion in the VE as a function of training and system characteristics (Lampton, Knerr, Goldberg, Bliss, Moshell, & Blau, 1994). The battery was designed to serve several important functions. First, the tasks provide a means to bring research participants to a basic level of proficiency in prerequisite VE skills (e.g. locomotion, object manipulation). Second, the tasks can be used as "behavioral benchmarks" for interface hardware and software comparisons, to determine quickly the effects of system changes (display resolution, update rate) on task performance. Third, task performance can provide statistical controls for the future. Fourth, the tasks provide an initial set of conditions for investigating consequences of use of VE such as simulator sickness and the sense of "presence". or immersion. Finally, the tasks can be used to compare performance between different populations.

Several goals of this research were to a) gather baseline performance measures on a locomotion task performed in the VE comparing older and younger adults, b) determine whether performance differences do exist, c) entertain possible explanations for any differences, d) ascertain whether older adults as well as younger adults will show performance improvements over trials, and e) compare the rates of improvement. Reported here are the results of a locomotion task referred to as doorways (Lampton et al., 1994).

METHOD

Participants

One group consisted of sixteen active healthy older adults, eight males and eight females, ranging in age from 65 to 76 years old, who participate in the LIFE at UCF continuing education program for senior citizens. The mean age of this group was 70 years old, all being active in a variety of activities. A second group consisted of sixteen college students, eight males and eight females, attending the University of Central Florida, whose ages ranged from 19 to 36 years old. The mean age of this younger group was 24 years old. All participants were required to be in good physical health, have had no previous experience with virtual environments, have at least 20/40 corrected or uncorrected far point static visual acuity, and have normal stereoscopic vision. Older participants received no incentive for participation in this study, while some of the younger participants were able to receive extra credit points in their class for their participation.

Apparatus

A Virtual Research VR-4 Helmet was used as the display device. The VR-4 has a horizontal field of view (FOV) of 48 degrees and a vertical FOV of 36 degrees, with 742 X 230 pixels to each eye and uses LCD technology. Head tracking was not used in this study. The VE images were generated by two 50 MHz IBM compatible PCs. Perspective changes and locomotion were controlled by a six-degree-of-freedom Gravis joystick.

Locomotion/Navigation Task: Doorways

The locomotion tasks require the participant to "walk" through the VE by using an input control device to direct the speed and direction of the participant's simulated body. Eye level is set during pre-testing activities, and corresponds to the participant's actual eye height in inches. There is no visible representation of the body. The body can move forward or backward, laterally, or rotate. Software settings, chosen to represent normal walking parameters, control the maximum speed and the rates of acceleration and rotation. The body interacts with the VE through collisions with walls and doorframes. Because collisions almost always stop forward progress, movement at a reasonable rate requires emphasis on both speed and accuracy.

The Doorways task represents a VE "road test" of the kind and difficulty of walking performance that might be required in a VE training application. The course is formed by a series of 10 rooms connected by 7 by 3-foot doorways offset from one another.

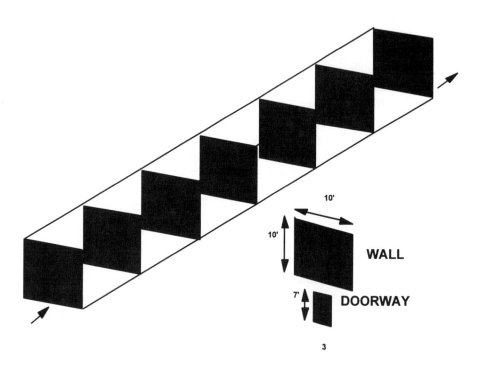

Figure 2. Doorways Task.

The positions of the doors in the walls vary so that a series of non-90 degree turns must be made to navigate the course efficiently as shown in Figure 2. Participants were asked to traverse this set of rooms as quickly as possible while trying to minimize their collisions with the VE. time to traverse and collision totals were collected for each of the ten rooms.

RESULTS

Video Game and Computer Experience

Participants were asked to report the number of hours they spend each week using computers and video games. The older participants reported spending and average of 5.50 hours using a computer, whereas the younger participants reported a mean of 17.19 hours of weekly computer use. Gender differences were also observed. Males, overall, reported using a computer 12.81 hours a week, and females indicated that they spent an average of 9.87 hours each week using a computer. A two factor between subjects ANOVA revealed that younger participants use computers more frequently than do the older participants $F(1,28)=9.139$, $p<.01$. It was also shown that males tended to use computers more frequently than did females $F(1,28)=4.623$, $p<.05$. No other effects or interactions were found involving computer use.

Mean weekly video game use for older participants was 3.78 minutes as compared to 1.38 hours for younger participants. Males overall reported playing video games on the average of .875 hours per week, whereas females reported .563 hours of video games per week. An ANOVA was not conducted to determine whether these means significantly differ because 15 of the 16 older participants reported playing video games zero hours per week. It was investigated whether computer use will be correlated with initial performance on the doorway locomotion task. A significant correlation between computer use and performance might suggest that experience with computers has an effect on how participants initially perform in the VE tasks, and begin to explain some of the differences in performance found between groups. A Pearson correlation of $r= -.566$, $p<.001$ indicates significant relationship between reported weekly computer use and mean traversal time of the average of the first two rooms. Higher reported computer use is related to better performance and faster initial traversal times.

Mean room traversal times were calculated for each of the ten rooms in the doorways task. A three-factor mixed ANOVA was used to determine whether there were any differences or interactions between age, gender or trials in mean time to traverse a room in the doorway task. The overall mean times across groups for traversing room one through ten are 18.75, 19.19, 17.27, 14.73, 14,52, 15.22, 14.23, 13.09, 13.02, 11.25 seconds respectively. As Figure 1 indicates, it appears that an individuals' performance as measured by time to traverse rooms improved with the number of trials. A Rao R test for multivariate equivalence showed a significant effect for practice on room traversal times. $R(9, 20)=8.09$, $p<.0001$. A trend analysis was conducted to determine if, in fact, traversal times decreased systematically with practice. It was revealed that room traversal time did decline linearly with the number of rooms traversed $F(1, 28)=36.02$, $p<.0000$.

The overall mean room traversal times for the main effect of age were 10.39 and 19.94 seconds for young and old. It was found that the mean overall room traversal times differ significantly as a function age, $F(1,28) = 38.84$, $p < .0000$. Overall, younger individuals moved through the rooms faster than older individuals.

The overall mean room traversal times for the main effect of gender were 13.39 and 16.93 seconds for male and female respectively. It was found that these overall mean room traversal times differ significantly, $F(1,28) = 5.02$, $p < .05$. Overall, male individuals traversed the rooms in the doorway task faster than female individuals.

An Age by Trial interaction was also found $F(9,252)= 2.06$, $p<.05$. This indicates that perhaps although older individuals' average room traversal times were slower, the discrepancy in traversal times between the two age groups grew significantly less with the number of trials.

A trend analysis was performed on both younger and older participants to determine whether a general trend toward improvement over trials resulted for each group. For the older group, trend analysis revealed a significant negative linear trend, $F(1,28) = 36.02$, $p<.0000$. There was no significant quadratic trend found. However, younger participants, whose overall performance was faster than older participants, showed no significant trends toward improvement.

DISCUSSION

There was an overall improvement in locomotion performance as a function of number of rooms traversed. Younger individuals were able to traverse the rooms more quickly than older individuals. Also, males tended to be faster than females. An interaction showed that although there was a disparity in performance between older and younger persons, this became significantly less over trials. It was also found

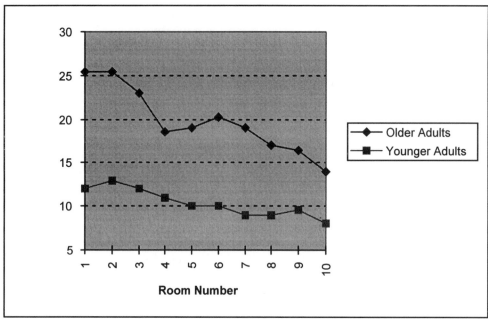

**Figure 1. Mean Room Traversal Times
for Both Older and Younger Adults.**

for this task, that there was a high correlation between computer use and traversal times. Interestingly, younger males reported the highest amounts of computer use of all subgroups, and computer use was shown to correlate highly with performance on this task. The interaction indicates that prior exposure to computers may have again, been a large predictor of initial performance, but less reliable of a predictor after some training. Performance in this task required skillful use of a control device in order to manipulate the object which can be developed by both young and old through minimal practice. Experience may not be the only factor contributing to overall slower traversal times, visual motor, and cognitive differences may also explain some of the variance in performance. Some aging issues related to performance in VE are discussed.

Visual Changes

The visual system shows a definite decline with age. The various optical media through which light must pass before encoding by retinal receptors can begin to deteriorate noticeably at around 40 years of age. The crystalline lens of the eye begins to yellow and thicken with age, along with other changes in the cornea. These changes result in a decrease in the amount of light being transmitted to the photoreceptors of the eye, and an increase in the amount of light that is scattered within the eye (Rinalducci, Smither, & Bowers, 1993). It is estimated that the amount of light reaching the retina of a 60 year old is approximately two-thirds of that of a 20 year old (Weale, 1973). Moreover, a decrease in ability to accommodate and to adjust the diameter of the pupil have significant effects on dark adaptation, resistance to glare, acuity and a host of other abilities (Hughes, 1981).

HMDs and lighting

Display brightness is a potential problem with the aging population. In short, as the optical media of the eye ages, the amount of light entering the eye decreases and glare sensitivity increases. A display that is sufficiently bright for a younger individual may not be bright enough for an elderly individual. An elderly individual, therefore, may require a greater amount of light to achieve the same level of retinal illuminance.

Consequently, the visual performance of an elderly individual may be significantly impaired in low light situations. As Rinalducci (1996) points out, there are limits to the levels of luminance obtainable in most simulator displays with such levels never approaching those found in actual daylight. Most HMDs have a luminance level comparable to that of night dawn/dusk.

The problem with low luminance may be further exacerbated by the recent push toward high resolution, smaller field of view (FOV) displays. Smaller FOV displays require a greater percentage of the visual scene to be presented in the central area of the retina. Information processed through the central portion of the retina and fovea is processed primarily through the parvocellular visual channel. The parvocellular visual channel is responsible primarily for object recognition and resolving increasingly finer detail. Visual information cast upon the more peripheral portions of the retina are primarily processed through the magnocellular channel. The magnocellular channel is more sensitive to such things as movement and spatial orientation. A characteristic of the parvocellular channel is that its performance degrades much sooner than the peripheral magnocellular channel as luminance levels are decreased. Degradation of this visual channel begins at about the luminance level equivalent dusk, whereas the magnocellular channel remains functioning at much lower luminance levels (Leibowitz & Owens, 1977). This phenomenon of selective degradation coupled with already reduced amounts of light through aged optical media may compound a visual display lighting problem for the older population. On the other hand, as the level of luminance increases, the amount of light scatter within the older eye also increases. Thus, since older individuals require increased levels of light to achieve the same level of retinal illuminance, brighter displays may result in a greater degree of glare also resulting in a reduction in visual effectiveness.

Accommodation and convergence

The focal point of the lenses for most HMDs are at a fixed distance. This means that the distance at which people focus, regardless of the intended distance portrayed in the VE, is constant. Fixed accommodative distance can present perceptual problems as well as eyestrain for all, and may present different problems for older adults who have degraded accommodative ability .

LCDs in HMDs are usually a fixed distance apart, with the amount of visual overlap being constant resulting in ocular convergence, which is unnatural. Convergence is the movement of the eyes inward in unison as an object moves closer in the real world. Convergence is directly related to accommodation in that as an object moves closer, the amount of accommodation should increase and the convergence of the two eyes should also increase. These two separate functions that occur in the eyes are normally coordinated and linked together in the real world. In virtual environments there appears to be an uncoupling of accommodation and convergence, which has been cited for problems with such things as distance perception (see Rolland, Burbeck, and Ariely, 1994). This problem may be compounded in the older adults with degraded abilities of both accommodation and vergence. Conversely, perhaps less difficulty may be experienced by older persons simply because degradation of these functions and their decoupling has already occurred naturally, and less emphasis is placed upon these cues to determine depth, and more emphasis on pictorial cues. Many questions remain unanswered.

Cognitive and motor responses

The most fundamental age-related change in cognitive and motor performance is the general slowing of behavior. Undoubtedly the human capacity to process information is limited, and with advancing age the rate of information processing becomes slower and less efficient (Rinalducci, Smither, & Bowers, 1993). Older persons appear to have more difficulty with tasks that are more complex. One reason for this is the limits older adults have in their short-term memory. This doorways task required participants to perform locomotion functions on the joystick for which they were trained prior to the actual trials.

267

Remembering control functions coupled with the motor demands of locomotion in a virtual environment made the doorways tasks complex enough to see initial performance differences.

For the doorway task, although it appears that experience with computers may have been a major contributing factor determining performance. However, it was demonstrated that older adults were indeed capable of learning how to use a joystick for locomotion and navigation in VE, thus indicating sufficient cognitive and motor skills. Czaja (1988) suggests that research regarding response latency and older adults implies that the design of displays, as well as response devices, and user software are critical for the effective implementation of computer systems. Studies using computer or video game experience as a covariate in future research might allow for researcher to isolate some of these other factors potentially contributing to performance differences. Therefore, further studies must be conducted to address the issues touched upon in this paper, as well as issues regarding older adults' susceptibility to simulator or Cyber sickness.

ACKNOWLEDGMENTS

We would like to thank Bruce Knerr and the U.S. Army Research Institute Simulator Systems Research Unit (ARI SSRU), Orlando Field Unit for all their support. We would also like to show our appreciation to the Institute for Simulation and Training (IST) at the University of Central Florida (UCF), especially Kimberly Parsons, Dwayne Nelson, and Greg Wiatrowski for all their technical support. We would also thank the Learning Institute for Elders (LIFE) at UCF and the individuals that participated in this study.

REFERENCES

Czaja, S. J. (1988). Microcomputers and the Elderly. In M. Helander (Ed.), *Handbook of Human-Computer Interaction*, North-Holland: Elsevier Science Publishers B. V., 543-568.

Ehrlich, J. A., Knerr, B. W., Lampton, D. R. and McDonald, D.P. (1997). *Team situational awareness training in virtual environments: Potential capabilities and research issues.* (Technical Report), US Army Research Institute for the Behavioral and Social Sciences: Alexandria, VA.

Hays, R. T. and Singer, M. J. (1989). Simulation Fidelity as an Organizing Concept. In R. T. Hays and M. J. Singer (Eds.), *Simulation Fidelity in Training System Design: Bridging the Gap Between Reality and Training.* (pp 47-67), New York: Springer-Verlag.

Harbin, T.J., (1991). Environmental toxicology and the aging visual system. In D. Armstrong, M.F. Marmor, and J.M. Ordy (Eds.), *The effects of aging and environment on vision,* 219 -224. New York: Plenum Press.

Hughes, P.C. (1981). Lighting for the Elderly: A psychobiological approach to lighting. *Human Factors, 23(1),* 65-85.

Kennedy, R.S., Lane, N.E., Lilientahl, K. S., Berbaum, K. S., and Hettinger, L. J. (1992) Profile analysis of simulator sickness symptoms: Application to virtual environment systems. *Presence: Teleoperators and Virtual Environments,* 1(3), 295-301.

Lampton, D. R., Knerr, B. W., Goldberg, S. L., Bliss, J. P., Moshell, J. M., & Blau, B.S. (1994). The virtual environment performance assessment battery (VEPAB). *Presence,* 3(2), 145-157.

Lampton, D. R., Gildea, J. P., McDonald, D.P. and Kolasinski, E. M. (Oct. 1996). *Effects of display type on performance in virtual environments.* (Technical Report 1049), US Army Research Institute for the Behavioral and Social Sciences: Alexandria, VA.

Leibowitz, H. W. and Owens, D. A. (1977). Nighttime Driving Accidents and Selective Degradation. *Science* 197, pp 422-423.

Levinson, W. H., and Pew, R. W. (1993). *Use of virtual environment training technology for individual combat simulation.* (Technical Report 971). US Army Research Institute for the Behavioral and

Social Sciences: Alexandria, VA.

Regian, J. W., Shebilske, W. L., and Monk, J. M. (1993). *A preliminary empirical evaluation of virtual reality as a training tool for visual-spatial tasks*. (Report No. AL-TR-1993-0004). Brooks AFB, TX: Armstrong Laboratory. (DTIC Report No. AD-A 266 110.

Rinalducci, E. J. (1996). Characteristic of Visual Fidelity in the Virtual Environment. Presence, 5(3), 330-345.

Rinalducci, E. J., Smither, J. A. and Bowers, C. (1993). The effects of age on vehicular control and other technological applications. In J. A. Wise (Ed.) *Verification and Validation of Complex Systems: Additional Human Factors Issues*. Embry-Riddle Aeronautical University Press.

Rolland, J. P, Burbeck, C. A., Gibson, W. & Ariely, D. (1994). *Towards Quantifying Depth a Size Perception in 3D Virtual Environments*. (Technical Report TR93-044), Dept. of Computer Science: University of North Carolina at Chapel Hill.

Singer, M.J., Ehrlich, J. A., Cinq-Mars, S. and Pappin, J. (1995). *Task performance in virtual environments: Stereoscopic vs. monoscopic displays*. (Technical Report 1034), US Army Research Institute for the Behavioral and Social Sciences: Alexandria, VA.

Singer M. J., Allen, R. C., McDonald, D. P. & Gildea, J. P. (1997). *Terrain Appreciation Virtual Environments: Spatial Knowledge Acquisition*. (Technical Report), US Army Research Institute for the Behavioral and Social Sciences: Alexandria, VA.

Weale, R.A. (1973). The effects of the aging lens on vision. *Ciba Foundation Symposium, 19*, 5-20.

Witmer, B. G., Bailey, J. H. and Knerr, B. W. (1996). Virtual spaces and real world places: transfer of route knowledge. *International Journal of Human-Computer Studies*, 45, 413-428.

Effects of Age and Automation Level on Attitudes Towards Use of an Automated System

James M. Hitt, II, Mustapha Mouloua, and Dennis A. Vincenzi
Center for Applied Human Factors in Aviation
University of Central Florida

INTRODUCTION

As automated systems continue to saturate our daily activities we as human factors specialists and design engineers must be concerned with the issues that may inhibit or promote the use of an automated system. Automated systems are found in many facets of daily life. It is difficult for a person not to see the many daily uses of automated systems. For example, automated teller machines (ATM's) can be used for deposits or withdrawals, transferring funds between accounts, purchasing stamps, making account inquiries and some ATM's can check current exchange prices. Aircraft can now take off, cruise, and land with the use of automated systems. Word processing software can now automatically correct spelling and grammar mistakes. Braking systems for automobiles can now detect the exact amount of pressure to put on the pads to stop the automobile without locking the brakes (often causing an uncontrolled spin). But our point here is not to belabor the existence of automation but to further understand the important design issues for automated systems. One important design issue is how does the age of the system user effect performance.

Aging and Automation Systems and Automation Systems

Several studies have addressed the issue of aging and its effects on human performance in automated systems. Research by Hardy, Mouloua, Molloy, Dwivedi, and Parasuraman (1995) examined age differences under dual task conditions using the MAT battery. Their results showed an equal degradation in performance on the two tasks between the young and elder participants. Another study by Hartley (1992), revealed that in dual task conditions there is a greater level of divided-attention cost for older adults when compared to younger adults. These studies report varied empirical evidence for an age effect when examining performance in automated systems.

Other studies have addressed such issues as how older adults view automation use. Issues such as reliance (Riley, 1995), trust (Muir, 1987), and safety have been cited as important deign considerations for automated systems. Research examining attitudes towards automation use has been conducted on a limited sample size (Sams, Sierra, Sahagian, Nichols, & Mouloua, 1997). Using a composite score from the Complacency-Potential Rating Scale (CPRS) these researchers revealed a significant difference between age groups with younger adults having a higher level of positive attitudes towards automated systems than older adults. Other studies have examined ATM use by older adults and in several cases the elderly population declared the desire for training programs to further understand the uses of the automated systems. All of these studies have taken different paths to examine possible age related differences when humans interact with automated systems. The purpose of this study is to determine if differences among any of the five components of the CPRS (overall automation, reliance, trust, safety, and confidence) are attributed to age differences and/or exposure to various levels of automation (single, dual, or multi-task). We hypothesize that the CPRS scores will be lower for the elderly population and as the task load increases (more manual control), the scores of CPRS will decrease.

METHOD

Participants

A total of 72 participants, 36 between the age of 18 and 25 (mean = 20.3), and 36 over the age of 65 (mean age = 65.4), were recruited from the general population of undergraduate students at the University of Central Florida and the surrounding Orlando community. Participants received either extra credit (undergraduates) or $25 (elderly non-student) for their participation in the experiment.

Materials

The task used for this experiment was the Multi-Attribute Task (MAT) Battery (Comstock, 1990). The MAT was designed to test workload levels in a multi-task environment. For this study, three of the tasks within the MAT were used, the tracking task, system monitoring task, and the fuel management task. The MAT was run using a 486DX, 66MHz IBM compatible computer, a 14" VGA color monitor and an analog joystick was used for the tracking task. The Complacency-Potential Rating Scale (CPRS) was used to measure user attitudes towards automation use. This scale, developed by Singh, Molloy, and Parasuraman (1993), is a 20-item, 5 point Likert-type scale (ranging from "strongly agree" to "strongly disagree") used to measure attitudes towards commonly used automation systems (banking, aviation, VCR's, etc). Previous factor analyses have determined five components measured by the scale. These factors relate to issues such as confidence, reliance, safety, and trust in the use of automated systems. The final component measures an overall measure of automation. These components were the focus of the study.

Design

The study used an Age (young and elderly) x Level of Automation (single, dual or multi-task) factorial design. Each participant entered a 10-minute training session before engaging in three 30 minute automated sessions. During the training session, all three tasks (tracking, system monitoring, and fuel management) were manually operated. Participants were required to achieve a minimum detection rate of 50% on the system-monitoring task (in the training session) before being allowed to proceed to the experimental sessions. This was done to insure a minimum proficiency level among all participants before data collection began. Each participant was randomly assigned to a single, dual, or multi-task condition. The measured variables in the study were the 5 components from the CPRS. The sum of the scores from related questions to each variable (overall automation, confidence, reliance, safety, and trust) from the CPRS were subjected to analysis via ANOVA methods.

Procedures

For the training session, each participant performed the three tasks involved in MAT battery. The compensatory tracking task is meant to simulate the stick control skills used in aviation flight. Participants were instructed to steer a target to the center of a large cross hair target and attempt to hold the target as close to the center as possible. Performance is measured as the RMSE (in pixels) from the center of the cross hair target. The monitoring task requires the participant to monitor four separate analog dials randomly fluctuating. At random intervals one of the dials will exceed the allowable boundaries and this is the cue for the participant to reset the particular dial. The performance measure for this task was the mean malfunction detection rate. The fuel management task requires the participant to maintain a certain level of fuel supply within various tanks. This is accomplished by activating or deactivating fuels tanks as necessary. Measures for this task were taken as the mean error from the desired fuel level. Total testing time for the study did not exceed 2 hours.

For this experiment, each participant was randomly assigned to one of the three automation levels. If assigned to the single task condition, the subject manually performed the tracking task while the system monitoring and fuel management tasks were automated. Those in the dual task condition performed the tracking and system monitoring tasks simultaneously while the fuel management task is automated. The multi-task condition required the participants to perform all three tasks in unison. Upon completion of the last 30 minute testing session, the CPRS was given to each participant for completion. After completion of the attitude questionnaire, participants were debriefed and excused from the testing site.

RESULTS

Reliance

A main effect for age was found between the two groups, $F(1, 66) = 9.79$, $p < .01$. Young participants held the attitude of automated systems being of greater reliance when compared to the scores of the elderly group (means = 9.64 and 8.30, respectively). Figure 1 graphically depicts the differences between the two groups.

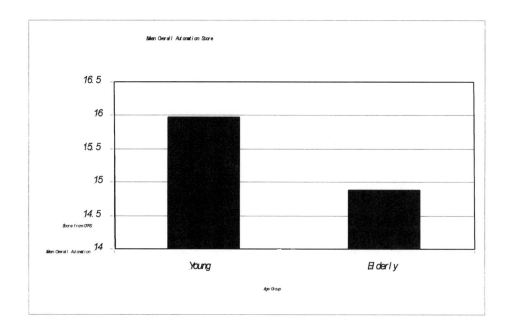

Figure 1. Mean reliance scores from the CPRS for both young and elderly age groups.

General Automation

Although not statistically significant ($p = 0.07$) differences were found between the two age groups for the component labeled general automation (see Figure 2). No main effect for level of automation or interaction effect was found.

272

Mean Reliance Score

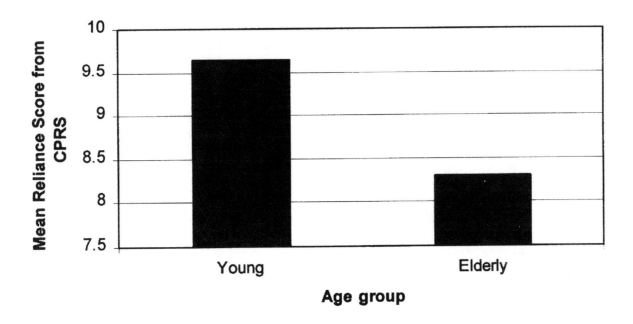

Figure 2. Mean Overall Automation Score from the CPRS for both young and elderly age groups.

Confidence

No main or interaction effects ($p > .05$) for the confidence component of the CPRS were found to exist between the age groups and the level of automation variable.

Safety

No effect ($p > .05$) was found to exist for the age variable or level of automation when examining to safety component of the scale.

Trust

Both age groups reported trusting automated systems equally ($p > .05$). No effect was also found for the level of automation and the interaction term.

DISCUSSION

The goal of the present study was to examine potential changes in attitudes towards automation use based on age and exposure to varied levels of automation. The data shows our original hypotheses received little support. Differences in the reliance component of the CPRS questionnaire were found between the age groups, with the young age group expressing a greater positive attitude towards the reliance in automated systems. As being the only significant effect found for all of the components, the reliance factor here seems to stand alone.

This is surprising based on the amount of research and model development which states the constructs of trust, confidence, and reliance as being inter-related (see Riley, 1989) as well as automation accuracy (Muir, 1989). Reliance in this study refers to the likelihood (probability) of the user to activate an automated system.

Although the results here are somewhat surprising, an examination of the literature might help to explain the results of this study. One potential problem associated with the use of automated systems is the over-reliance on automation often exhibited by pilots. This over-reliance is defined as the reluctance of pilots to take over control from an automated system. Evidence of this type of problem is evidenced by the number of reports in the Aviation Safety Reporting System (ASRS) linking monitoring failures and over-reliance in automated systems (Mosier, Skitka, & Korte, 1994). Future studies should use a more valid pre/post design for determining the effects of exposure to automation levels. More detailed efforts should be made to determine the task parameters in which performance of older adults begins to decrease. The various literature studies cited, in conjunction with the current findings, have only supplied partial support for their proposed hypotheses.

REFERENCES

Comstock, R. (1990). *Multi-attribute task battery*. (Preliminary Technical Report). Hampton, VA: NASA Langley Research Center.

Hardy, D.J., Mouloua, M., Molloy, R., Dwivedi, C.B., & Parasuraman, R. (1995). Monitoring of automation failures by young and old adults. *Proceedings of the Eighth International Symposium of Aviation Psychology* (pp. 1382-1387). Columbus, OH: OSU Press.

Hartley, A.A. (1992). Attention. In F.I.M. Craik & T.A. Salthouse (Eds.), *The Handbook of aging and cognition* (pp. 3-49). Hillsdale, NJ: LEA.

Mosier, K., Skitka, L., & Korte, K.J. (1994). Cognitive and social psychological issues in flight crew/automation interaction. In M. Mouloua & R. Parasuraman (Eds.), *Human Performance in automated systems: Current research and trends* (pp. 191-197). Hillsdale, NJ: LEA.

Muir, B. (1987). Trust between human and machines, and the design of decision aids. *International Journal of Man-Machine Studies, 27*, 527-539.

Muir, B. (1989). *Operators' trust in and use of automatic controllers in a supervisory process control task.* Unpublished doctoral thesis, University of Toronto.

Riley, V. (1989). A general model of mixed-initiative human machine systems. Paper presentation at the Human Factors Society, 33rd Annual Meeting, Denver, CO.

Riley, V. (1995). Factors influencing operator reliance on automation. *Proceedings of the Eighth International Symposium of Aviation Psychology* (pp. 270-275). Columbus, OH: OSU Press.

Sams, K., Sierra, C., Sahagian, J., Nichols, B., & Mouloua, M. (1997). Attitudes Towards Automation Usage by Younger and Older Adults. Unpublished manuscript, University of Central Florida.

Singh, I.L., Molloy, R., & Parasuraman, R. (1993). Automation-induced "complacency": Development of the Complacency-Potential Rating Scale. *International Journal of Aviation Psychology, 3(2),* 111-122.

TRUST

IN

AUTOMATION

Human Trust in Aided Adversarial Decision-Making Systems

Younho Seong   James Llinas   Colin G. Drury   Ann M. Bisantz
Center for Multisource Information Fusion
State University of New York at Buffalo

INTRODUCTION

The study of trust has a long history in sociological literature: however, that history is not rich in empirical studies. Some representative, recent studies of human trust in automation have been performed on a continuous chemical process control simulation (Lee & Moray, 1992, 1994; Muir & Moray, 1996). This means that the objects which human operators dealt with were machines or displays representing the behavior of machines. Thus, the induced characteristics of human trust in automation were essentially concerned with predictability, dependability, and faith, whose attributes could be easily captured from the behavior of machines. Current automation, which in the context of this study is technology using a data fusion process, produces estimates of situational conditions which are, ideally, of reasonable but imperfect quality, i.e., there is some uncertainty in the estimates. However, this study extends the studies of trust to situations where, additionally, the automation system is open to deliberate manipulation by adversaries. Having another person or a group of people *at the other end* changes the central scheme of the problem. In this environment, called *Information Warfare* (IW), human operators must deal with faults resulting from both mechanistic automation failure (imperfect data fusion) and premeditated deception or misguidance manipulated by an adversary. Human operators should have the ability to distinguish the faults perpetrated by the foe, to calibrate their trust in decision-making aids and to eventually accomplish their mission efficiently. Without knowing the system's vulnerability from an adversary, human operators may regard faults of corrupted information by an adversary as automation failures in the data fusion process or malfunction of displays. In this paradigm, some characteristics of trust, which were identified in previous studies, such as fiduciary responsibility, may not be applicable because of the hostile environment. Simultaneously, the new paradigm should be able to include the simple relationship between human operators and decision-making aids.

POTENTIAL CHARACTERISTICS OF HUMAN TRUST AND IMPLICATIONS FOR IW

Among the sociological studies defining trust in the interpersonal relationships, Rempel, Holmes, & Zanna (1985) characterized trust as a multi-faceted construct having three dimensions. While this represents one classification of trust characteristics, Sheridan (1980) suggested a more comprehensive set of seven possible characteristics of human trust in the human-machine systems. As we are dealing in IW with trust of machines, and the data fusion processes, Sheridan's classification is a better starting point. We will consider the applicability of Sheridan's characteristics in turn where the domain is IW.

The first of Sheridan's aspect of trust in automation seems to relate to notions of *reliability*. This implies a system of reliable, predictable, and consistent functioning. In other words, a sense of reliability in a decision aid (DA) is established when it can be observed to create the same output repeatedly under a particular set of circumstances. There are three fundamental points of vulnerability from the real world situation to the display for operators. These are true input, observable input and displayed output. True input is susceptible to deception and corruption of operations and actions by artificially creating a false circumstance. Observable input is susceptible to parametric or algorithmic corruption. This is creating false output from internal processes in spite of a *correct* circumstance in the real world. Displayed output is susceptible to algorithmic corruption. This is creating a false display in spite of both (a) an unperturbed real

276

world and (b) unperturbed (non-display related) processing operations.

The second aspect of trust is a sense of *competence* or *robustness* in the DA. That is, robustness supports expectations of future performance based on capabilities and knowledge not strictly associated with specific circumstances that have occurred before. In some sense, this represents a *calibration* of the decision-aiding software processes or algorithms. Corruption of a operator's trust in the DA through IW techniques places a requirement on the deceiver to know the nature of adaptivity in the DA internal workings. However, if the operator is astute in the likely technology for the DA, a reasonable approximation to the inner workings may be feasible. Alternately, if the deceiver is aware of the limits of capability of the DA, the operator could create circumstances which are in those fringe operating regions and reduce the operators impressions of the competence of the DA.

Familiarity is the third feature or aspect of trust. Often a person confronts a situation or an object with a high degree of novelty, but still feels familiar with and able to deal with the situation. Either from a naturalistic or inherent cultural expectation, familiarity may prevent any exploratory risk-taking behavior to diagnose the situations, or to identify objects whether new or familiar. Consequently, it may induce biased decision-making. Because of the fact that familiarity is not based on any scientific knowledge or expertise and tends to be inherited from those who have cultural similarity with us, the person who is confronting an unfamiliar or unanticipated situation or object will be very vulnerable to deception. Unlike other industrial settings where unanticipated, and so unfamiliar events are sometimes confronted by operators, operators in military command, control, communication and information system (C^3I) may not have been exposed to or trained in unanticipated events. For automated DA's, this can involve the use of common or perhaps military-standard symbology, nomenclature, etc.

Sheridan's fourth characteristic is *understandability*. The construct of understandability is equivalent to developing an appropriate mental model, possibly with the aid of familiarity. In designing a machine to aid an operator, understandability usually is affected by the degree of transparency of the system which the operator can *see* through the interface to the underlying system. Opaque machines or interface media will not only prevent the operator from trusting the machines, but also from engaging in problem-solving activities in cases of warnings or mishaps. Thus, any means by which an adversary could corrupt the graphic user interface or other interface functions, in order to confound the operators' ability to understand a system would lead to distrust. This is particularly important if the operators do not participate in the development process. If this is the case, then the degree of understandability governs the operators' ability to form his own mental model of the inner workings of the DA, which also lends itself to predictability. Here the deceiver can simply have the goal of generating randomness in his or her attack, since irregular patterns between input and output will aid in loss of trust.

Next we consider *Explication of Intent*. Instead of leaving a person in a position where the covert meanings have to be discovered and understood from the systems' behavior, this attribute allows people trust others over those who just perform tasks. However, current technological improvements in designing intelligent computers are not well enough developed to allow operators to communicate using higher level intentions. Unless we develop intelligent machines, which can specify their intentions of future actions outright, we have to rely on the current available technologies, e.g. in the form of symbols, short statements, or a combination of both which are pre-programmed by system designers. Therefore, we are often forced to trust (or not to trust) based on a symbolic medium through which one produces effects and on the basis of which one derives an interpretation of "what is happening". Estimating intent is one of the more difficult things to do in military situations but is also, if done correctly, one of the highest payoff areas. Intent is approximately an explicit indication of a future planned action to which the actor is committed; the difficulty is in defining and observing those indicators. An actor develops intent based in part on his value system; that is, the operator will plan actions that have a sense of payoff in the context of a value system. The value system is, however, multidimensional; it has relatedness to military goals and objectives but also to notions of personal value and the societal notions of value we all grow up with; these are the harder elements to judge. The development of an intended action is dependent on a particular or sequence of particular outputs (from a DA) in the context of judged value;

$$Action(T + \Delta T) \approx (\ DA\ Output\ |T, DA\ Output\ |(T-1), ...and\ implied/ca\ lculated\ value\ of\ action,\ given\ the\ DA\ Output\ )$$

By creating any false output of the DA (by any means), the deceiver corrupts this relationship and leads the deceived toward taking an alternative action.

Usefulness is Sheridan's sixth trait associated with trust. Usefulness of data or machines means responding in a useful way to create something of value for operators, eventually developing into trust. In fact, one branch of decision theory is explicitly based on such values: "Utility theory". This, however, raises a question: Does usefulness of data ensure the quality of decision-making, or make operators dependent on the DAs? In other words, do notions of data values help decision performance, induce trust, or both? Studies indicated that humans tend to behave in different ways rather than using the estimated utility (e.g., Klein, 1997; Pulford & Colman, 1996; Tversky & Kahneman, 1974). Usefulness relates to the ability of a DA to respond in a *useful* or *responsible* way and to notions of value and utility of the DA itself in a local or specific sense. By this latter remark is meant that if the DA performs and produces output that is consistent with its planned concept of employment, then it will be judged *useful* in this sense. However, Usefulness also takes us back to the notions of value; DA's can be useful but not valuable. Most deceivers would probably not have insight into notions of usefulness since this can mean many things, but they may have some notions of value which can be exploited as described above.

The final aspect of trust in Sheridan's classification is *Dependency*. Trust is not a useful concept unless an operator is willing to depend on a machine. Dependency is one aspect of trust accessible to empirical measurement, i.e., by the fraction of time an operator behaves as if the machine were trustworthy. From Wickens (1992), we see that likelihood of use of automation is commensurate with level of trust in automation and inversely related with level of self-confidence of operator which implies a number of things. For instance, if the operators are known to be poorly trained then their self-confidence would presumably be low and their reliance on automated DA's high, perhaps to an unwarranted degree. Also, if the DA has only been built to handle easy cases and has never confronted the fringe-condition where it can fail, the operator may develop an unwarranted trust in its competence.

STRUCTURING EXPERIMENTS FOR INVESTIGATING TRUST IN AN IW DOMAIN

Given the tools for measuring trust that were presented in human factors literature, it is possible to consider the types of empirical studies which could be performed to investigate human trust in IW domains. Such controlled investigations would provide a better understanding of what situation characteristics influence trust, as measured by the either or both of the psychophysical ratings and performance and process measures, and also how changes in an operator's trust in system components affect ultimate system performance. In order to develop possible scenarios for investigating aspects of trust in an IW, it is instructive to consider how trust has been investigated in other automated systems.

Previous Investigation of Trust in Automated Support Systems

As described before, empirical work in the area of human trust in an automated support (decision-aided) system is limited, and has concentrated primarily on investigating trust in simulated, semi-automated process control environments. Moreover, and importantly as regards our concerns for IW environments, these studies have been in *non-adversarial* domains (i.e., Lee & Moray, 1994; Muir & Moray, 1996). Different system aspects were altered to see how participants' trust in systems components, such as the automated controller, was affected. In particular, Muir and Moray (1996) altered the quality of the pump systems by introducing either *random* or *constant errors* in its ability to maintain a set-point, introduced errors into the pump's *display* of its pump rate (although actual pump rate was error-free), and the performance of the automated controller in setting and maintaining appropriate settings for the pump. Lee and Moray introduced *faults into pump performance* (Lee & Moray, 1992) or *faults into either automatic or manual controllers* (Lee & Moray, 1994). These conditions are not unlike the type conditions that may arise in IW environments. Trust was measured both subjectively, using rating scales, and objectively, by logging participants' actions (e.g., hypothesizing that more or less use of an automated control system implied more or less trust in that automated system). Because faults were introduced into different components, these experiments investigated trust in a particular system aspect rather than trust in automation generally.

Designing Experimental Scenarios for Studies of Trust in AADM Environments

In the AADM environments of interest to this study, pictured in Figure 2, data fusion techniques are used to aid the decision-maker by synthesizing data from numerous sources into a form useful for the decision-maker. Because the environments of interest are ones involving adversaries, the possibility of corruption in either or all of the data, fusion algorithms, and displays involved in such decision-aiding systems can be introduced by *Information Operations* manipulated by the hostile forces. It is in such environments that we would like to perform human-in-the-loop experiments to study various hypotheses related to human trust under IW conditions and in AADM environments. As mentioned before, the literature is not very helpful in this regard. The multi-faceted manner in which trust was investigated in the above experiments suggests two dimensions along which studies of human trust in complex environments, such as an AADM environment, could vary: we called these the *system dimension* and the *surface-depth level*.

System Dimension. In the pasteurization experiments (Lee & Moray 1994; Muir & Moray, 1996) the quality of system performance was manipulated at what could be called different *system* levels. Faults or random errors were introduced at the *level of the (system) environment;* the process control system itself (i.e., the pumps), and at the *level of a (system) control intervention* (i.e., the automated controller). There are analogous levels in an AADM environment. The physical component level in the pasteurization experiments–the pumps and heaters–corresponds to the *actual tactical situation* that is taking place. Just as the states of pumps and heaters can be observed and controlled, the states of hostile and friendly assets can be assessed, and actions related to the situation can be taken. The next level, *data fusion systems and algorithms*, which automatically combine and synthesize information obtained form the tactical environment, can be considered analogous to the automated controller in the pasteurization experiments, which used information from the physical control system to automatically take control actions. Finally, in an AADM environment, one can consider a third level, the *interface level*. At this level, the results of the data fusion algorithms are displayed to the operator, in order to aid decision making.

Surface-Depth Level. Another dimension along which investigations of trust can vary is a *surface-depth level*. The *surface level* corresponds to the information available about the environment (as formalized in Brunswik's Lens Model; Cooksey, 1996), whereas the *depth level* corresponds to the actual state of the environment. The manipulations performed by Muir and Moray (1996) can be described in terms of these dimensions. Muir and Moray (1996) manipulated both the characteristics of the pump itself (depth level) and the display of the pump rate (surface level). This surface-depth dimension can be applied at all three of the system dimension levels described above, resulting in six combinations (see Figure 1).

Further Categories of Corruption. Within each cell of upper portion of Figure 1, it is possible to identify various types of malfunction, or causes of information degradation or corruption.

- Degradation. The *quality* of the system component can be degraded through constant, random errors, or discrete failures.

- Failure. System components can *fail* completely resulting in a loss of data. Different causal factors for the corrupting processes can also be considered:
- Non-intentional. System components can degrade due to non-intentional malfunction.
- Sabotage. An enemy can take *intentional action* to interfere with a system component.
- Subterfuge. An enemy can take intentional action both to interfere with a system component, *and to disguise that sabotage*.

System Dimension	Surface-Depth Dimension	
	Surface Level	Depth Level
Environment: Tactical Situation	Sensed and Observed Data	Evolving Tactical Situation
Intervention: DF Algorithms	Result of Algorithms	Data Fusion Algorithms
Interface: Decision Aid	Display Format	Information to be Displayed

Level of Malfunction
(Degradation / Failure)

Causal Factors
Non-intentional,
Sabotage, Subterfuge

Figure 1. Components of an AADM Environment described along a system and a surface-depth dimension. Potential experimental scenarios and manipulations, organized by system and surface-depth dimensions and levels of malfunction, causal factors.

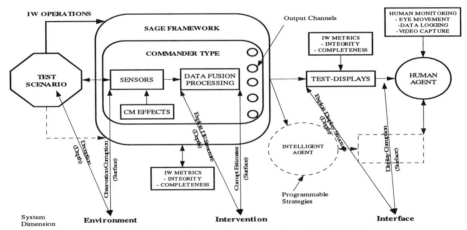

Figure 2. Experimental concept for AADM IW

Given a particular experimental context, the surface-depth and system dimensions, along with the levels of malfunction, and causal factors, can be used to systematically define a series of experimental manipulations which can be used to investigate issues of trust in AADM environments (Figure 1).

Design of Experimentation. For studies of AADM, a possible experimental context would be an interactive battle simulation in which people must make interpretations and/or decisions (e.g. identification of unknowns, decisions to engage hostile forces) based on information gathered and fused into decision-aiding estimates about the situation, such as those related to electronic emissions, weapons profiles, and locations and movements of various agents. The simulation would include data fusion modules which could synthesize environmental information in order to aid the participants decisions.

In either cases of having single or multiple participants, it is also possible to consider that some agents or participants would be synthetic; that is, created in software as so-called *intelligent agents*. This would add a dimension of experimental control, since the behavior of the agent would be fully controllable, or at least controllable within known limits. If one is to study such environments experimentally, a synthetic environment of this type needs to be created in a controlled way. In developing such simulation environments one of the primary capabilities to establish is that of the *problem space* or the framework from which problem information, data, and parameters evolve. Some call this type of capability a *scenario generator*, in which simulated observable cues are produced as inputs to processes under study. In our case, we are proposing the use of an existing capability for this, the so-called "SAGE" (Semi-Automated Ground Environment) software system. SAGE can also provide representation of two adversarial commanders, or at least the information environments in which their decision-making is being conducted.

SUMMARY AND FUTURE RESEARCH

Applicability of the Sheridan's classification into AADM systems has been investigated. Because the classification offers a variety of characteristics of trust in supervisory control systems, it forms a good basis for the conceptualization of trust in AADM environment and further development of experimental framework. Due to the particular characteristic of the domain which adversaries, and multiple processes are involved, however, the experimental framework should consider the possible points of attack which could be varied from the real world to human operators. Based on the interpretation of the classification from the AADM perspective, then, three factors, and scenarios are discussed for an experimental framework in AADM environment. These factors include the dimension of the point of attack, degree of attack, and position of attack which may affect human operator's role of monitoring activities and decision-making actions, which consequently may have an impact on operators' calibration of trust.

ACKNOWLEDGMENTS

This work was supported by the Air Force Armstrong Laboratory; the support of Mr. Gil Kupperman is gratefully acknowledged. SAGE was made available by Ball Aerospace Corp., whose support is gratefully acknowledged.

REFERENCES

Cooksey, R. W. (1996). *Judgment analysis: Theory, methods and applications*, New York: Academic Press.

Klein, G. (1997). The Recognition-Primed Decision (RPD) Model: Looking back, Looking forward. In C. E. Zsambok and G. Klein (Eds.), *Naturalistic decision making* (pp. 285-292). Mahwah, NJ: Lawrence Erlbaum.

Lee, J. D., & Moray, N. (1992). Trust, control strategies and allocation of function in human-machine systems. *Ergonomics, 35*(10), 1243-1270.

Lee, J. D., & Moray, N. (1994). Trust, self-confidence, and operators' adaptation to automation. *International Journal of Human-Computer Studies, 40*, 153-184.

Muir, B. M., & Moray, N. (1996). Trust in automation: Part II. Experimental studies of trust and human intervention in a process control simulation. *Ergonomics, 39*(3), 429-460.

Pulford, B. D., Colman, A. M. (1996). Overconfidence, base rates and outcome positivity/negativity of predicted events. *British Journal of Psychology, 87*(3), 431-445.

Sheridan, T. B. (1980). Computer control and human alienation. *Technology Review*, October, 61-73.

Tversky, A., & Kahneman, D. (1974). Judgment under uncertainty: Heuristics and biases. *Science, 185*, 1124-1131.

Wickens, C. (1992). *Engineering psychology and human performance*. 2nd Ed., New York: Harper-Collins.

Modeling Human Trust in Complex, Automated Systems Using a Lens Model Approach

Younho Seong   Ann M. Bisantz
State University of New York at Buffalo

INTRODUCTION

Automation has played an important role in supporting human and system performance in complex modern systems, such as aviation and process control. The advent of automation has changed the role of the human operator from performing direct manual control to the management of different levels of computer control. Human operators assume roles as supervisory controllers, interacting with the system through different levels of manual and automatic control (Sheridan & Johannsen, 1976). Therefore, the human operator must understand how to interact with the automated system, how the automation works, how to respond to system outputs, and how and when to intervene in the process, if the process fails. One factor affecting this interaction is the operator's trust in the automated system. Sheridan (1980) emphasizes the importance of human trust in automation as playing a key role in determining the level of a human operator's reliance on and the degree of intervention in automation and appropriate use of automation. Trust has been studied mainly from a sociological perspective which focused on interpersonal relationship between individuals. Following the sociological definitions of trust, more recent studies (Lee & Moray, 1992; Muir & Moray, 1996) have constructed models of human operator's trust in automated systems and shown how human trust in automated process control systems may affect system performance. These studies have focused on determining the extent to which human operator's trust in machines might affect system performance, and if so, identifying potential factors affecting the level of the operator's trust. An important concept regarding human trust is the notion of calibration: operators must have an appropriate level of trust in the information or automated system, given the characteristics of the situation. As Muir (1994) indicated, "well-calibrated" operators are better able to utilize automated systems. In case of aided adversarial decision-making systems, understanding how well an operator judges the level of data integrity, based on the observable characteristics of the situations becomes a very critical issue. We propose that Brunswik's Lens Model of human judgments (Brunswik, 1955; Hammond, Stewart, Brehmer, & Steinmann, 1975) may be useful in formalizing the study of trust. The Lens Model provides dual models of a human judge and the environment to be judged, and allow the extent to which an individual's judgment behavior captures the structure of the environment to be assessed. This extent can provide a description of how well an operator's trust in the information, is calibrated to the actual environmental situation, as described by the relationship between those characteristics and the actual integrity of the information.

PREVIOUS MODELS OF HUMAN TRUST

Sociological Models of Human Trust

Rempel, Holmes, and Zanna's (1985) definition of trust contains critical aspects of trust which can be used to examine human trust in automation from the human factors perspective. They emphasized not only components of interpersonal trust, but also the dynamic characteristics of trust toward a partner, regarding trust as a generalized expectation related to the subjective probability which an individual assigns to the occurrence of some set of future events (Rempel, *et al.*, 1985). That is, the study suggested that humans evaluated their partners based on the characteristics they observed. Therefore, these characteristics served as cues to determine the level of human trust. The study is valuable in that it allows us to understand the importance of the role of human trust in the sociological domain and to identify certain characteristics of trust. Other research has also suggested that trust is a multi-factorial concept (Barber, 1983; Zuboff, 1988).

Human Factors Models of Human Trust

Based on Barber's (1983) study of trust, Muir (1994) constructed a hypothetical model of trust in machines, consisting of a linear combination of the characteristics identified by Barber. That is, the model represents human trust as a combination of persistence, technically competent performance, fiduciary responsibility, and interaction effects between these characteristics. In addition, Muir produced an integrated framework or model by crossing Barber's (1983) dimensions of trust, with Rempel *et al.*'s (1985) framework of trust as a process of hierarchical stages, developing over time. Lee and Moray (1992) extended the Muir's work and established a dynamic, mathematical model of trust based on a series of experiments. The model reflected dynamic characteristics, in that the current level of trust was affected by the previous level of trust and system-oriented factors such as the presence of automation faults and level of joint system performance. Both models, however, failed to explicitly consider the calibration of trust and the true state of automation trustworthiness, although the need to assist human operators in calibrating their trust was recognized. We propose that Brunwik's Lens Model may be valuable in describing human trust in automated systems, since it provides a mechanism for capturing this notion of calibration as well as the true state of automation trustworthiness.

LENS MODEL

Brunswik's Lens model, shown in Figure 1, is a symmetrical framework which describes how both the *environmental structure* and patterns of *cue utilization* collectively contribute to judgment performance. In this model, the judge combines cue information (X_i) about the environment to make a judgment (Y_s). The model represents the classical notion of information transformation from stimulus (information presentation) to response (judgment) in which humans process information internally to yield some functional response based on cues observed, which in turn are representations of the environmental state. Thus, the model includes not only a classical decision concept, i.e., how humans sample and combine cues presented to them, but also the relationship between available cues and the true state of the environment. By analyzing a judge's cue utilization policy, therefore, we may be able to understand how that judge has adapted to the structure of the environment. The predictability of the environment, given a set of cues (the ecological validity of the cues) can also be assessed. Therefore, this model allows us to assess and evaluate how well the true environment structure is represented via a set of cues. Additionally, achievement, denoted as r_a, represents how well human judgments correspond to the actual values of the environmental criterion to be judged. Achievement is shown in Figure 1 as a line connecting judgments to criterion values. Because the Lens Model provides the means for considering the judge's adaptation to the environment, and the degree of achievement, both of which relate to the calibration of human trust, it seems that the use of the Lens Model approach to model human trust in automated systems is reasonable.

APPLYING THE LENS MODEL INTO HUMAN TRUST IN COMPLEX, AUTOMATED SYSTEMS

Conceptually, modeling human trust in automated systems using the Lens Model is relatively straightforward. The judgment modeled in this case is the operator's judgment of the trustworthiness of some system component or output. That is, the operator decides whether or not a system component is to be trusted. In Lens Model terms, then, the environmental criterion is the actual trustworthiness of the component. The judgment is the operators' assessment of that trustworthiness. To make this judgment, the

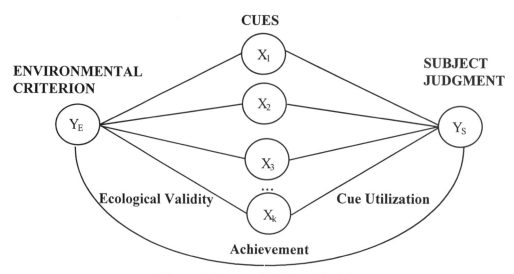

Figure 1. Brunswik's Lens Model

operator must rely on a set of observable cues which have some relationship to the components trustworthiness. In this paradigm, the concept of calibration is explicitly measured by achievement (r_a) – the extent to which the operator's assessment of trustworthiness matches the true state of the environment. One can also consider calibration to include operator's adaptation to the structure of the environment, in terms of the relationship between the cues and actual integrity of information.

Further specification and experimental verification of this model of trust in automation beyond the general level noted above presents certain difficulties, however. First, there is no clear, objective measurement of the true state of environment, in terms of its trustworthiness. Generally, trust as a state in itself has been measured only subjectively. This is problematic in terms of the Lens Model formulation, since application of the Lens Model and evaluation of the model parameters requires knowledge of the *true* environmental state. To circumvent this difficulty, we propose transforming the judgment from one of an assessment of trustworthiness to one that is more performance oriented. From an engineering standpoint, we are interested in human trust in a system to the extent to which that trust affects system performance. For instance, we are interested in whether or not operators utilize an automated controller, or obtain certain data, given their trust in that controller or information source. The true state of the environment, in terms of the adequacy of the controller, or the integrity of the data source, can be objectively determined. For these examples, the operator judgment would be whether to use the controller or the data. More generally, the operator judgment is one of component utilization, and true state of the environmental criterion is whether or not the component should have been used. In terms of trust, this assume that an operator's behavior in utilizing a system component reflects their trust in that component.

Second, to implement a Lens Model description of human trust in automation, it is necessary to specify what cues might be available for an operator to make a judgment about whether to use a system component. Candidate cues include the components of trust identified by previous studies of trust (e.g., Barber, 1983; Rempel, *et al.*, 1985; Zuboff, 1988). For instance, cues could include such factors as predictability, dependability, faith, reliability, competence or robustness. To be included in a quantitative Lens Model, these cues would be both measurable, and available to the operator. The availability of these candidate cues to the operator depends to some extent on how information is displayed to operators. However, the consideration of how to measure these cues must be addressed. For example, consider predictability. If we define the environment to be judged in terms of a subsystem, or set of systems, we can represent predictability in terms of the degrees of freedom in performance that were designed into the system. That is, predictability could be measured in terms of allowed error or performance variance. The smaller the degree of freedom, or allowable error, the more predictable the system is. If predictability is one component of trust, as Barber claimed, then trust will be negatively impacted by a large degree of performance variability. Additionally, the reliability of a system or component could be measured in terms of past performance (e.g., breakdowns, errors, etc.).

Instantiating the Model

To evaluate the model, an experimental framework has been established in an Information Warfare (IW) domain (Seong, Llinas, Drury, & Bisantz, 1998) in which one can consider trust in the context of aided adversarial decision making, where military officers must assess the integrity of information which may be intentionally altered or degraded by an enemy. In this domain, the points of attack by an enemy can be the real battle situation, data gathering or fusion algorithms, or a data transfer network. By changing the points of simulated attack, we may able to observe how operators successfully calibrate their trust in terms of accurately pinpointing the point of attack, and changing the level of trust. In the IW domain, studying human trust is important for several reasons. For example, forces might be vulnerable to information attacks which diminish their trust in data fusion or other decision aids, rendering these assets less useful, or to deceptive attacks, in which an inappropriately high level of trust in the aid is maintained. In terms of the Lens Model approach, data, fusion algorithm outputs, would be judged as usable or not (e.g., trustworthy or not), based on operators understanding of the predictability, reliability, etc. of the information displayed to them.

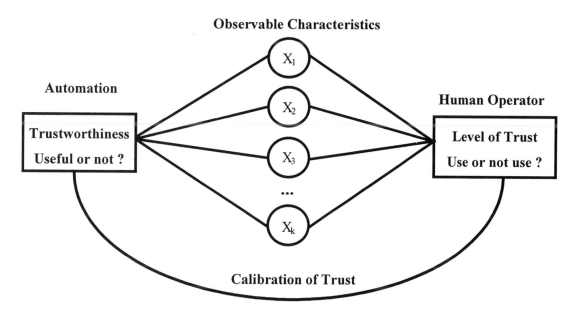

Figure 2. Model of human trust in automation using the Lens model

SUMMARY AND FUTURE RESEARCH

A Lens Model approach for modeling human trust in automated systems has been proposed. Because the Lens Model provides the means for modeling both human judgment policy and the actual structure of the environment, it allows operator calibration to the actual trustworthiness of a system to be explicitly considered. Conceptual solutions for addressing certain difficulties with this approach, such as the objective determination of the true state of system trustworthiness, and the identification and measure of cues which reflect system trustworthiness, were discussed. Finally, an experimental framework in the domain of

Information Warfare was described, which may provide the means for further instantiating and evaluating the effectiveness of this model of human trust in automation.

REFERENCES

Barber, B. (1983). *The logic and limits of trust*, New Brunswick, NJ: Rutgers University Press.

Brunswik, E. (1952). *The conceptual framework of psychology*. Chicago, IL: Univ. Chicago Press.

Cooksey, R. W. (1996). *Judgment analysis: Theory, methods and applications*, New York: Academic Press.

Hammond, K. R., Stewart, T. R., Brehmer, B., & Steinmann, D. O. (1975). Social judgment theory. In M. F. Kaplan & S. Schwartz (Eds.), *Human judgment and decision processes*. New York: Academic Press.

Muir, B. M. (1994). Trust in automation: Part I. Theoretical issues in the study of trust and human intervention in automated systems. *Ergonomics*, 37(11), 1905-1922.

Muir, B. M., & Moray, N. (1996). Trust in automation: Part II. Experimental studies of trust and human intervention in a process control simulation. *Ergonomics*, 39(3), 429-460.

Lee, J. D., & Moray, N. (1992). Trust, control strategies and allocation of function in human-machine systems. *Ergonomics*, 35(10), 1243-1270.

Rempel, J. K., Holmes, J. G., & Zanna, M. P. (1985). Trust in close relationships. *Journal of Personality and Social Psychology*, 49(1), 95-112.

Seong, Y., Llinas, J., Drury, C. G., & Bisantz, A. M. (1998). Human trust in aided adversarial decision-making systems. In M.W. Scerbo & M. Mouloua (Eds.), *Automation technology and human performance: Current research and trends*. Mahwah, NJ: Lawrence Erlbaum.

Sheridan, T. B. (1980). Computer control and human alienation. *Technology review*, October, 61-73.

Sheridan, T. B., & Johannsen, G. (1976). *Monitoring behavior and supervisory control*. New York: Plenum.

Zuboff, S. (1988). *In the age of the smart machine: The future of work and power*. New York: Basic Books.

HUMAN FACTORS ISSUES

Flight Deck Automation Issues

Beth Lyall
Research Integrations, Inc.

Ken Funk
Oregon State University

Hundreds of articles and presentations have addressed problems and concerns with flight deck automation. Many have raised general concerns about the approaches taken to automation use philosophies and automation design. Others have addressed specific problems or concerns identified with particular designs or implementations of automation.

The first phase of our research identified 114 human factors problems and concerns with flight deck automation (Funk, Lyall, & Riley, 1995). In the second phase of our project we used a wide variety of sources to locate and record evidence related to these problems and concerns. Because an *issue* is "[a] point of discussion, debate, or dispute ..." (Morris, 1969), we decided to change our terminology and refer to these problems and concerns as *flight deck automation issues*. This is what will be used for the remainder of this paper.

The sources we reviewed for evidence included accident reports, documents describing incident report studies, and documents describing scientific experiments, surveys and other studies. We also conducted a survey of individuals with broad expertise related to human factors and flight deck automation (human factors scientists, aviation safety professionals, pilots, and others), and evaluated a related set of ASRS incident reports. We reviewed these sources for data and other objective information related to the issues. For each instance of evidence we qualitatively assessed the extent to which it supported one side of the issue or the other, and assigned a numeric strength rating between -5 and +5. We assigned a positive strength rating to evidence supporting that side of the issue suggested by its issue statement (supportive evidence) and a negative strength rating to evidence supporting the other side (contradictory evidence).

For example, consider the statement of *issue065: Pilots may lose psychomotor and cognitive skills required for flying manually, or for flying non-automated aircraft, due to extensive use of automation.* If we found evidence in a source indicating that pilots lose manual flying skills due to extensive use of automation (at least under some circumstances), we recorded the related excerpt from the source document and assigned this supportive evidence a positive rating, perhaps as great as +5. If we found evidence in a source indicating that pilots can and do maintain manual proficiency even with extensive use of automation (at least under some circumstances), we recorded the related excerpt and assigned this contradictory evidence a negative rating, perhaps as great as -5.

We developed detailed strength assignment guidelines for evidence from each type of information source. For example, in pilot surveys of automation issues, if at least 90 per cent of the respondents were in agreement with a statement consistent with an issue statement, we assigned a strength rating of +5. If at least 90 per cent were reported as disagreeing with a statement consistent with an issue statement, we assigned a strength rating of -5. During the process of collecting and recording evidence, we revised, updated, consolidated, and organized the issues, finally yielding 92 flight deck automation issues.

THE FLIGHT DECK AUTOMATION ISSUES WEBSITE

Our work has yielded a body of information consisting of flight deck automation issues, unsubstantiated citations of those issues, supportive and contradictory evidence for the issues, and a bibliography of documents related to the issues and flight deck automation in general. To disseminate this information, we created a World Wide Web site accessible by the following Universal Resource Locator:

http://flightdeck.ie.orst.edu/

The website includes details about all aspects of our project. In particular, the methodology for each component (including strength rating assignment) is fully described, the list of flight deck automation issues is presented, several taxonomies organizing the issues under different sets of categories are included, and access to a searchable, read-only version of our database is provided. Though we hope to continue adding to it, the database currently contains the following:

- a bibliography of more than 500 automation-related documents,
- more than 2,000 unsubstantiated citations of possible problems and concerns (from Phase 1),
- 92 flight deck automation issues,
- more than 100 studies reviewed for evidence related to those issues,
- more than 700 records of supportive and contradictory evidence related to the issues, and
- 282 ASRS incident reports reviewed for evidence in our own incident study.

EVIDENCE APPLICATION

The following example is given to show how the searchable database on the website can be used. A search of the current database asking for evidence associated with the problem "Pilots may not understand the structure and function of automation or the interaction of automation devices well enough to safely perform their duties (Issue105)," would identify 51 evidence records. These 51 records are from 14 references - five accident reports, two experiments, five survey studies, one observation study, and one incident study. Thirty-seven of the evidence records were assigned positive strength ratings (i.e. provide information consistent with the issue statement about how pilots may not understand automation), and 14 evidence records were assigned negative strength ratings (provide information refuting the issue statement about how pilots have been shown to understand automation). Each of the evidence records includes the strength of the evidence, the bibliographic reference for the source, an excerpt from the document stating the information upon which the evidence was based, and the type of source it comes from (e.g. accident report, survey, etc.) A representative subset of this evidence query is presented in Table 1.

The evidence information can be used to summarize the perspectives and nuances that have been documented about this issue, as well as provide a starting point for identifying and evaluating potential solutions. This type of query can be accomplished for any of the issues by using the database search capabilities included on the website. It is our goal to keep the database current to allow researchers and practitioners to have a common base of information from which to work, and to provide a means for coordinating future efforts to address flight deck automation issues. The database can also be queried by keyword or for evidence records related to other factors like a particular reference source or study type.

DATA SUMMARY

Besides looking at each issue individually, it is valuable to look across all of the issues to identify those that may need more attention or may not be considered in need of attention at this time. We have summarized the information using several methods; two will be presented here. The first is a representation of the top 10 issues with overall positive evidence as measured by the sum of the values of the strength ratings (positive and negative) for each issue. These issues are presented in Table 2. It can be argued that for these issues we now need to focus on solutions, not more research to confirm that they are problems.

Table 1

Sample evidence records for "Pilots may not understand the structure and function of automation or the interaction of automation devices well enough to safely perform their duties (Issue105)."

Strength	Evidence Excerpt	Source Reference
-4	"Another problem related to mode engagement was the attempt to activate a mode without the prerequisites for this activation being met. Fifty percent of the transitioning pilots and 1 of the 14 experienced pilots tried to engage VORLOC without being in the manual radio mode as required." That is, 16 of 20 (80%) did it correctly.	Sarter, N.B., & Woods, D.D. (1994)
-3	Statement 34: "There are still modes and features of the B-757 FMS that I don't understand." From the histogram of the responses in Phase 1 of the study, 34% of the pilots agreed or strongly agreed with the statement and in Phase 2 of the study, 21% of the pilots agreed or strongly agreed with the statement while 53% disagreed or strongly disagreed in Phase 1, and 64% disagreed or strongly disagreed in Phase 2. The neutral responses were 13% in Phase 1 and 15% in Phase 2.	Wiener, E.L. (1989)
-2	"... when asked to intercept the LAX 248 [degree] radial, all 6 of the transition pilots had difficulties carrying out the task using LNAV, as compared to only 7 of the 14 experienced pilots." That is, 7 out of 20 (35%) did not have difficulties.	Sarter, N.B., & Woods, D.D. (1994)
+2	Statement 34: "There are still modes and features of the B-757 FMS that I don't understand." From the histogram of the responses in Phase 1 of the study, only 34% of the pilots agreed or strongly agreed with the statement and in Phase 2 of the study, only 21% of the pilots agreed or strongly agreed with the statement while 53% disagreed or strongly disagreed in Phase 1, and 64% disagreed or strongly disagreed in Phase 2. The neutral responses were 13% in Phase 1 and 15% in Phase 2.	Wiener, E.L. (1989)
+3	"... the training observations indicate that pilots do not perceive the FMS as one large integrated system consisting of a variety of closely related, interacting subsystems such as the MCP or the CDU. ... Our data show that pilots think of and operationally use the MCP and CDU as, at least two different systems."	Sarter, N.B., & Woods, D.D. (1992)
+4	"The GA mode becomes available when descending below 2,000 ft radio altitude with autothrottles armed. Out of 20 pilots, only 5 [25%] recalled the altitude at which this occurs. Eight pilots (40%) knew that the availability of the mode depends on reaching a certain altitude, but they did not remember the actual height."	Sarter, N.B., & Woods, D.D. (1994)
+5	"2.2 Conclusions (a) Findings ... 7. The autopilot was utilized in basic CWS. 8. The flightcrew was unaware of the low force gradient input required to effect a change in aircraft attitude while in CWS. ..."	NTSB (1972)

We have also done this summary based only on the difference between the number of positive evidence records and the number of negative evidence records. The 10 issues remain the same, although the rank order changes somewhat.

Table 2

Top 10 issues with overall positive evidence measured and ranked by the sum of the strength ratings (positive and negative).

Issue ID	Abbreviated Issue Statement	# pos evid records	# neg evid records	sum of pos evid strength	sum of neg evid strength	sum pos + sum neg	Rank
105	understanding of automation may be inadequate	37	14	94	-31	63	1
83	behavior of automation may not be apparent	18	4	40	-5	35	2
131	pilots may be overconfident in automation	16	4	38	-5	33	3
92	displays (visual and aural) may be poorly designed	32	7	48	-16	32	4
133	training may be inadequate	25	12	48	-17	31	5
106	pilots may over-rely on automation	10	1	28	-1	27	6
40	automation may be too complex	15	3	26	-5	21	7
108	automation behavior may be unexpected and unexplained	16	5	29	-8	21	8
99	insufficient information may be displayed	16	3	26	-6	20	9
65	manual skills may be lost	14	12	37	-17	20	10

It can be seen in Table 2 that Issue105 (also presented in Table 1) is far beyond the other issues on this measure. Besides Issue105, it is interesting to note that the other three issues in the database related to the understanding of automation are also in the top 10: Issue40, Issue108, and Issue133. This suggests that it is clear that problems exist with pilot understanding of automation. Resources should now be committed to identifying and evaluating solutions to these related problems in design, operations, and training. Another area that is represented by two issues in the top 10 list is pilot over-reliance on automation represented by Issue106 and Issue131. Once again, this suggests that problems have been clearly supported in this area and now resources should be committed to developing solutions, although this may be more difficult than those related to the understanding of the automation.

Table 3 presents the other end of the spectrum: those issues that have overall negative evidence based on the sum of the strength ratings. The table represents all the issues with a sum less than zero. The overall weight of available evidence suggests that resources would be better used elsewhere than in the development of solutions to problems suggested by these issues. Further summary of all the issues can be found in Funk, et al. (in press). We also plan to add these to the website.

CONCLUSION

We believe that the work we have done will provide valuable information to those who would like to conduct research on the most pertinent issues related to flight deck automation. Our searchable database can also be used to identify work that has already been accomplished that may lead to possible solutions to these problems.

Table 3

Issues with overall negative evidence measured and ranked by the sum of the strength ratings (positive and negative).

Issue ID	Abbreviated Issue Statement	# pos evid records	# neg evid records	sum of pos evid strength	sum of neg evid strength	sum pos + sum neg
166	company automation policies and procedures may be inappropriate or inadequate	6	4	8	-9	-1
123	inadvertent autopilot disengagement may be too easy	2	1	2	-3	-1
115	testing may be inadequate	1	1	1	-2	-1
84	crew coordination problems may occur	10	6	11	-14	-3
156	fatigue may be induced	3	4	4	-10	-6
46	pilots may lack confidence in automation	16	12	21	-30	-9
139	inter-pilot communication may be reduced	5	5	7	-16	-9
79	automation may adversely affect pilot workload	19	25	32	-43	-11
13	job satisfaction may be reduced	4	6	4	-18	-14

REFERENCES

Funk, K., Lyall, B., & Riley, V. (1995). Perceived Human Factors Problems of Flightdeck Automation (Phase 1 Final Report for Federal Aviation Administration Grant 93-G-039). Corvallis, OR: Dept. of Industrial and Manufacturing Engineering, Oregon State University. Internet address: **http://www.engr.orst.edu/~funkk/Auto/autoprob.html**

Funk, K., Lyall, B., Wilson, J., Vint, R., Niemczyk, M., Suretogeh, C., Owen, G. (in press). Flight Deck Automation Issues. *International Journal of Aviation Psychology.*

Morris, W. (Ed.) (1969). *The American heritage dictionary of the English language.* Boston: Mifflin.

NTSB Aircraft Accident Report: Eastern Airlines, Incorporated, L-1011, N31OEA, Miami, FL, December 29, 1972. NTSB-AAR-73-14

Sarter, N.B., & Woods, D.D. (1992) Pilot interaction with cockpit automation: Operational experiences with the Flight Management System. *International Journal of Aviation Psychology, 2*(4), 303-321

Sarter, N.B., & Woods, D.D. (1994) Pilot interaction with cockpit automation II: an experimental study of pilot's model and awareness of the Flight Management System. *International Journal of Aviation Psychology, 4*(1), 1-28.

Wiener, E.L. (1989) Human Factors of Advanced Technology ("Glass Cockpit") Transport Aircraft (NASA CR 177528). Moffet Field, CA: NASA Ames Research Center.

Injury Reduction Ergonomics in the Development of an Automated Squeezer Molder Line in a Gray and Ductile Iron Casting Facility

Angela Galinsky, *University of South Dakota*
Carryl Baldwin, *Western Iowa Technical Community College*
T. Dell, Joanne Benedetto, J. Berkhout, *University of South Dakota*

INTRODUCTION

In the course of a company-wide ergonomic evaluation of workstations, three squeezer-molder stations in the casting division of the foundry were found to generate an unacceptably high rate of musculoskeletal injuries. OSHA 200 logs for these workstations provided documentation of strains, tears and back injuries dating back to 1981.

We conducted a detailed inventory of all discrete operator tasks performed at these stations. Each task was further analyzed for its effect on operator safety, productivity, machine integrity, and for the derivation of design requirements for possible automation of the squeezer-molder stations.

Extensive automation of these workstations was determined to be the most cost-effective way of reducing the physical hazards to the operators.

This foundry produces gray and ductile iron castings of up to 1,500 pounds. Approximately 150 employees work in the casting facility. Most of the workstations require considerable physical effort. Forklifts move pallets of molds, cores and castings. An overhead conveyor moves crucibles of molten metal. Other lifts and displacements are human powered, some using wheeled carts.

The company is known for its low scrap rate (less than 2%), and its aggressive quality control procedures. The company has an active safety committee, and its safety policies and procedures are well documented and implemented.

Ergonomic hazard inventory

Seventeen years of accident records were reviewed. Workstations were flagged for extensive analysis if they met any of the following tests:

1) a frequency of accidents above norms for the metal working industry;
2) any severe accidents involving lost time;
3) any injuries related to repetitive stress or chronic musculoskeletal disorders;
4) any upper extremity disorders;
5) any accidents suggesting the presence of noxious environmental conditions.

Minor injuries tended to occur to the most recent hires, while serious injuries were distributed among employees without regard to tenure. This suggests that experience reduced exposure to minor injuries, but that major workstation changes would be needed to reduce the more serious incidents.

Tables 1 and 2 below are excerpted from Dell and Berkhout (1998), who published an extensive review of injuries by work category at this same metal working facility.

One startling statistic is that less than one-fourth of all the molders working at the three sand mold stations escaped injury. This was due to their low turnover rate and long tenure at the shop, as well as the nature of the tasks they performed.

Table 1.
Percentage of workers escaping injury by job classification.

	All incidents	OSHA recordable
Coremakers	62%	84%
Hot metal workers	59	75
Grinders	50	69
Detail workers	47	67
Movers	33	54
Molders	**23%**	**43%**

Table 2.
Incident rate per 100 full time employees per year.

Grinders	101.7
Molders	**70.2**
Hot metal workers	65.0
Movers	58.3
Detail workers	53.8
Coremakers	42.7

Most of the injuries to grinders involved foreign particles in the eye and minor abrasions. Workstation design did not seem applicable to this group of minor injuries. The molders were identified as those most at risk for musculoskeletal disorders, and most in need of a thorough workstation analysis. In a subsequent analysis which includes the last three years (1995-1997), 82% of all lost time injuries were identified as having occurred at the molder workstations. Of those injuries, 97% were related to musculoskeletal strain. Also, the majority of days lost in this three year period were attributable to two workers who are in litigation proceedings with the workmen's compensation carrier and the company as a result of their injuries. Both workers have been with the company for several years and were at risk for musculoskeletal injury. Both workers presented with symptoms relating to work-related low back disorders (WRLBDs).

ANALYSIS OF SAND-MOLD WORKSTATIONS

Each worker at the squeezer-molder stations was videotaped while performing normal tasks. Open-ended interviews were conducted with all the employees doing this job. The employees were asked to define their tasks themselves, and comment on what they felt worked or did not work at their particular station.

Physical evaluations of the squeezer-molder stations included documentation of lighting conditions. Illumination intensity maps were prepared, showing both levels of illumination at all material handling points, and showing contrast glare levels. Sound and vibration levels were also documented. Anthropometric fit-to-equipment was assessed, with an emphasis on the heights of work tables and roller tables, sizing of tool grips, tool vibration, tool storage and retrieval, foot-stool use, manual assists, and both static and dynamic postures while working.

INPUTS TO DECISIONS ON AUTOMATION PROCEDURES

Given the videotapes of worker performance, employee interview data, and detailed accident records, the molder workstations were given the highest priority for hazard correction. An analysis of all operator tasks was undertaken using a systems approach. This included quantification of workload at different points in the sand molding procedure, and implementation of the Job Hazard Analysis (JHA) systems safety technique.

Hazards identified by the JHA procedure included chemical and particulate exposure, trip hazards, and musculoskeletal overload under certain limiting conditions. Sand overflow and other debris accumulate at the workstation and need to be periodically cleared. The weight lifting requirements at this workstation exceed NIOSH lifting limits. Under revised limits the company would still need to select new hires from men in the 90th centile of ability to lift heavy loads.

The molders' activities were performed within the following sequence: after positioning the mold an overhead hyperextension motion is required to release the gravity flow of molding sand from an overhead buffer. The operator is required to mechanically tamp the sand into the mold shells, turn the mold over after it is charged with sand on both sides, and place the completed mold assembly on a conveyor. Any of these motions can cause an overexertion injury or a twisting injury. Fatigued workers may substitute explosive motions for steady strength motions, further increasing the risk of a lower back injury, since overexertion of the upper torso in a twisted orientation puts a large strain on the muscles fastened to the lumbar vertebrae.

Cost / benefit ratios for justifying the automation of the molder stations included net company costs of $60,000 in a single year from expenses related to the two lower back injuries. These costs included actual sums paid out as a consequence of injuries; workers' compensation premium increases, worker replacement costs and litigation expenses.

Two station upgrades were considered. A partial station modification included a lift table with powered rotation to move the molds between work tables, and a redesigned sand release mechanism to eliminate the high arm reach currently needed.

This solution, while costing less than a single year's exposure to injury expenses, was felt to be inadequate because it didn't address the sand accumulation problem, and the addition of the table and redesigned sand bunker controls encroached on operator space in ways we felt could lead to unpredictable problems subsequent to installation.

A fully automated installation eliminates the three operators at the molding stations and replaces them with a process control line. The time spent shoveling sand overflow is, approximately, 20 hours a week. This cost is eliminated in the new installation. Mold shells will be loaded onto the line by human operators, but the sand filling, tamping and sealing operations will be done by mechanical devices monitored by a single observer. Although the full automation solution has a higher cost, it still promises a three to five year payback period, and is still justified by the potential savings in injury related outlays.

ACKNOWLEDGEMENTS

The authors would like to thank all the employees in the Foundry for their effort and insight into industrial ergonomics and safety.

REFERENCES

Dell, T., & Berkhout, J. (1998). Injuries at a metal foundry as a function of job classification, length of employment, and drug screening. *Journal of Safety Research, 29*, 9-14.

Human Factors and Ergonomics Society. (1997). Ergonomics and musculoskeletal disorders. In Karwowski, W., Wogalter, M., and Dempsey, P., Editors. Santa Monica, CA.: Human Factors and Ergonomics Society.

Pure-Tone Threshold Shifts During Moderate Workload Conditions

Carryl L. Baldwin
Western Iowa Tech Community College

Angela M. Galinsky
University of South Dakota

AUDITORY DISPLAYS

In today's complex systems, heavy demands may be placed on the operator both in terms of mental workload and human perceptual abilities. Automation may decrease operator demands in some instances; however, as pointed out by Parasuraman and Mouloua (1996) automation may also merely redistribute demands without reducing mental workload.

Because automated systems are not limited in terms of the amount of information which can be processed at any given time, designers have tended to make automated systems increasingly complex, thus shifting the operator's role to one of supervisory control or monitoring (Wickens, 1992). The more complex the system, the greater the likelihood that some part of the system will fail. Wickens points out that failures can occur in terms of: a) failures of the initial system (like the aircraft engine); b) failure of the automation (like the auto pilot); or, c) failure of the monitoring system either in terms of failing to indicate system failure or indicating system failure when in fact there is none. Auditory displays can be used to enhance the human performance capabilities of operators in complex environments and are therefore frequently used to convey essential information to the operator both supplementing visual information and providing early warning of system failure.

Due to the essential nature of auditory displays in many complex environments, operators are frequently required to demonstrate minimum auditory capabilities as measured by audiometric testing. However, in light of recent research demonstrating performance changes across decibel (dB) levels under conditions of moderate mental workload, which are not found in low mental workload conditions, the validity of standard audiometric tests for predicting actual performance in complex environments is called into question (Baldwin, 1996; 1997).

Older individuals experiencing decrements in auditory abilities may be particularly affected by the intensity of auditory displays. As our workforce ages, and the human role shifts increasingly to one of supervisory control in automated environments, strategies aimed at optimizing the performance capabilities of older workers become essential. Further investigation of the role of signal intensity and its relation to perceptual and cognitive processing of auditory displays has tremendous potential toward this end.

In the present investigation, the nature of the relationship between sensory abilities and the cognitive environment in which sensory abilities are examined was explored. It was hypothesized that pure-tone auditory thresholds would increase when participants were tested under moderate mental workload conditions (when participants had to perform a secondary task). Specifically, it was predicted that participants would require higher decibel levels to detect pure-tone frequencies when performing the detection test simultaneously with a cognitively challenging task as compared to detection thresholds obtained under standard assessment methods.

METHODS

Forty volunteers from a Midwestern college, with a mean age of 22.2 years, participated in the experiment. Each participant completed two audiometric exams testing pure-tone threshold levels at 500, 1000, 2000, 3000, 4000, 6000, and 8000 Hz for both the left and right ears. In one of the audiometric exams, participants were also required to perform a secondary task which required them to stack falling geometric shapes, as quickly as possible, into completed rows leaving no gaps between the various shapes. The secondary task was presented via a commercially-available, hand-held game device. All participants were allowed to practice the task prior to the experimental trials. The presentation order of the two audiometric exams was randomized across participants.

RESULTS

Analysis of threshold levels across all frequencies revealed a significant difference between decibel levels required for detection during standard versus cognitive workload trials, (F (1,39) = 30.78, p <.000). The mean threshold level for detection at each frequency for right and left ears obtained with standard and cognitive load methods is presented in Table 1 and Table 2 respectively, along with the difference in dB

Table1: **Mean Threshold and (Standard Deviation) in Decibels (dB) Obtained under Standard and Cognitive Load Methods for the Right Ear**

Frequency in Hertz	Standard Method Mean (SD) in dB	Cognitive Load Mean (SD) in dB	Difference in Threshold
500	10.9 (6.0)	13.4 (7.3)	2.5
1000	5.5 (4.2)	7.2 (4.9)	1.7
2000	4.5 (4.4)	6.9 (5.0)	2.4
3000	5.5 (6.6)	8.2 (6.6)	2.7
4000	6.9 (6.4)	10.2 (10.3)	3.3
6000	17.4 (9.1)	18.0 (9.0)	.6
8000	9.6 (8.7)	11.6 (8.9)	2

Table 2: **Mean Threshold and (Standard Deviation) in Decibels (dB) Obtained under Standard and Cognitive Load Methods for the Left Ear**

Frequency in Hertz	Standard Method Mean (SD) in dB	Cognitive Load Mean (SD) in dB	Difference in Threshold
500	12.4 (12.1)	14.5 (12.5)	2.1
1000	6.9 (11.4)	8.8 (10.4)	1.9
2000	5.5 (9.5)	9.3 (13.2)	3.8
3000	5.4 (9.0)	7.8 (10.3)	2.4
4000	7.1 (9.0)	10.1 (11.0)	3
6000	18.0 (9.9)	19.0 (9.3)	1
8000	7.9 (8.3)	10.6 (8.6)	2.7

between the two methods. Note that for all frequencies, threshold levels are higher when obtained under cognitive load conditions. The threshold shift is particularly dramatic for the 2000 Hz tone, a frequency critical for speech perception. These data are presented graphically in Figure 1.

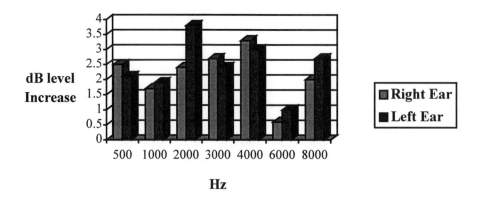

Figure 1. Increase in Signal Amplitude Required for Detection During Workload
as Compared to Standard Assessment Procedures

DISCUSSION

These results provide support for the proposed relationship between mental workload and sensory performance. Participants required higher signal amplitude to detect pure-tone frequencies when simultaneously performing a cognitively challenging task. These results call into question the validity of standard audiometric testing procedures for predicting performance in occupational settings. Standards for auditory presentation levels should take into account the cognitive as well as the sensory load imposed on the operator.

Older individuals in particular, may be affected by the intensity of auditory displays. The increased incidence of hearing impairment with advanced age is well documented (Corso, 1963; Gelfand, Piper, & Silman, 1985; Schieber and Baldwin, 1996). The presence of hearing deficit has the potential to exacerbate the performance decrements observed in young, normal-hearing-hearing listeners under low presentation intensities. In fact, Lindenberger and Baltes (1994) have found empirical evidence which indicates that visual and auditory acuity account for nearly half of the age-related reliable variance in intelligence and are a strong predictor of individual differences in intellectual functioning.

Previous research with speech stimuli presented to young, normal-hearing listeners in dual task trials has indicated that performance for simple speech processing tasks improves as auditory signal amplitude increases from 45 to 70 dB (See Baldwin, 1996; 1997). Although the precise nature of this intensity-performance relationship is still under investigation, the range in question is clearly audible and does not result in performance differences in single task trials. In other words, participants make no more errors and respond just as quickly to speech stimuli presented at 55 dB as to speech stimuli presented at 65 dB under ideal conditions where full concentration can be given to the speech processing task. However, when participants are required to share mental resources with a second task, this same dB presentation range results in performance decrements at the lower dB levels. Of interest is the fact that the performance decrements observed in young listeners resemble the magnitude of decrements sometimes observed in older individuals which have previously been attributed to cognitive impairment (Baldwin & Schieber, 1996).

What remains to be ascertained is whether this performance enhancement is due to a decrease in the amount of processing resources required. (Increasing amplitude might decrease the resources required for signal extraction and encoding.) Or, whether increasing amplitude leads to a stronger auditory trace or "echo" in sensory storage. If increasing the amplitude of an auditory signal strengthened the auditory trace, it might increase the duration of the auditory trace analogous to Averbach and Sperling's (1961) finding in the visual arena. Averbach and Sperling found that for visual stimuli, increasing the contrast between the stimuli and the background upon which the stimuli are displayed dramatically increased the length of the visual trace, a difference of 250 ms using low contrast (light fields) versus 2 seconds using high contrast (dark fields).

It can be reasoned that increasing the duration of the auditory trace would increase the accessibility of the auditory information simply by increasing the likelihood that the information would remain long enough to be encoded, a characteristic which would be particularly valuable in situations involving somewhat fluctuating but relatively high cognitive load. If for instance, the operator was temporarily engaged in a maneuver requiring the upper limits of his or her cognitive resources, then a competing auditory signal with brief duration might not get encoded. However, in the same situation, an auditory signal which persisted longer would have a greater chance of being available if a transient temporary reduction in cognitive demand occurred. Even slight increases in the duration of an echoic trace would have the potential to greatly increase the processing abilities of operators. This potential might be highest for populations experiencing perceptual-cognitive difficulties resulting from changes in peripheral or central processing mechanisms.

Auditory displays are advantageous in numerous situations, including settings where the visual system is overburdened or when a verbal response is required. Additionally, auditory displays can facilitate visual displays. For example, Fujawa and Strybel (1997) have shown that auditory indicators can significantly reduce the time taken to locate a visual target. Sanders and McCormick (1993) recommend that tonal signals be presented 40-50 dB above threshold for at least 500 ms and suggest that frequencies between 200 and 3000 Hz be used. Patterson (1983) recommends that alarm intensity be a minimum of 15 dB above background noise but less than 25 dB above background noise. Although providing valuable insights, current guidelines fail to take into account the cognitive complexity of the operating environment. Fujawa and Strybel (1997) found that the performance enhancement achieved with auditory indicators increased as the amplitude of the indicator increased from 40 dB to 70 dB. Results of the current investigation indicate that as the cognitive complexity of the operating environment increases, the signal amplitude necessary for optimum performance needs to increase. These findings have implications for establishing new guidelines for auditory displays.

Beringer and Harris (1997) cite an interesting example of failure to detect auditory signals which would normally be at or above recommended intensity levels. According to Beringer et. al., a report from the FAA Aircraft Certification Service describes instances where pilots in the cockpit fail to detect an auditory alarm which is clearly audible to the test administrator. In these instances, the test administrator at ground level has no trouble hearing the alarm which indicates that the auto pilot has disengaged. In fact, the alarm is intense enough to be annoying to the test administrator and yet the pilot fails to notice the alarm.

Beringer reports that this phenomena is most prevalent among older pilots despite the fact that the older pilots have successfully passed at least an informal "whisper test" of auditory abilities. Essentially, the anecdotal test is described as consisting of whispering a message to the pilot. If the pilot can demonstrate comprehension of the message then the pilot's hearing is deemed adequate for operation. If the pilot fails to demonstrate comprehension of the message, then the pilot is referred for standard audiometric testing. Although the whisper test is far from scientific, it would seem to indicate that one's hearing thresholds were within reasonable levels. And yet under simulated flight conditions, these pilots are failing to detect an auditory warning considerably higher than threshold.

This could be interpreted as severe auditory deficit on the part of the pilot; however, an alternative explanation is that while experiencing periods of high cognitive demand, the pilot may have lacked sufficient available mental resources with which to process the auditory signal. Results of the current investigation indicate that detection thresholds may be higher during periods of high mental workload.

The current results provide support for the interaction between mental workload and sensory performance. Future research exploring the nature of this relationship among aging populations is warranted. Increasing the presentation intensity of auditory displays has the potential of facilitating optimal performance in complex automated environments. This faciliatory effect may prove particularly advantageous to older workers. As our population as a whole ages, strategies aimed at optimizing the performance capabilities of older individuals become increasingly imperative.

REFERENCES

Averbach, E., & Sperling, G. (1961). Short-term storage of information in vision. In C. Cherry (Ed.), *Information theory: Proceedings of the fourth London symposium.* London: Butterworth.

Baldwin, C. L. (1997). Speech Amplitude and Cognitive Processing Requirements. *Proceedings of the Human Factors and Ergonomics Society 41st Annual Meeting,* (Abstract p. 1406). Santa Monica, CA: Human Factors/Ergonomics Society.

Baldwin, C. L. (1996). Speech Intensity and Cognitive Processing Requirements During Concurrent Visual and Auditory Task Performance. Unpublished Dissertation. Vermillion, SD: University of South Dakota.

Baldwin, C. L., & Schieber, F. (1995). Age Differences in Mental Workload with Implications for Driving. *Proceedings of the Human Factors and Ergonomics Society 39th Annual Meeting,* (pp. 167-171). Santa Monica, CA: Human Factors/Ergonomics Society.

Beringer, D. B., & Harris, H. C. (1997). Automation in general aviation part II: Four ways to reach ground zero feet AGL unintentionally-Autopilot and pitch trim malfunctions. *Proceedings of the Human Factors and Ergonomics Society 41st Annual Meeting,* (pp. 75-79). Santa Monica, CA: Human Factors/Ergonomics Society.

Corso, J. F. (1963). Aging and auditory thresholds in men and women. *Archives of Environmental Health, 6,* 350-356.

Fujawa, G. E., & Strybel, T. Z. (1997). The effects of cue informativeness and signal amplitude on auditory spatial facilitation of visual performance. *Proceedings of the Human Factors and Ergonomics Society 41st Annual Meeting,* (pp. 556-560). Santa Monica, CA: Human Factors/Ergonomics Society.

Gelfand, S. A., Piper, N., & Silman, S. (1985). Consonant recognition in quiet as a function of aging among normal hearing subjects. *Journal of the Acoustical Society of America, 78,* 1198-1206.

Lindenberger, U., & Baltes, P. B. (1994). Sensory functioning and intelligence in old age: A strong connection. *Psychology and Aging, 9,* 339-355.

Patterson, D. (1983). Guidelines for auditory warning systems on civil aircraft: A summary and a prototype. In G. Rossi, (Ed.), *Noise as a public health problem.* Milan, Centro Ricerche e Studi Amplifon.

Parasuraman, R., & Mouloua, M., Eds. (1996). *Automation and Human Performance: Theory and Applications.* Mahwah, NJ: Lawrence Erlbaum.

Sanders, M. S., & McCormick, E. J. (1993). *Human Factors in Engineering and Design.* New York: McGraw Hill.

Schieber, F., & Baldwin, C. L. (1996). Vision, Audition, and Aging Research. In F. Blanchard-Fields and T. M. Hess (Eds.), *Perspectives on Cognitive Change in Adulthood and Aging.* New York: McGraw-Hill.

Wickens, C. D. (1992). *Engineering Psychology and Human Performance, 2nd edition.* New York: HarperCollins.

Contextual and Cognitive Influences on Multiple Alarm Heuristics

Daniel P. McDonald, Richard D. Gilson, Mustapha Mouloua,
Jeremy Dorman, and Patricia Fouts
University of Central Florida

INTRODUCTION

Trained operators and the general public alike, are increasingly faced with alarms in their environment. Alarms must be salient enough to capture attention with the intent to redirect it toward desired cues in the environment. Second, alarm should be easily understood. The alarm should be designed to clearly communicate the intended information to the operator. The question is how do we facilitate this communication.

Alarms are often intended to indicate conditions not immediately detectable by the senses. In a sense, alarm systems provide replacement stimuli, designed to represent to us an undetected condition of interest. For example, a fire alarm informs people of the presence of a fire, before detection by human senses. This is desirable to promote safe exit from a building or home. Alarms often take advantage of cue salience to alert people. Highly salient auditory or visual cues can increase arousal, calling for immediate action. Many of these cues may share a natural association with potential danger for the observer. An alarm's association with urgency can be useful to motivate evasive behaviors. However, excessive salience can also prove to be detrimental. A response in haste can prove to be fatal. Some alarm, such as in complex systems, may call for consideration of several situational factors before making appropriate responses.

In addition to capturing attention, alarms are also used to convey information regarding the situation. For example, on a fire alarm, a flashing display indicating "fire" can help indicate the cause of the alarm. Previous experience with fire alarms may be sufficient for identifying the alarm, through association. Similarly, experience may also prescribe appropriate responses such as "exit the building". Alarms convey their messages through these associations made in the past, which may be accurate or inaccurate, depending upon many factors, some related to alarm context. In order to effectively communicate messages through alarms, we must understand better, how natural and learned associations, as well as contextual factors play a role in determining behaviors associated with alarms.

One characteristic associated with man-made alarms, dramatically affecting the message, is the potential for false alarms to occur. False alarms have been a concern in a number of applied contexts. For example Tyler, Shilling, and Gilson (1995) found that 5% (1450) of all aircraft hazard reports submitted to the Naval Safety Center from 1984 through 1994 indicated that false or erratic indications were at least partially responsible for a mishap. Anecdotally, experienced pilots may learn to ignore or even disengage an alarm which tends to be false or is perceived as irrelevant within the context (Gonos, Shilling, Deaton, Gilson, & Mouloua, 1996). The plausibility of a false alarm reduces both the confidence in and responses to that alarm (Bliss, Gilson & Deaton, 1995; McDonald, Gilson, Deaton, & Mouloua, 1995). This phenomenon, sometimes referred to as the "cry-wolf" or false alarm effect, may exert a powerful influence on how system anomalies are handled in the present and in future situations. Breznitz (1984) maintains that a result of false alarms is the loss of credibility of the warning system. Bliss, Gilson and Deaton (1995) found that an alarm's expected reliability influences the likelihood of a response to that alarm. Specifically, most people tended to probability match their responses to the expected probability of an alarm being true. These results appear similar to other matching behavior found with animals, where responses are made in direct proportion to the frequency of the reinforcement (see Herrnstein, 1970). Notably, probability matching is not optimal behavior from the standpoint of individual success.

During high workload, when faced with time pressure, presented with insufficient or ambiguous information, or placed under conditions containing uncertainty, all of which can accompany alarm conditions, decision shortcuts or heuristics are often used. For example, an operator may not entertain all possible hypotheses regarding the nature of a failure, but may focus on one which comes to mind first, or on one which resembles most the current situation, given the context (Tversky & Kahneman, 1973). For unique circumstances, operators may draw from their past experiences, even when similarities between the past and current experiences is minimal, or their mental model is incorrect. Moreover, contextual factors may help to frame the situation, thus affecting the selection and relation of contextual information with respect to an alarm.

The current study was conducted to first, determine whether participants will employ a decision strategy based upon cognitive information presented in the context. Specifically, it was investigated whether the degree of understood functional dependency between two implicated components, one presented on a given alarm, would affect decisions regarding that alarm. It was predicted that response rates to a 60 percent true alarm would increase with an increase in the degree of perceived functional dependency between two implicated automobile components. Second, it was investigated whether responses to this alarm are subject to perceptual influences within the context, even in this more cognitive-rich environment. Specifically, will increases in the number of alarms affect responses to the alarm in question? Third, it was attempted to ascertain whether employed strategies were stable across experience, or if changes occurred between a first and second block of alarm trials. Fourth, it was asked whether different response strategies were employed by those who are more knowledgeable of the system as compared to the less knowledgeable, and whether they differ in terms of response confidence.

METHOD

Participants

Forty eight undergraduate psychology students from the University of Central Florida volunteered for the experiment. They were all recruited from psychology classes and received course extra credit for their participation. Participants were required to have a current drivers license.

Materials

One Macintosh Centris 650 25 MHz computer, and one IBM clone 486SX 66 MHz computer were used. The alarm test bed used in this experiment was created by the experimenters using Aldus Supercard 1.6 and run on the Macintosh computer. A central tracking task was implemented using that function from the Multi-Attribute Task (MAT) Battery (Comstock and Arnegard, 1992). Participants used a joystick to control the compensatory tracking task, and a single button mouse for their responses to the alarms. The CRT displaying the tracking task was placed directly in front of the participant, with the alarm screen to each participant's right.

Procedure

All participants were asked to perform a task which represented the operation of an unmanned vehicle. They were informed that an added difficulty associated with operating an unmanned vehicle is that the operator may not be exposed to essential environmental cues. Participants were informed that their task was composed of a central navigation compensatory tracking task, and an alarm management task presented on the right hand screen. They were instructed to keep the tracking cursor in the center box area of the screen. They were told that satisfactory performance was required for successful navigation of the remote vehicle, since substandard performance could constitute crashing the vehicle.

Participants were also told that the display to the right contained an alarm panel that could present various indications regarding the vehicle's condition. They were told that throughout their mission the center alarm may activate on the display panel. The panel contained five alarms in all, the center alarm plus four other alarms. The four peripheral alarms were all equidistant from the center alarm, with one alarm above, one below, one to the right, and one to the left of the center alarm. Participants were told that the sensitivity to the center alarm could be set. It was explained that increases in sensitivity for that alarm, while improving the likelihood of detecting a malfunction, also increases the likelihood of a false center alarm. Participants were instructed, for this experiment, to set the center alarm sensitivity to 60 percent true. They were told that this meant that the center alarm was true 60 percent of the time, and that, consequently, it was false 40 percent of the time. This number was displayed as a setting in the upper left hand corner of the display in a green box. Participants were told that their job was to determine whether the center alarm was true or false each time it was active, and to respond appropriately by either initiating a response or resetting the alarm.

On most trials, the activated center alarm also displayed component information selected from the pairs of components rank ordered earlier. On these trials a second peripheral alarm was also activated. This alarm displayed the second component of a given ranked pair. During each of two blocks, participants were presented the center alarm along with a peripheral alarm, collectively displaying a component pair ranging from highly dependent to highly independent. Also, during both trial blocks, the same conditions were presented along with two additional undefined alarms, for a total of four active alarms.

Throughout the experiment the center alarm was activated, while participants performed the central navigation task. Participants were instructed to determine, in each case, whether the alarm was true or false, and then to respond appropriately. If they believed that the center alarm was true, they were to use the mouse to place the cursor on a button labeled "true" and select. If they believed the center alarm to be false then they were to select the button labeled "false". They were instructed that responding "true" would initiate a response sequence, and responding "false" would reset the alarm. They were told that they would not receive feedback regarding their performance until after the experiment. They were also told that, following the experiment, their overall performance would be scored and compared to other performances. They were also informed that substandard performance on either the tracking or alarm task would result in having to perform another block of trials. This was done to add realism and to motivate participants to perform adequately. They were instructed to make their responses as accurately and quickly as possible. Participants were informed that both the central navigation and the alarm management tasks were equally important to the success of the mission, and that improper response to any central alarm could result in catastrophic failure, regardless of the implied component importance to the overall functioning of the vehicle. Response type, response times, and response confidence was recorded and analyzed.

RESULTS

No effects were found for any of the factors on response times. Therefore, they will not be considered here. However, it appears from viewing the Figure 1, that response rates steadily increased as the assumed functional dependency between component pairs increased. A trend analysis revealed that this was, in fact, a positive linear relationship between functional dependency and response rates. $F(1,9)=11.48$, $p<.001$. Moreover, a four factor mixed ANOVA revealed a significant increase in response rates as the total number of active alarms increased from two to four $F(1, 8)=11.29, p<.01$. The graph shows clearly that the situations containing four active alarms had overall higher response rates than analogous cognitive situations containing only the two alarms. During events of low cognitive proximity, an increase in the overall number of alarms from two to four resulted in a response rate of around 60 percent. Interestingly, this can be compared to about 37 percent for the same component pair in a two alarm condition. This suggests that even though participants "knew" that two components were highly independent or disconfirming, additional

303

alarms resulted in increased response rates to the central alarm. From the graph it appears that, in fact, there may also be an interaction, where there are even larger differences during occasions of lower cognitive relatedness rather than higher. However, a significant interaction was not found. Results indicate that alarm number may overcome the confirming strategy used during situations of low functional component dependency. Although people knew the components were functionally independent, an increase in overall alarm number resulted in a deviation in this cognitive strategy. No other effects were found for responses made.

From viewing Figure 2, it appears that response confidence steadily increased as the assumed functional dependency between component pairs increased. A trend analysis revealed that there was, in fact, a positive linear relationship between functional dependency and response confidence $F(1,22)<14.78$, $p<.001$. Moreover, a four-factor mixed ANOVA revealed reveal a significant effect for experience on response confidence $F(1,20)=5.27, p<.05$. As participants gained experience responding to the alarm, they became more confident in their adopted strategy. Response confidence increased in spite of no performance feedback being provided to participants. Also, even though no differences in response strategies were found between automobile experts and novices, a difference between these groups existed in terms of their response confidence, where those who were more knowledgeable in mechanics tended to be more confident in their responses $F(1,20)=5.70, p<.03$.

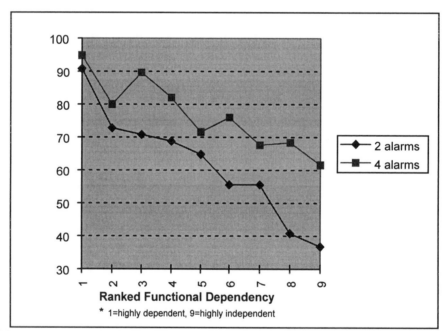

Figure 1. Mean response rates to the 60 percent true center alarm plotted as a function of perceived functional dependency between components, for two and four alarms.

CONCLUSIONS

It was found that participants' response rates to a 60 percent true alarm increased as the degree of functional dependency between two implicated components increased. Response rates ranged from well below to well above 60 percent in a linear fashion as the degree of agreement between components increased, demonstrating a systematic heuristic based upon cognitive relationships for all who participated regardless of expertise in automobile mechanics. This means that although participants may have ranked the same component pairs as having a different level of dependency, they all tended to use the degree of dependency

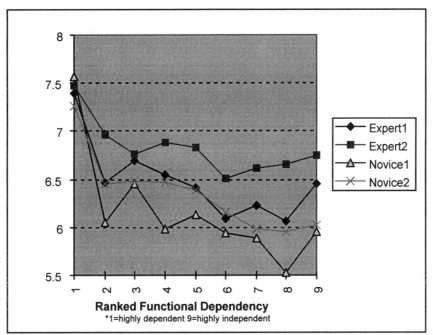

Figure 2. Mean response confidence to the 60 percent true center alarm plotted as a function of perceived functional dependency between components, for self-rated novices and experts in automobile mechanics, on both their first and second exposure.

as a factor to consider their response. Confidence in response also increased with increased functional dependency. Moreover, increased alarm numbers from two to four resulted in increased response rates. This increase in response rate occurred even when implicated system components were of low functional agreement. Results suggest that increased alarm numbers may override strategies that are based on their understanding of the system, and this consideration should be taken when designing complex alarm displays. Perhaps a confirmation bias is occurring, where during occasions containing low component agreement, participants tended to disregard that information in the presence of other active confirming alarms. This influence of increased alarm number on response further suggests a contextual influence, one which is perceptually based. Experience with responding was also shown to affect confidence in responses made, in that confidence increased between the first and second block of trials. This indicates that they were adopting a strategy or "understanding" of the system. This has training implications in that elevated confidence as a function of experience, regardless of receiving any meaningful feedback, could interfere with future performance. Results also revealed that those who rated themselves as more knowledgeable in automobile mechanics tended to have higher overall response confidence than the less knowledgeable. However, no differences in response strategies were found between more and less knowledgeable persons. These results clearly demonstrate the importance of understanding alarm context as well as cognitive influences for design of alarm displays as well as training application.

REFERENCES

Bliss, J. P., Gilson, R. D., & Deaton, J. E. (1995). Human probability matching behavior in response to alarms of varying reliability. *Ergonomics*, 38, 2300-2312.

Breznitz, S. (1984). *Cry wolf: The psychology of false alarms*. London: Lawrence Erlbaum Associates, Publishers.

Comstock, J.R., & Arnegard, R.J. (1992). *Multi-attribute task battery*. (NASA Technical Memorandum

104174). Hampton, VA: NASA Langley Research Center.

Gonos, G.H., Shilling, R.D., Deaton, J.E., Gilson, R.D., & Mouloua, M. (1996). *Cockpit alarm diagnostic and management systems: Current status and future needs* (Special Rep. No. 96-001). Orlando, FL: Naval Air Warfare Center, Training System Division.

Herrnstein, R. J. (1970). On the law of effect. *Journal of the Experimental Analysis of Behavior, 13,* 243-266.

McDonald, D. P., Gilson, R. D., Deaton, J. E., & Mouloua, M. (1995) The effect of collateral alarms on primary response behavior. *Proceedings of the Human Factors and Ergonomics Society 39th Annual Meeting,* (pp. 1020-1024) Santa Monica, CA: Human Factors and Ergonomics Society.

Tversky, A., & Kahneman, D. (1973). Availability: A heuristic for judging frequency and probability. *Cognitive Psychology, 5,* 207-232.

Tyler, R. R., Shilling, R. D., & Gilson, R. D. False alarms in Naval aircraft: a review of Naval Safety Center mishap data. *Special Report 95-003. Naval Air Warfare Center Training Systems Division,* Orlando, Florida.

Ramifications of Contemporary Research on Stimulus-Response Compatibility for Design of Human-Machine Interfaces

Chen-Hui Lu
Providence University, Taiwan

Robert W. Proctor
Purdue University

INTRODUCTION

Stimulus-response (S-R) compatibility effects are differences in reaction time (RT) and accuracy of responding as a function of the relation or mapping between stimulus and response sets. Such effects typically are attributed to response-selection processes, that is, to processes that intervene between stimulus identification and execution of a specific motor action. Most well known are spatial compatibility effects, which were first demonstrated by Fitts and colleagues. Fitts and Deininger (1954) varied the S-R mapping for circular stimulus and response arrays and showed that responses were fastest and most accurate when each stimulus location was mapped to the corresponding response location. Mapping effects of this type are what are examined most commonly in compatibility studies.

Spatial compatibility is described in most textbooks and handbooks on Human Factors and Ergonomics (e.g., Sanders & McCormick, 1993), and other types of compatibility are often mentioned as well, for example, conceptual compatibility, modality compatibility, and so on. Consequently, virtually all ergonomists are familiar with the principle of using compatible display-control sets and mappings when possible. However, the treatments given S-R compatibility typically are rudimentary and do not reflect the level of empirical and theoretical knowledge on the topic that exists at present.

The purpose of this paper is to provide a brief survey of current theoretical and conceptual views regarding the determinants of spatial compatibility effects. We will also examine implications of these views and of pertinent empirical findings for display-control design.

SPATIAL CODING

Central to most accounts of S-R compatibility is the concept of coding. In the case of spatial compatibility effects, the emphasis is on spatial codes that mediate between stimuli and responses. Evidence for spatial codes is most apparent in two-choice reaction tasks, where a left or right keypress response is to be made to a left or right stimulus. The spatially direct mapping is faster than the indirect mapping, even when the arms are crossed so that the left response key is pressed by the right hand and the right response key with the left hand (e.g., Dutta & Proctor, 1992). In most situations, the spatial codes are based on relative location, that is, regardless of whether the display or response keys are directly in front of the performer or to the side, the mapping of left-to-left and right-to-right is superior to the reversed mapping (e.g., Proctor, Van Zandt, Lu, & Weeks, 1993). It is worth pointing out that spatial compatibility effects also occur when the spatial information is conveyed by words (e.g., LEFT), symbols (e.g., a left-pointing arrow), and direction of motion (e.g., leftward movement), as well as for vocal responses of a spatial nature (e.g., the utterances LEFT and RIGHT; Wang & Proctor, 1996).

One interesting fact is that stimulus location affects performance even when it is irrelevant to the task. Thus, spatial compatibility cannot be ignored by a designer even when the relevant stimulus information is nonspatial. For example, if you are to respond to a red stimulus with a left response and a green stimulus with a right response, responses are faster when the stimulus and response locations correspond than when they do not This phenomenon, first observed by Simon in the late 60s, is now called the Simon effect (see Simon, 1990, and Lu & Proctor, 1995, for reviews). It indicates that a spatial code corresponding to the stimulus location is formed even when location is irrelevant to the task. Because the issue of why spatial codes are formed when irrelevant is both intriguing and interesting theoretically, the Simon effect has been the focus of much of the recent research on compatibility effects.

Two proposals are currently being debated about the nature of spatial coding. Some authors link the formation of spatial stimulus codes to shifts of attention involved in the control of saccadic eye movements (e.g., Stoffer & Umiltà, 1997). Others propose that the spatial codes are generated automatically with respect to referent objects or frames (e.g., Hommel, 1993c). Tests of these hypotheses have yet to produce unequivocal evidence regarding what role, if any, attention shifts play in spatial coding. However, the results from several studies imply that spatial codes are formed with respect to multiple frames of reference when more than one is present.

Studies by Lamberts, Tavernier, and d'Ydewalle (1992) and Roswarski and Proctor (1996) illustrate this point. They used displays with eight locations, grouped as in Figure 1. Each trial involved presentation of the left or right fixation cross, followed by presentation of two boxes to the left or right of it, one of which contained the imperative stimulus. The relevant stimulus attribute was nonspatial, for example, left keypress to red stimulus and right keypress to green stimulus. As can be seen in the figure, left and right can be defined with respect to three different aspects of the display. Spatial correspondence effects were evident in the RT data for all three reference frames, hemifield, relative position, and hemispace. In other words, responses were fastest when the spatial codes with respect to all frames of reference corresponded to the response that was to be made and slowest when they all conflicted.

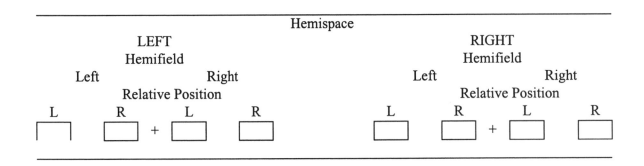

Figure 1. Displays Used by Lamberts et al. (1992) and Roswarski and Proctor (1996).

The importance of multiple frames of reference in the design of display and control panels is apparent in the crash of a British Midland Airways Boeing 737-400 aircraft in January, 1989 (Learmount & Norris, 1990). The primary cause of the accident was that the pilot shut down the wrong engine following fan-blade failure. As shown in the left half of Figure 2, the instrument layout was one with multiple frames of reference. The engine instruments were set in two side-by-side blocks; one frame of reference was thus left block versus right block. Within each block, the instruments were arranged in two columns, one for the left engine and one for the right engine, with the left block containing the primary instruments and the right block the secondary instruments. In terms of check reading, this organization is reasonable because a discrepancy between the two engines will pop out. However, the layout is confusing from the perspective of left-right codes due to the distinct frames of reference. More specifically, if a primary display for the right engine is deviant, an incorrect "left" spatial code with respect to the blocks frame conflicts with the correct spatial code of "right" with respect to the primary block. The same occurs if the secondary display for the left engine is deviant. Thus, it is evident that confusion may occur as to which engine to shut off when a rapid decision must be made, which is what apparently occurred in the accident of the 747. The conflict can be eliminated by having all displays for an engine in a single column. Alternatively, if the global frame is the more dominant one, then a grouping such as that shown in right half of Figure 2 might be preferable.

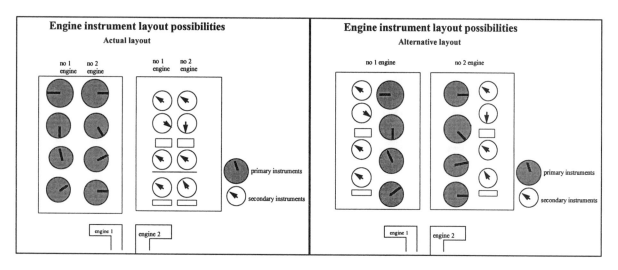

Figure 2. Engine Instrument Layout Possibilities.

DUAL RESPONSE-SELECTION ROUTES

Beginning with Fitts, compatibility effects have been attributed to a response-selection route that usually is characterized as involving intentional "translation" of the stimulus information into a response (e.g., Proctor et al., 1993). Translation is considered to be fastest when an identity rule is applicable, that is, respond with the spatially corresponding response, and slowest when a search or look-up process must be used, as when there is no systematic relation between stimuli and their assigned responses. Translation also benefits from other rule-based relations, such as respond with the mirror opposite location.

According to Kornblum et al. (1990), mapping effects due to translation differences will occur whenever there is dimensional overlap, or similarity, between the stimulus and response sets. For example, performance is better with a mapping of a high pitch tone to an "up" response key and a low pitch tone to a "down" response key than with the opposite mapping. The Nissan CD changer shown in Figure 3 uses the preferred mapping for selecting the desired disc number, but it uses the nonpreferred mapping for changing the track number.

An interesting finding that appears to be a function of translation processes is the occurrence of compatibility effects for orthogonal dimensions. Weeks and Proctor (1990) showed that for two-choice tasks, the mapping of "up" with "right" and "down" with "left" is preferred for a variety of stimulus types (physical locations, words, arrows) and response sets (aimed movements, keypresses, vocal words). Their explanation is that "up" with "right" are the salient polar referents for the two dimensions and that less translation is required when the salient referents correspond than when they do not. Compatibility effects for orthogonal dimensions are important because there are many situations in which they exist for control panels. The Nissan cassette receiver shown in Figure 3 uses mappings of controls that are in agreement with the up-right/down-left principle. The CD changer follows this relation for the labels on the track-select control but violated the principle in putting the right pointing arrow below the left pointing one.

Although many authors, including the present ones, have favored single-route translation models in the past, there is growing consensus that a second response-selection route, characterized as "direct" or "automatic," contributes to compatibility effects. The general idea is that a stimulus will automatically activate its natural, corresponding response via this route. The most prominent model currently, that of

Kornblum et al. (1990), is an example of a dual-route model of this type.

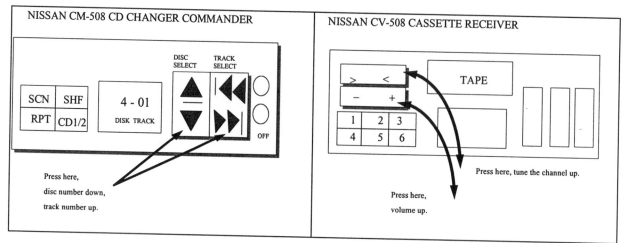

Figure 3. CD Changer and Cassette Receiver Interfaces.

One reason for proposing an automatic activation route in addition to the translation route is that, as described earlier, stimulus location affects performance even when it is irrelevant to the task. The effects of irrelevant location information are assumed to arise because the stimulus automatically activates the corresponding response. This activation is presumed to be due to links that are hardwired genetically or learned through a lifetime of experience with spatially compatible relations (Umiltà & Zorzi, 1997). However, this pattern of activation can be altered with relatively little experience (Proctor & Lu, in press). Subjects who practiced with a spatially incompatible mapping of stimulus locations to response locations in a two-choice task showed a reversed Simon effect, that is, faster responses when stimulus location and response location did not correspond, when stimulus location was subsequently made irrelevant. This reversal of the typical correspondence effect persisted over 600 trials, so it was not extremely short-lived.

When compatible and incompatible mappings are mixed, there is relatively little effect on the response latencies for the incompatibly mapped responses relative to a pure incompatible block, but the latencies for the compatibly mapped responses are slowed substantially (e.g., Van Duren & Sanders, 1988). The favored interpretation for this finding is that responses are selected by way of the automatic route when all relations are known to be compatible. However, when not all are compatible, this route must be inhibited and the slower translation route used. The point for designers is that the benefit of compatible mappings may be lost if other information in the environment is mapped incompatibly.

RELATIVE STRENGTH OF ASSOCIATION AND TIMING OF ACTIVATION

It is well established that effects of irrelevant information are asymmetric (Lu, 1997). That is, if an irrelevant stimulus dimension produces a correspondence effect when it is irrelevant, a correspondence effect often is not obtained when the relevant and irrelevant dimensions are switched. Stimulus locations are more strongly associated with manual responses than are arrows, and arrows more strongly than spatial words. Thus, when responses are manual, little if any spatial correspondence effect will occur from irrelevant spatial words, but irrelevant stimulus location may produce strong effects.

Effects of irrelevant stimulus location on manual responses are reduced by factors that slow the response to the relevant stimulus information. The activation of the spatially corresponding response appears to occur quickly and then decay. If the activation to the irrelevant information has decayed by the time that the relevant information is available for response selection, then it has little impact on performance. The magnitude of the effect of irrelevant information is a function of the temporal overlap of the activation produced by it with that produced by the relevant stimulus information (Hommel, 1993b).

TASK GOALS AND THE PERFORMER'S INTENTIONS

Most people think of compatibility effects as reflecting relatively fixed perceptual-motor relations. However, compatibility effects vary as a function of the goal of an action. Guiard (1983) had subjects move a cursor to the left or right, in response to a high or low pitch tone, by turning a steering wheel. This produces a standard Simon effect, responses are faster when the tone occurs to the side to which the cursor must be moved. Of most interest is a task in which subjects were required to hold the wheel at the bottom. In this case, the hands move left to shift the cursor to the right. Yet, a standard spatial correspondence effect was obtained such that responses were faster when the pitch assigned to the rightward cursor movement occurred in the right ear rather than in the left ear, and vice versa.

Hommel (1993a) conducted a standard Simon task with auditory stimuli (high and low pitch tones) in which the keypress lit a light on the opposite side. When subjects were told to ignore the light, responses were faster when the stimulus and response locations corresponded. However, when subjects were told to trigger the assigned light as quickly as possible in response to the stimulus, responses were faster when the stimulus location and response light location corresponded than when they did not. The important point from Guiard's (1983) and Hommel's studies for system designers is that very different spatial compatibility effects can be obtained as a function of the task goals of the performers.

PERSISTENCE OF COMPATIBILITY EFFECTS

One of the adages in Human Factors is that training cannot overcome the consequences of a bad design. That is clearly illustrated in studies of S-R compatibility. Even in two-choice reaction tasks, subjects who practice for up to 2,400 trials with an incompatible mapping never catch up with subjects who practice the same amount with a compatible mapping (Dutta & Proctor, 1992). This is true both for the standard left-right task and for the task in which the stimuli are up and down locations and the responses are left and right locations. It also holds for tasks in which stimulus location is irrelevant. Contrary to what one might think, even the simplest compatibility effects are not transient phenomena.

SUMMARY

In this paper we have only touched on several theoretical issues and findings of contemporary research on S-R compatibility. We have provided some illustrations of implications for Human Factors, and there are likely many more that could be derived. The compatibility literature contains a wealth of information about which anyone who is interested in the design of displays and controls should be aware.

REFERENCES

Dutta, A., & Proctor, R. W. (1992). Persistence of stimulus-response compatibility effects with extended practice. *Journal of Experimental Psychology: Learning, Memory, and Cognition, 18,* 801-809.

Fitts, P. M., & Deininger, R. L. (1954). S-R compatibility: Correspondence among paired elements within stimulus and response codes. *Journal of Experimental Psychology, 48,* 483-491.

Guiard, Y. (1983). The lateral coding of rotations: A study of the Simon effect with wheel-rotation responses. *Journal of Motor Behavior, 15,* 331-342.

Hommel, B. (1993a). Inverting the Simon effect by intention: Determinants of direction and extent of effects of irrelevant spatial information. *Psychological Research, 55,* 270-279.

Hommel, B. (1993b). The relationship between stimulus processing and response selection in the Simon task: Evidence for a temporal overlap. *Psychological Research, 55,* 280-290.

Hommel, B. (1993c). The role of attention for the Simon effect. *Psychological Research, 55,* 208-222.

Kornblum, S., Hasbroucq, T., & Osman, A. (1990). Dimensional overlap: Cognitive basis for stimulusresponse compatibility -- A model and taxonomy. *Psychological Review, 97,* 253-270.

Lamberts, K, Tavernier, G., & d'Ydewalle, G. (1992). Effects of multiple reference points in spatial stimulus-response compatibility. *Acta Psychologica, 79,* 115-130.

Learmount, D., & Norris, G. (1990). Lessons to be learned. *Flight International, 31 October - 6 November,* 24-26.

Lu, C.-H. (1997). Correspondence effects for irrelevant information in choice-reaction tasks: Characterizing the S-R relations and the processing dynamics. In B. Hommel & W. Prinz (Eds.), *Theoretical issues in S-R compatibility* (pp. 85-117). Amsterdam: North-Holland.

Lu, C.-H., & Proctor, R. W. (1995). The influence of irrelevant location information on performance: A review of the Simon and spatial Stroop effects. *Psychonomic Bulletin & Review, 2,* 174-207.

Proctor, R. W., & Lu, C.-H. (in press). Processing irrelevant location information: Practice and transfer effects. *Memory & Cognition.*

Proctor, R. W., Van Zandt, T., Lu, C.-H., & Weeks, D. J. (1993). Stimulus-response compatibility for moving stimuli: Perception of affordances or directional coding? *Journal of Experimental Psychology: Human Perception and Performance, 19,* 81-91.

Roswarski, T. E., & Proctor, R. W. (1996). Multiple spatial codes and temporal overlap in choicereaction tasks. *Psychological Research, 59,* 196-211.

Sanders, M. S., & McCormick, E. J. (1993). *Human factors in engineering and design.* New York: McGraw-Hill.

Simon, J. R. (1990). The effects of an irrelevant directional cue on human information processing. In R. W. Proctor & T. G. Reeve (Eds.), *Stimulus-response compatibility: An integrated perspective* (pp. 31-86). Amsterdam: North-Holland.

Stoffer, T., & Umiltà, C. (1997). Spatial stimulus coding and the focus of attention in S-R compatibility and the Simon effect. In B. Hommel & W. Prinz (Eds.), *Theoretical issues in stimulus-response compatibility* (pp. 181 -208). Amsterdam: North-Holland.

Umiltà, C., & Zorzi, M. (1997). Commentary on Barber and O'Leary: Learning and attention in S-R compatibility. In B. Hommel and W. Prinz (Eds.), *Theoretical issues in stimulus-response compatibility* (pp. 173- 178). Amsterdam: North-Holland.

Van Duren, L., & Sanders, A. F. (1988). On the robustness of the additive factors stage structure in blocked and mixed choice reaction designs. *Acta Psychologica, 69,* 83-94.

Wang, H., & Proctor, R. W. (1996). Stimulus-response compatibility as a function of stimulus code and response modality. *Journal of Experimental Psychology: Human Perception and Performance, 22,* 1201-1217.

Weeks, D. J., & Proctor, R. W. (1990). Salient-features coding in the translation between orthogonal stimulus and response dimensions. *Journal of Experimental Psychology: General, 119,* 355-366.

PANEL SESSIONS

Automation and Human Performance:
Considerations in Future Military Forces

Thomas Mastaglio
Virginia Modeling, Analysis, and Simulation Center

The United States Military is committed to applying information system technology to enhance the effectiveness of its combat systems and aid the performance of its soldiers, sailors, marines and airmen. The Department of Defense has undertaken several experimental and demonstration programs to determine the best approach to applying and integrating information systems to achieve this goal. These advanced technology demonstrations include the Army's Force XXI experiments, its Army After Next war games, the Navy's Smartship initiative and its efforts to build an Arsenal Ship. Some of these programs are ongoing, others completed, and some discontinued.

The lessons the military is learning in terms of how to design their systems, automate existing operational procedures, and when to replace them with new ones can be of use to the nondefense sector confronting similar challenges. A panel of experts who have been involved in these technology development and demonstration programs will address the challenges in dealing with these automated information and command and control systems. Topics will include: information overload, decision making using information presented using virtual displays, interacting with computers under stressful conditions, team cohesion, data visualization, and determining a suitable organization structure with clearly defined roles for all its members.

The central issue for this panel is the military's new awareness that the demands and stress of a combat environment impact human performance in ways that have not been predicted or considered in the past. Automation brings new capabilities, but associated new challenges for managing combat operations and training operators to work effectively. We will also allow time for interaction with the conference audience for questions and comments regarding ways the military can take advantage of knowledge gained through research and industry efforts to automate their operations.

U.S. ARMY FORCE XXI PROGRAM: SOME HUMAN DIMENSIONS

Richard E. Christ
U.S. Army Research Institute

During the 1990s the Army dedicated a massive effort toward beginning the development of a digitized land combat force. This effort was fostered by anticipated leaps in the capabilities of communication and information technology. Its application to the domain of military operations was further stimulated directly by a pamphlet published by the U.S. Army Training and Doctrine Command, Force XXI Operations. The developmental effort was supported by large-scale experimentation with digital technology, focused largely on issues related to the operation of a wireless tactical internet, and on the integration and performance of various hardware and software systems. As experimentation continues, a knowledge gap has been recognized in areas related to the human dimensions of digitization.

My presentation will provide a very brief description of the objectives of and the methods used by the Army during a series of Advanced Warfighting Experiments and an equally brief summary of the official results of those experiments. Then, as a stimulus for expanding the scope of issues that are considered by this panel and the conference, I will provide a summary description of three quite different research programs with which I have been associated. These research programs, identified below, taken together with others described at this conference, underscore clearly the need for more research designed to identify and to recommend responses to the new capabilities and challenges brought to the domains of human and organizational performance by emerging information technologies.

Human dimensions assessment of a warfighting experiment investigating the potential for digitizing land combat forces. Three successive surveys were conducted of up to 8,000 soldiers and leaders. The results show that in spite of dramatic increases in work hours, perceptions of increased work-family conflict, and a general belief that the new technology would increase the complexity of their jobs and the amount of training required, respondents felt they were able to adequately cope with new work demands caused by their participation in the experiment. Factors that might account for high levels of job satisfaction and commitment include the belief that their missions were important and that the new technology would benefit their own and their unit's performance.

The relationship among factors affecting the effectiveness of Span Of Command and Control (SOCC) in Army organizations. Fifty-five Army officers from Captain to Full General were interviewed regarding factors that have been identified as affecting the effectiveness of SOCC and the difficulty of command and control. The results show interacting effects of SOCC factors, level in the hierarchy, and type of unit. These results have several implications for designing future high technology forces.

Lessons for inserting information technology from the business and management science literature. The results of an analysis of the literature on possible effects of information technology insertion were synthesized with the Army's digitization objectives and experiences. Two major points are developed: (a) factors other than the potential of the technology interact with one another and the technology itself to determine the resultant nature, form, and functionality of the "digitized" organization; and (b) the most immediate impact will not be quantum increases in operational performance but, rather, the insertion process itself, with sustained performance gains possible only after a process of mutual adaptation and convergence between the new technology and existing organizational and contextual forces.

ARSENAL SHIP: SHIP AND MISSION OPERATIONS CENTER (SMOC) OPERATOR WORKLOAD MODELING

Kelly O'Brien
Rome Research Corporation

Arsenal Ship, a forerunner to the Navy's SC21 initiative, was a joint Navy/DARPA program to acquire a new capability for delivery of large quantities of ordnance (approximately 500 Vertical Launch System cells) in support of land and littoral engagements. Arsenal Ship was not to be fitted with long range surveillance or fire control sensors, but was to be remotely controlled via robust datalinks to other assets, such as Aegis and AWACS. The objectives of this project to develop high leverage technologies and improve acquisition methods have far-reaching implications for the future of surface ships for the US Navy. However, development of the virtually unsinkable ship with massive firepower capabilities was a challenge due to the design goal of a very small crew (less than fifty). In January 1997, three defense contractors, including the Northrop Grumman team, were selected to compete for a contract to design, construct, and test one Arsenal Ship demonstrator. During Phase II Functional Design, a preliminary workload analysis was conducted to verify the Ship and Mission Operations Center (SMOC) manning allocations. The workload analysis was developed to examine the potential impact of defining operator roles non-traditionally and introducing automation in an effort to reduce manning for various operating conditions. An interactive PC-based model was developed using functions identified and data collected from domain experts. Tasks specific to the functions were assigned to candidate SMOC operators. Each task was categorized by type (such as monitoring or decision/action) and the average time to complete the task was used to arrive at a certain workload value. Variables specific to the operator position, such as the number of missiles launched over a time period for the weapons control operator, were introduced. From here, the analysis was run and an overall workload value was compared with a workload threshold value to determine if the operator could perform the tasks in the allocated time period.

This model is unique in the sense that it is specifically designed to identify potential workload concerns at an early stage in the development process. While robust, the model is easy to use and the summary data analysis is easy to interpret. Results can be given in a graphical format for each operator or can be shown in a summary graphical format. The model can also be used as a sensitivity tool to conduct "what-if" analyses.

This discussion will describe the operator workload model, the procedure used to conduct the analysis, and the results that were generated to support Phase II Arsenal Ship development. The results confirm that during peace time cruising, the SMOC could be manned by a minimum crew of two. During war time cruising, the SMOC could be manned by a minimum crew of four.

THE NEED TO MODEL HUMAN AGENTS AND HUMAN PERFORMANCE IN DISTRIBUTED COMMUNICATION SYSTEMS

Joseph Psotka
Army Research Institute

Human behavioral characteristics are critically important to determining the outcome of land combat but there are few ways of inserting them into modern simulations and models. In particular, current army doctrine depends upon cognitive, motivational, and moral variables such as leadership, level of training, unit morale and cohesion. Distributed interactive simulations using the High Level Architecture (HLA) and virtual training technologies have matured steadily for use in training diverse skills at dispersed or remote locations, or for refreshing previously learned skills or knowledge. Yet, these technologies still do not include the impact of human behavior in warfighting analyses or simulations. Current models of constructive simulations use attrition-based algorithms and do not include important variables such as information flow management, fear, training proficiency, artillery suppression, and fatigue. These are difficult issues to capture in computational models and require new avenues of research in this fundamentally important area for the Army, to correlate and leverage diverse efforts. Significant progress might be possible if research efforts took advantage of the advances in modeling human agents and human performance in distributed communications of intelligent agents within financial, economic, industrial, and entertainment industries; using architectures arising from artificial intelligence and virtual reality research, such as production rules, neural networks, and semantic network systems. The enormous cost of computers and networking systems can only be controlled by developing systems that work more efficiently with humans in the ways that humans work and think. Computers must carry more of the load. They are much too passive, and because they are deaf, dumb, and blind they often fail to take users' needs into account. Automated search technologies to produce latent semantic indexing appear to have a novel role. The potential gain from even small improvements in our understanding and ability to model human behavioral factors would be significant. Novel frameworks for modeling emotion and personality based on human cognitive modeling have arisen in the past years. Exploiting these theoretical developments and automated search and indexing technologies could fundamentally improve human behavior and decision making representation in CGF.

SMART SHIP

Commander R.T. Rushton
United States Navy

SMART SHIP was initiated in late 1995 in response to the results of an NRAC summer study, tasked to look at emerging technologies for use in reducing crew sizes for new ship classes being planned for construction . The project began an aggressive search of technologies with a call to industry and to government laboratories for mature technologies that had potential in streamlining crew operations. While awaiting the results of this search, YORKTOWN began an in-depth look at her own internal operations. There was early recognition that technologies existed in YORKTOWN that had never been fully utilized: like the "autopilot" in the steering control system and the unmanned engine room design in the propulsion plant. There was also a realization that many naval customs, fleet policies, and operational procedures had become suspect with changes in the world operating environment and improvement of "off ship" supporting technologies.

In the spring of 1996 the Naval Sea Systems Command proposed installing distributed architecture based ship control technologies centered around a robust fiber optic network. The proposal included "Windows NT" based software for an "Integrated Bridge" (IBS), Integrated Condition Assessment System (ICAS), Damage Control System (DCS) and Standard Machinery Control System (SMCS). All four of these software technologies were either commercial "off the shelf" systems or had been developed for the U.S. Navy under other projects. These innovations were coupled with a wireless communications system called HRDRA.

The LAN is a 16 fiber, fiber optic mesh connected to five control hubs with "asynchronous transfer mode" (ATM) capability. All software operates in a dual channel configuration for redundant control system signal paths. Each of the five hubs were placed in a different watertight section of the ship for battle survivability.

The Integrated Bridge (IBS) is a triad of ship maneuvering and control software. It provides: a near real time navigation picture using military precision Global Positioning System (GPS) data, a collision avoidance capability that allows for operator alertment of potentially dangerous surface targets and maneuvering decision aids, and helm control software that allows for steering and throttle control via the LAN, to multiple stations.

The Damage Control System (DCS) is a set of software, to automate the monitoring, command, and control requirements of shipboard damage control. It includes an embedded main engineering space fire doctrine as well as the capability to place damage control status information on the LAN. The Standard Machinery Control System (SMCS) replaced the aging computer main frame supported propulsion, electrical, and auxiliary control system. It provides the ability to take existing machinery monitoring and control nodes, multiplex their signals via Remote Terminal Units (RTUs) and place them on the LAN. The use of "WindowsNT" technology and ability to consolidate control features, shed them off, or move them anywhere with LAN access, revolutionized the watch architecture in engineering.

The Integrated Condition and Assessment System (ICAS) provides engineering leadership with clear concise data recording, trend analysis, and on-line system monitoring for the first time. ICAS has become a key ingredient in shifting the surface combatant force over to a condition based maintenance philosophy.

HYDRA is a multi-channel wireless communications system that provides complete mobile communications between primary command and control personnel and stations throughout the ship.

Why Humans are Necessary on the Flight Deck:
Supporting the Requisite Role of the Pilot

Paul C. Schutte,
NASA Langley Research Center

Many papers and discussions in Aviation Human Factors take the pilot's physical presence on the commercial flight deck as a given. Statements such as, "Until significant advances in automation are made, the human is required..." or "For the foreseeable future, pilots will be necessary ..." are made without much analysis. Once this premise is stated, we often move on to defining appropriate allocation of functions between humans and automation on the flight deck. Most of these allocation decisions address overcoming human limitations and improving situation awareness and workload. However, it seems that these decisions fail to return to the basic question. Why is the human necessary? We need to explicitly address how function allocations and design directly support the requisite role of the human.

The purpose of the panel session is to define why the pilot is necessary in the flight deck. It is not to define the reasons that humans do not perform well at some tasks nor to directly debate their presence. It is hoped that after making these required duties explicit, the discussion can then turn to how best to enable and support the pilot in this requisite role.

THE NECESSITY OF THE PILOT

Capt. Richard B. (Skeet) Gifford (Ret.)
NASA Langley Research Center

Airplanes are much more reliable and therefore much safer than they were when I began my career in aviation. While the pilot does not encounter serious emergencies with the frequency that he once did, vigilance is still required to operate the flight safely. The first airplane I flew with the Air Force, a C-119, had 14 procedures in the Emergency checklist. Those 14 procedures took care of just about anything that would happen to the airplane. In contrast, the B-777 has procedures for over 180 non-normal procedures. Several examples are included which describe incidents where the pilot on the flight deck corrected malfunctions of automation or made other inputs to preserve the safety of the flight. The examples are drawn from my personal experience and from incidents that occurred while I was a Flight Manager. These examples include map shift, uncommanded thrust reduction, spurious calculations by the Flight Management Computer, failure of both Flight Management Computers and database errors.

We must recognize that complacency can be the unintended consequence of reliability and automation. Forty-five years ago, failures were predictable and often catastrophic. Today, failures are infrequent but can be subtle, unpredictable, and difficult to analyze. In automating human error out of the cockpit, we have sometimes merely traded it for human error on the part of the engineer.

Flight crew members should have better training in problem recognition and analysis. Engineers should design systems to keep the pilots in the loop and provide sufficient operational flexibility to enable the pilot to effectively deal with unexpected situations. The NTSB should provide a more sophisticated analysis of accidents so that the industry can correct the root-cause of the failure.

WHY ELIMINATING THE HUMAN OPERATOR DOES NOT ELIMINATE HUMAN ERROR

Victor Riley
Honeywell Technology Center

When an airplane crashes, the first thing we do is ask whether the pilot or the controller made an error. If not, we start looking for some type of system failure. If we find such a failure, we then try to determine whether it was due to a maintenance shortcoming or, if not, a design flaw. If the airplane breaks, our ultimate purpose is not to blame the airplane, but rather to figure out what type of human activity (design, maintenance, operation) led to the failure.

People who would like to get rid of human error by getting rid of the human operator may not recognize that automation is the product of a human activity. In fact, automation simply replaces the operator with the designer; to the extent that a system is made less vulnerable to operator error, it is made more vulnerable to designer error. This is well demonstrated by incidents where a landing aircraft failed to trigger the weight on wheels switch, causing the ground spoilers, brakes, and thrust reversers to be locked out of pilot control.

Defenders of the "automate, don't operate" position may appeal to the fact that the designer has considerably more time and tools at his or her disposal when confronting an operational problem than the pilot has, and that there do exist methods of validating and verifying system software. Leaving aside the arguments over how well these methods work, validation and verification is a means of measuring the software product against the specification. The hard, possibly intractable, part is making sure that the specification is complete and valid for all possible situations and conditions. Until this is possible, design will always have a large subjective element to it, and systems will work according to either the operator's judgment, or the designer's judgment.

This brings us to the matter of trust. Is the public more likely to trust a designer to foresee all possibilities, when he or she is probably safely on the ground somewhere when their system gets into trouble, or a pilot who is on board the aircraft and wants to live (presumably) just as much as they do? If we automate everything and get rid of the operator, should we make sure that a member of the design team is on every flight of that series of aircraft so the other passengers don't feel that they're taking more of a risk than the designer? That the designer is backing up his or her fiduciary responsibility?

Consider the evidence so far, that despite the sophisticated level of technology we have now, the only completely automated civil transportation system is the elevator and its horizontal counterpart, the airport tram.

HUMAN AUTOMATION INTERACTION: AN AIRLINE PILOT'S PERSPECTIVE

Richard B. Stone
Bountiful, UT

With the introduction of the B767 in the early 1980's a dramatic change was made. Automation of systems, flight controls, and navigation were bundled together in an integrated fashion. Pilots immediately embraced the new displays but at the same time spoke of the feeling of being "out of the loop" or the added burden of programming during times of high workload.

Over the years a training philosophy has been adopted to limit the pilots knowledge to those technical details that the pilot could change, particularly in regards to the Flight Management System (FMS). Now some 18 years later that philosophy is being reexamined.

When viewed over the period since introduction the new automated aircraft appear to have successfully:
1. Improved situational awareness through the active display of present position
2. Relieved pilots of boring and repetitive tasks such as optimizing aircraft performance, maintaining aircraft track, and automatic downloading of maintenance information.
3. Improved navigational accuracy.
4. Improved aircraft performance i.e. fuel economy and time enroute.

On the contrary some problems have been identified:

1. Workload may be shifted to times of high activity such as reprogramming a new landing approach and runway during descent and landing preparation.
2. Command strategies are different in a two person, highly automated cockpit.
3. Training has not kept pace with the changes in cockpit operations. A need to know philosophy characterizes the present aircraft training systems.
4. Procedures have not kept pace with the changes in cockpit operations. Changes in procedures continue to be piecemeal.

Generally, the automation of airline aircraft has progressed in a stilted fashion. The new aircraft have highly automated systems by 1980 standards but they have remained that way while the personal computer has been through many new iterations. The planned enhanced GPWS will take years to implement. In contrast, many older aircraft still carry a significant number of passengers. The challenge for all is adopt and quickly install new safer equipment and procedures.

Medical Automation: Help or Hazard?

Marilyn Sue Bogner, Ph.D.
Institute for the Study of Medical Error

The computer chip and resultant automation as well as technology in general have revolutionized equipment across domains. Early in this technological revolution, there were expressions of concern about the unquestioning acceptance of automation — acceptance that contributes to problems. Among the early cautions are lack of user oriented system linking (Chapanis, Garner, & Morgan, 1949) and lack of operator knowledge about the state of the system when human intervention is necessary (Bainbridge, 1983). These concerns and others are increasingly expressed for aviation as well as process control and nuclear power (Parasuraman & Riley, 1997). However, automation in medicine, a domain in which technology is pervasive, has received a paucity of attention (Bogner, 1994). It is particularly important to address medical automation concerns because the current fiscal constraints of managed care incorporate the implicit if not explicit assumption that technology can compensate for fiscally induced changes in health care delivery.

In health care facilities, the workload of the clinician is increasing and the work environment is complicated by greater cognitive demands and accelerated work tempo afforded by automation (Cook, 1994). Such system complexity contributes to an increased number of problems while making it more difficult to identify and correct the source of a problem. These conditions give rise to the inappropriate attribution of a problem to human error. The complications, however, reflect well documented problems with technology such as lack of feedback (Norman, 1992); a need for simplicity, transparency, comprehensibility, and predictability in system behavior (Billings, 1997); and information overload (Stokes & Wickens, 1988).

The environments in which complications are most evident, those in which fast and accurate decisions and treatment are most critical and stress and fatigue the highest, are the emergency room (Xiao et al., 1996), the operating room (Finley & Cohen, 1991), and intensive care unit (Donchin et al., 1995). Health care delivery is further impacted by the presumed cost-saving personnel changes mandated by managed care such as replacing experienced registered nurses with lower-salaried, less experienced licenced practical nurses. This is occurring as demand is increasing for clinical experience and skill to safely use technologically sophisticated medical devices. Other manifestations of fiscal constraints are discharging a patient in a sub-acute state with attendant medical devices to home care provided by inexperienced, stressed friends or relatives as lay care givers, and involving the patient in self-care using sophisticated and difficult to use devices (Obradovich & Woods, 1996). Concurrent with increasing the role of lay care givers in providing care using medical technology is the reduction in assistance to those care givers by restricting reimbursement for home visits by nurses. These scenarios that mismatch inexperienced users with sophisticated devices establish the conditions for incidents with medical technology that have adverse outcomes. Such outcomes often require further medical treatment with additional trauma and expense, and affect the quality of life of the patient, the lay care giver, and the clinician through the specter of litigation.

For medical automation to be a help rather than a hazard, its design must be user-driven (Bogner, 1996), complex displays must be configured with regard for the users' needs (Chapanis, 1996), and situations comparable to computer controlled flight in aviation that might invite large blunders while eliminating small manual system errors (Wiener, 1993) must be avoided. In short, to be a help rather than a hazard, medical automation and technology must be socialized from being dictators that demand conformance to their terms forcing users to persevere in tailoring their behavior to accommodate those terms however inappropriate they may be in the operational context, to being colleagues with the users *on the users' terms*.

REFERENCES

Bainbridge, L. (1983). Ironies of automation. *Automatica, 19*, 775-779.

Billings, C.E. (1997). *Aviation automation.* Mahwah, NJ: Lawrence Erlbaum Associates, Inc.

Bogner, M.S. (1994). Human error in medicine: A frontier for change. In M.S. Bogner (Ed.), *Human error in Medicine* (pp. 373-383). Mahwah, NJ: Lawrence Erlbaum Associates, Inc.

Bogner, M.S. (1996). Special Section Preface. *Human Factors, 38*, 551-555.

Chapanis, A. (1996). *Human factors in systems engineering.* NY: John Wiley & Sons, Inc.

Chapanis, A., Garner, W.R., & Morgan, C.T. (1949). *Applied experimental psychology.* NY: John Wiley & Sons.

Cook, R.I. (1994). Shifting the burden to the users: Clever tricks of the design masters. Notes for presentation to the Human Error/Human Factors course of the US Food and Drug Administration, Rockville, MD.

Donchin,Y., Gopher, D., Olin, M., Badihi, Y., Biesky, M., Sprung, C.L., Pizov, R., & Cotev, S. (1995). A look into the nature and causes of human errors in the intensive care unit. *Critical Care Medicine, 23*, 294-300.

Finley, G.A. & Cohen, A.J. (1991). Perceived urgency and the anaesthetist: Responses to common operating room monitor alarms. *Canadian Journal of Anaesthesiology 38*, 958-64.

Norman, D.A. (1992). *Turn signals are the facial expressions of automobiles.* NY: Addison-Wesley.

Obradovich, J.H. & Woods, D.D. (1996). Users as designers: How people cope with poor HCI design in computer-based medical devices. *Human Factors, 38*, 574-592.

Parasuraman, R. & Riley, V. (1997). Humans and automation: use, misuse, abuse. *Human Factors, 39*, 230-253.

Stokes, A.F., & Wickens, C.D. (1988). Aviation displays. In E.L. Wiener & D.C. Nagel (Eds.), *Human factors in aviation* (pp. 387-430). San Diego: Academic Press, Inc.

Wiener, E.L., (1993). *Intervention strategies for the management of human error.* (NASA Contractor Report 4547). Moffett Field, CA: NASA Ames Research Center.

Xaio, Y., Hunter, W.A., Mackenzie, C.F., Jefferies, N.J., Horst, R.L, & the LOTAS Group (1996). Task complexity in emergency medical care and its implications for team coordination. *Human Factors, 38*, 636-645.

AUTOMATION AND SURGICAL PERFORMANCE IN A BIOPSY PROCEDURE

Corinna E. Lathan
Biomedical Engineering, The Catholic University of America

Biopsy procedures are often needed to acquire tissue samples for pathological analysis so a treatment plan can be developed. Computed Tomography (CT) directed needle biopsies are routinely performed by interventional radiologists to gather tissue samples near the spine. As currently practiced, this procedure requires a great deal of spatial reasoning, skill, and training on the part of the radiologist. Image-guided systems that could assist the radiologist in needle placement and alignment would be a great improvement. Even further gains could be realized by the development of a semi-autonomous robotic biopsy system that could place the needle under the guidance of the radiologist.

Three critical task components of the spine biopsy procedure are as follows: *Select* the best CT slice for viewing the lesion; *Plan* the path between the skin entry point and the end-point; *Move* the needle to the biopsy location. This procedure can be time consuming and tedious since the biopsy needle must be advanced slowly and its position checked several times to ensure vital organs are not damaged. While this technique is generally effective, there is a time-accuracy tradeoff. The error tracking the planned path may be large, resulting in an increase in time, however, the error in end-point accuracy is usually low since the procedure continues until the desired endpoint is reached. The goal of an automated procedure would be to both decrease path tracking error and increase precision.

The current biopsy procedure, as represented by Figure 1(a), shows that the human operator carries most of the perceptual (*select image*), cognitive (*path planning*), and motor (*move needle*) workload. The radiologist controls the insertion of the needle directly and only uses occasional "snapshots" from the CT scanner for visual feedback. The system in Figure 1(b) is still manually controlled, but the visual feedback has been enhanced greatly to give the radiologist real-time image guidance during the insertion. Real-time 3-D image guidance would relieve some of the operator workload as well as reduce path planning error.

Figure 1(c) depicts a semi-autonomous system in which a haptic master arm provides force feedback to help guide the radiologist and drive a robotic system that actually performs the task. In this scenario, the functions handled by the human are significantly reduced, allowing the radiologist to concentrate on the parts of the procedure where human judgment is required most. Shared control will reallocate some of the motor function from the human to the robot, reducing total error and improving accuracy. Finally, note that the progression from Figures 1(a) to 1(c) is downward compatible; it does not preclude the semi-autonomous system from being operated in manual mode.

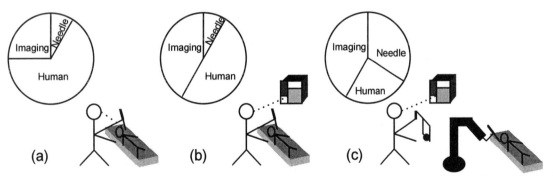

Figure 1. Relative function allocation of system components: the human operator, imaging information, and needle insertion.

TECHNOLOGY AND NURSING: ELECTRONIC FETAL MONITORS

Sonia C. Swayze
Center for Devices and Radiological Health, Food and Drug Administration

Fetal deaths associated with electronic fetal monitors have been reported to the Food and Drug Administration (FDA). The electronic fetal monitor is a microprocessor based medical device that uses ultrasound transducers to measure and translate signals from the fetal heart. The transducers are positioned on the maternal abdomen and/or thigh and connected by cables to a monitor. Fetal heart information translated from the transducer is displayed on the monitor panel and can be recorded and printed on a strip or trace. Some monitors provide alarm surveillance for fetal heart rates (FHR) that are outside of the limits set. The fetal monitor is used primarily by nurses to monitor the heart rate of single or twin fetuses throughout labor and delivery. The monitor is limited in evaluating FHR because of possible signal loss due to fetal movement, misreading due to maternal movement, and possible recording of maternal heart rate (MHR) instead of FHR. The fetal deaths reported to FDA involved single and twin fetuses and have raised concern about the monitor and whether it detects and records the FHR as it should.

One report involved a woman who was about to deliver twins. Her nurse reported that an external and internal fetal monitor recorded two normal baseline heart rates. During a forceps delivery, an external fetal monitor was used to monitor both twins. The first twin had died before birth (in utero) 48 hours earlier. When a twin dies in utero, the monitor automatically picks up the MHR. If the MHR and FHR coincide, the monitor is not able to differentiate between the two. There are other distinct limitations in detecting FHR: artifact may be mistaken for normal variability of the FHR and conceal a change in fetal status and with some monitors only one fetal heartbeat can be heard on the loudspeaker at any time. The channel monitoring the other fetal heartbeat for sound can only be accessed if the nurse presses a key. In this case, the nurse assumed the monitor was recording the twins' heart rates; however, it was apparently monitoring the mother's and the second twin's heart rate. The second twin was delivered alive and without incidence. No technical failure of the monitor was identified following the event.

Information on the cause of fetal death was not reported. Death may have been caused by lack of observation or misinterpretation of relevant information, or because the monitor failed to detect a change in the first twin's FHR. This underscores the importance of human monitoring technology. Nursing practice as well as literature on use of electronic fetal monitors emphasize, that monitoring should not replace a complete nursing assessment since the monitor may not detect and record the FHR.

PROBLEM IDENTIFICATION IN COMPLEX AUTOMATED MEDICAL SYSTEMS

Debbie Blum
Center for Devices and Radiological Health, Food and Drug Administration

Many of today's medical devices are "system" devices. They are comprised of multiple components and or accessories (devices) which must each work independently and feed into the system to result in one unified outcome. Understanding of the intended uses of the system are critical, but equally important is a knowledge and understanding of the mechanics of the system. Very often minicomputers are incorporated into the medical device system, such as those found in permanent implanted pacemakers. Also standard in automated medical device systems are various switches to permit the user to initiate new programming, override internal standard programming, turn off annoying warnings and alarms, and buttons to shut off everything at once. If the power supply fails, many systems, such as some IV pumps, even have battery backup to assure that the device will continue to function. Many complex systems, such as a fetal vacuum extractor, actually necessitate multiple persons be involved with the device system to assure proper and safe assembly, use and maintenance.

Problem identification of reported medical device adverse events is strongly based on the information provided in the product labeling, including patient population and clinical application(s) for which the product was cleared for marketing. This information is a limited resource. The reality of device use in a constantly advancing medical milieu may be quite different from the information available in the product labeling. As a result, when problems occur, it becomes very challenging to determine what error occurred, why it occurred, and how to prevent it from recurring in the future. For example, insertion of a coronary stem(s) involves elements such as: use of a special procedures room, specially trained staff, the stent delivery system(s), imaging equipment and a myriad of auxiliary medications, and supportive machinery. Failure in any one of these elements may cause or contribute to an adverse event. Yet when an adverse event does occur, it may be extremely difficult to reconstruct the scenario as it occurred, in order to try to determine which element(s) may have contributed to the event. This demonstrates how critical it is to identify all that is involved in the use of automated medical device systems. The user/operator is one variable, but we must consider the patient and the environment as equally important variables.

Not only are the device systems themselves complex, but the scenario in which they are used is also complex. We need to recognize that there are multiple variables which contribute to the success or failure of a system. We must also recognize that with complex system devices, the part that *appears* to malfunction or fail *may not* be the part that is *causing* the malfunction or failure. For example, a monitor may give faulty output because of a disruption in a circuit in an accessory device. Evaluation of the monitor itself may reveal that it is fully functional, within all specifications, and the problem cannot be duplicated. With only this information, the nature of the problem remains a mystery, and more investigation will be required before any plan can be developed to minimize recurrence of similar problems. When there are more questions than answers, it is often the device operators and support personnel who can provide critical information which can contribute to a better understanding of complex medical devices, their use, and problem resolutions.

USE ERROR WITH INFUSION PUMPS: MISLOADING

Christine M. Parmentier
Center for Devices and Radiological Health, Food and Drug Administration

Infusion pumps are microprocessor equipped devices used to regulate flow rate of hydration solutions, blood products, and medications delivered to patients in a variety of settings including hospitals, outpatient clinics, and even the home. The literature related to infusion pumps often cites programming errors as typifying the kind of use error associated with this medical device. While this type of error is one that is often reported through the FDA's MedWatch adverse event reporting system, it is not the only use error associated with infusion pumps. Numerous adverse events of inaccurate fluid delivery occurring with infusion pumps are causally linked to an incorrect connection between the infusion pump and the administration set and tubing that delivers the solution to the patient. This incorrect tubing placement is also referred to as misloading.

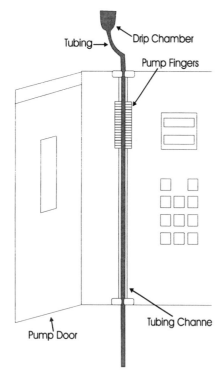

In many settings where the infusion pump is used, with the exception of the home setting perhaps, the nurse is often the health care provider who has the primary responsibility of administering infusion therapy for the patient. The nurse connects the container with the solution to be delivered to the infusion administration set and tubing, attaches this administration set to the pump and the patient, programs the pump, monitors the delivery of solution to the patient, and discontinues the delivery of solution when the treatment is completed.

Many infusion pumps use standardized administration sets and tubing. A linear peristaltic pumping mechanism propels fluid through the tubing. The tubing is stretched across the pump head where the pumping "fingers" come in contact with the tubing (see illustration) and as each successive pumping finger squeezes the tubing against the backplate, the fluid contained in the tubing is pushed along in a peristaltic fashion. When tubing is not correctly in place within the tubing channel, the pumping fingers are not able to compress the tubing, or do so only partially, resulting in altered fluid delivery.

One type of pump, for example, is associated with two different types of misloadings which result in different fluid delivery outcomes: a siphoning of fluid in the line and blood rom the patient backwards through the tubing, or inaccurate fluid delivery.

To minimize the potential for infusion pump misloading, all of the factors associated with the device, the environment and the user, as well as their interactions, must be considered during the design of the device.

Conference Participants

Dr. Anthony D. Andre
Interface Analysis Associates
1135 S. De Anza Blvd.
San Jose, CA 95129
Phone: (408) 342-9050
Fax: (408) 342-9059
andre@interface-analysis.com

Dr. Carryl L. Baldwin
WIT
P.O. Box 5199
Siox City, IA 51102-5199
Phone: (712) 274-8733 x1425
baldwic@witcc.cc.ia.us

Debbie Blum
Div. of Postmarket Surveillance
Office of Surveillance & Biometrics
CDRH, FDA
1350 Piccard HFZ-520
Rockville, MD 20850

Dr. Marilyn Sue Bogner
Institute for the Study of Medical Error
9322 Friars Rd.
Bethesda, MD 29817
Phone: (301) 571-0078
Fax: (301) 530-0679
msbogner@erols.com

Dr. Cristina Bubb-Lewis
Lucent Technologies
189 West Front Street
Red Bank, NJ 07701
Phone: (732) 345-7133
Fax: (732) 817-4606
cbubb@lucent.com

Dr. Peter Burns
Volvo Technology Development Corporation
Dept. 6930, PVH 32
Phone: +46 31 59 88 72
Fax: +46 31 59 54 15
peterb@VTD.VOLVO.SE

Dr. Evan Byrne
NTSB 490 L'Enfant Plaza S.W.
Washinton, DC 20594
Phone: (202) 314-6352
byrnee@NTSB.gov

Dr. Barrett S. Caldwell
Department of Industrial Engineering
University of Wisconsin-Madison
1513 University Ave., Rm 393
Madison, WI 53706-1572
ph: (608) 262-2414; fax: (608) 262-8454
caldwell@engr.wisc.edu

Diego Castano
Cognitive Science Laboratory
The Catholic University of America
Washington, DC 20064
Phone: (202) 319-5825
Fax: (202)319-4456
06castano@cua.edu

Cindy R. Ching
414 South Fountain Green Rd.
Bel Air, MD 21015-4755
Phone: (410) 836-0662
cindy_ching@chppm-ccmail.apgea.army.mil

Dr. Richard Christ
US Army Research Institute
3910 El Capitan Dr.
Temple, TX 76502-1117
(254) 286-6946
ChristR@hood-emh3.army.mil

Dr. J. Raymond Comstock, Jr.
NASA Langley Research Center
Hampton, VA 23681-0001
phone: (757)864-8262
fax: (757) 864-7793
j.r.comstock@larc.nasa.gov

Katherine Cook
Centre for Human Services
D30 East Court
DRA Portsdown West
Fareham,
Hants PO17 6AD
England

Mark S. Creasap
HQ USAF
4143-7 Falcon Place
Waldorf, MD 20603
(301) 932-8113
creasap.mark@andrews.af.mil

Eleanor Davy
DERA
Air Human Factors
Rm 2012, A5 Bldg.
DERA Farnborough, Hants
GU140LX, UK
Phone: +44 (0)1252 395382
Fax: +44(0)1252 394700
ecdavy@dera.gov.uk

Dr. John Deaton
297 Saxony Ct.
Winter Springs, FL 32708
Phone: (407) 699-5017
Fax: (215) 542-1412
john_deaton@chiinc.com

Dr. William Dember
Dept. of Psychology
University of Cincinnati
Cincinnati, OH 45221-0376
Phone: (513) 556-5537
william.dember@uc.edu

Dr. Dick de Waard
Centre for Environmental and Traffic Psychology
University of Groningen
Grote Kruisstraat 2/1
9712 TS Groningen
The Netherlands
Phone: +31 50 363 6758
d.de.waard@ppsw.rug.nl

Dr. Cynthia O. Dominguez
Air Force Research Laboratory/Human Engr.
Division
2255 H Street
Wright-Patterson AFB, OH 45433
Phone: (937) 255-7559
Fax: (937) 255-0879
email: cdominguez@al.wpafb.af.mil

Mica Endsley
SA Technologies
4731 East Forest Park
Marietta, GA 30066
Phone: (770) 565-9859
mica@satechnologies.com

Jacqueline A. Duley
Cognitive Science Laboratory
The Catholic University of America
Washington, D.C. 20064
Phone: (202) 319-5825
Fax: (202) 319-4456
21duley@cua.edu

Todd Eischeid
Dept. of Psychology
Old Dominion University
Norfolk, VA 23529
tme300g@oduvm.cc.odu.edu

Stephen Fairclough
Research Institute
The Elms, Elms Grove
Loughborough University
Leicestershire
LE11 1RG UK
Phone: +44 1509 611088
s.u.fairclough@lboro.ac.uk

Alexis Fink
Dept. of Psychology
Old Dominion University
Norfolk, VA 23529
aaf200g@oduvm.cc.odu.edu

Angela Galinsky
Sioux City Founding Co.
801 Division St.
Sioux City, IA 51104-3067
Phone: (712) 252-4181
Fax: (712) 252-4197
agalinzky@aol.com

Scott Galster
Cognitive Science Laboratory
The Catholic University of America
Washington, D.C. 20064
Phone: (202) 319-5825
Fax: (202) 319-4456
59galster@cua.edu

Capt. Richard Gifford
Lockheed Martin
MS 389
NASA Langley Research Center
Hampton, VA 23681-0001
r.b.gifford@larc.nasa.gov

329

Simon Green
DERA
Air Human Factors
DERA Farnborough, Hants
GU140LX, UK
Fax: +44 (0)1252 394700

Tessa Gorton
DERA
Air Human Factors
DERA Farnborough, Hants
GU140LX, UK
Phone: +44 (0)1252 394784
Fax: +44 (0)1252 394700
tgorton@dera.gov.uk

Gerald Hadley
Dept. of Psychology
Old Dominion University
Norfolk, VA 23529
gah400z@oduvm.cc.odu.edu

Dr. Lawrence Hettinger
Logicon Technical Services, Inc.
PO Box 317258
Dayton, OH 45431-7258
Phone: (937) 255-8770
Fax: (937) 255-8778
lhettinger@al.wpafb.af.mil

Dr. Brian Hilburn
Man Machine Integration Division (VE)
National Aerospace Laboratory (NLR)
Anthony Fokkerweg 2
1059 CM Amsterdam, The Netherlands
Tel: +31 20 511 36 42
Fax: +31 20 511 32 10
hilburn@nlr.nl

Edward M. Hitchcock
Department of Psychology, ML 376
University of Cincinnati
Cincinnati, OH 45221-0376
hitchcem@email.uc.edu

James M. Hitt
Center for Applied Human Factors in Aviation
University of Central Florida
Orlando, FL 32816-1390
jmh19074@pegasus.cc.ucf.edu

Tom Hughes
Veridian
5200 Springfield Park
Dayton, OH 45431
Phone: (937) 255-8848
thughes@al.wpatb.af.mil

Dr. Toshiyuki Inagaki
Insttitute of Information Sciences
and Center for TARA
University of Tsukuba
Tsukuba 305 Japan
inagaki@is.tsukuba.ac.jp

Matt Jackson
 DERA
Air Human Factors
DERA Farnborough, Hants
GU 140LX, UK
Phone: +44 (0)1252-394790
Fax: +44 (0)1252 394700

Keith S. Jones
Department of Psychology, ML 376
University of Cincinnati
Cincinnati, OH. 45221-0376
joneks@email.uc.edu

Peter G.A.M. Jorna
National Aerospace Lab (NLR)
Anthony Fokkerweg 2
1059 CM Amsterdam
The Netherlands
Phone: +31 20 5113638
jorna@nlr.nl

Dr. Elaine Justice
Dept. of Psychology
Old Dominion University
Norfolk, VA 23529

Dr. David B. Kaber
Department of Industrial Engineering
Mississippi State University
PO Box 9542
125 McCain Building
Mississippi State, MS 39762-7400
Phone: (601) 325-8609
Fax: (601) 325-7618
kaber@engr.msstate.edu

Hyon Kim
Dept. of Preventative Medicine
University of Wisconsin-Madison
Madison, WI 53706-1572
hyonkim@students.wisc.edu

Kirsten Kite
Florida Institute of Technology
949 Flotilla Club Drive
Indian Harbour Beach, FL 32937
Phone: (407) 779-3287

Dr. Jefferson Koonce
CHAFA
P.O. Box 1780
University of Central Florida
Orlando, FL 32816-1780
(407) 823-3577
(407) 823-5862
koonce@pegasus.cc.ucf.edu

Kristin Krahl
Dept. of Psychology
Old Dominion University
Norfolk, VA 23529
kxk100g@oduvm.cc.odu.edu

Corrina Lathan
Dept. of Biomedical Engineering
Catholic University of America
Washington, DC 20064

Kara Latorella
NASA Langley
MS 152
Hampton, VA 23681
Phone: (757) 864-2030
k.a.latorella@larc.nasa.gov

Dr. Mark Lehto
School of Industrial Engineering
Purdue University
Grissam Hall
Room 265
West Layfayette, IN 47907
lehto@ecn.purdue.edu

Dr. Elizabeth A. Lyall
Research Integrations, Inc.
PO Box 25405
Tempe, AZ 85285-5405
Phone: (602) 777-8292
b_lyall@ix.netcom.com

Anthony Macera
Dept. of Psychology
Old Dominion University
Norfolk, VA 23529

Dr. Debra A. Major
Dept. of Psychology
Old Dominion University
Norfolk, VA 23529
dmajor@odu.edu

Anthony Masalonis
Cognitive Science Laboratory
The Catholic University of America
Washington, DC 20064
masalonis@cua.edu

Thomas W. Mastaglio
VMASC
7000 College Drive
Suffolk, VA 23435
Phone: (757)686-6201
Fax: (757) 686-6214
mastaglio@VMASC.odu.edu

Gerald Matthews
Department of Psychology
University of Dundee
Dundee DD1 4HN
Scotland
Phone: +44 1382 344612
Fax: +44 1382 229993
G.MATTHEWS@dundee.ac.uk

David W. Mayleben
Department of Psychology, ML 376
University of Cincinnati
Cincinnati, OH. 45221-0376

Ulla Metzger
Cognitive Science Laboratory
The Catholic University of America
Washington, DC 20064
Phone: (202) 319-5825
Fax: (202) 319-4456
metzgeru@cua.edu

Dr. Peter J. Mikulka
Dept. of Psychology
Old Dominion University
Norfolk, VA 23529

Louis Miller
Dept. of Psychology, ML 376
University of Cincinnati
Cincinnati, OH 45221-0376
millerlu@email.uc.edu

Robert Molloy
NTSB
490 L'Enfant Plaza East SW
Washington, DC 20594
(202) 314-6516
molloyr@ntsb.gov

Catherine Morgan
DERA Centre for Human Sciences
DERA Portsdown West
D30 East Court
Portsdown Hill Road
Fareham, Hants
PO17 6AD UK
Phone: +44 1705 336346

Dr. Mustapha Mouloua
Center for Applied Human Factors in Aviation
University of Central Florida
Orlando, FL 32816-1390
mouloua@pegasus.cc.ucf.edu
Fax: (407) 823-5862

Mary Niemczyk
Research Integrations, Inc.
P.O. Box 25405
Tempe, AZ 85285-5405
(602) 777-8292
niemczyk@ix.netcom.com

Dr. Raymond Nickerson
5 Gleason Road
Bedford, MA 01730

Dr. Kent Norman
Dept. of Psychology
University of Maryland
College Park, MD 20742-4411

Dr. Kelly O'Brien
883 Greenwood Manor Circle
Melbourne, FL 32904

Dr. Raja Parasuraman
Cognitive Science Laboratory
The Catholic University of America
Washington, D.C. 20064
Phone: (202) 319-5825
Fax: (202) 319-6263
parasuraman@cua.edu

Christine Parentier
Div. of Postmarket Surveillance
Office of Surveillance & Biometrics
CDRH, FDA
1350 Piccard HFZ-520
Rockville, MD 20850

Dr. David Pick
Dept of Behavioral Sciences
Purdue University Calumet
Hammond, IN 46323
Phone: (219) 989-2622
pick@calumet.purdue.edu

Dr. Alan Pope
NASA Langley Research Center
MS 152
Hampton, VA 23681-0001

Lawrence J. Prinzel, III
Dept. of Psychology
Old Dominion University
Norfolk, VA 23529

Dr. Robert Proctor
Dept. of Psychology
Purdue University
W. Lafayette, IN 47907-1364
Phone: (765) 494-0784
proctor@psych.purdue.edu

Dr. Joseph Psotka
US Army Research Institute
ATTN: TAPSC-ARI-BR
5001 Eisenhower Ave
Alexandria, VA, 22333-5600
(703)617-5572
FAX: 617-5162

Paul Rau
NHTSA
400 Seventh Street SW
Washington, DC 20590
(202) 366-0418
paul.rau@nhtsa.dot.gov

Dr. Victor Riley
Honeywell, Inc. HTC
3660 Technology Drive
MN65-2600
Minneapolis, MN 55418
Phone: (612) 951-7405
riley-vic@htc.honeywell.com

Bert Ruitenberg
Appelgaarde 4
2771PK Boskoop
The Netherlands
B_Ruitenberg@compuserve.com

Dr. Nadine B. Sarter
Institute of Aviation - Aviation Research Laboratory
University of Illinois at Urbana-Champaign
1 Airport Rd.
Savoy, IL 61874
Phone: (217) 244-8657
Fax: (217) 244-8647
nsarter@s.psych.uiuc.edu

Dr. Mark W. Scerbo
Dept. of Psychology
Old Dominion University
Norfolk, VA 23529
Phone (757) 683-4217
Fax (757) 683-5087
mws100f@oduvm.cc.odu.edu

Dr. Brooke Schaab
974 Kelso Ct.
Va Beach, VA 23464-3058
phone: (757) 467-2745

Victoria S. Schoenfeld
Dept. of Psychology
Old Dominion University
Norfolk, VA 23529

Paul C. Schutte
Mail Stop 152
NASA Langley Research Center
Hampton, VA   23681-0001
Phone: 1+757-864-2019
Fax: 1+757-864-7793
P.C.SCHUTTE@LaRC.NASA.GOV

Younho Seong
801 Robin Road
Amherst, NY 14228
(716) 688-5734

John O. Simon
Dept. of Psychology
University of Cincinnati
Cincinnati, OH 45221-0376
simonjo@uc.edu

Dr. Philip J. Smith
Cognitive Systems Engineering Laboratory
The Ohio State University
210 Baker Systems, 1971 Neil Ave.
Columbus, OH 43210
phone: (614) 292-4120
fax: (614) 292-7852
phil+@osu.edu

Julie Stark
Dept. of Psychology
Old Dominion University
Norfolk, VA 23529

Dr. Alan Stokes
Florida Institute of Technology
949 Flotilla Club Drive
Indian Harbour Beach, FL 32937
Phone: (407) 779-3287

Richard Stone
944 South Fremont Rd
Bountiful, VT 84010
Phone: (801) 299-1561
richbstone@classic.msn.com

Sonia Swayze
Div. of Postmarket Surveillance
Office of Surveillance & Biometrics
CDRH, FDA
1350 Piccard HFZ-520
Rockville, MD 20850

James L. Szalma
Department of Psychology, ML 376
University of Cincinnati
Cincinnati, OH. 45221-0376
szalmaj@email.uc.edu

Mona L. Toms
Veridian Veda Operations
5200 Springfield Pike, Suite 200
Dayton, OH 45431-1289
Phone: (937) 255-5550
Fax: (937) 476-2900
mtoms@dytn.veridian.com

Anna C. Trujillo
NASA Langley Research Center, MS 152
Hampton, VA 23861-0001

Dr. Robert Tyler
The National Aviation and Transportation Center
Dowling College
1300 William Floyd Parkway, Suite 200
Long Island, NY 11967-1822
Phone: (516) 244-1331
Fax: (516) 395-2996
tylerr@dowling.edu

Frank Vetesi
Lockheed Martin
1 Federal St., A&E 3W
Camden, NJ 08102
Phone: (609) 338-3923
Fax: (609) 338-4122
fvetesi@atl.lmco.com

Dennis Vincenzi
Center for Applied Human Factors in Aviation
University of Central Florida
Orlando, FL 32816-1390

Feng Wang
The Catholic University of America
Cognitive Science Laboratory
Washington, DC 20064
Phone: (202) 319-5825
49wang@cua.edu

Dr. Joel S. Warm
Department of Psychology, ML 376
University of Cincinnati
Cincinnati, OH. 45221-0376

Dr. Harold Warner
22987 Butternut Lane
California, OH 20619

Donald O. Weitzman
TRW Government Information Systems
One Federal Systems Park Drive
Fairfax, VA 22033
phone: (202) 651-2280
don_weitzman_at_seta@mail.hq.faa.gov

Dr. Christopher Wickens
Aviation Research Laboratory
University of Illinois
ARL #1 Airport Rd.
Savoy, IL 61874
phone: (217) 244-8617
fax: (217) 244-8647
email: cwickens@psych.uiuc.edu

Dr. Kelli Willshire
NASA Langley Research Center, MS 152
Hampton, VA 23681-0001

Jennifer Wilson
Research Integrations, Inc.
P.O. Box 25405
Tempe, AZ 85285-5405
Phone: (602) 777-8292
wilson_j@ix.netcom.com

David Wing
NASA Langley Research Center
MS 499
Hampton, VA 23681
Phone: (757) 864-3006
d.j.wing@larc.nasa.gov

Rebekah Vint
Research Integrations, Inc.
P.O. Box 25405
Tempe, AZ 85285-5405
Phone: (910) 841-1302
rlyall@ix.netcom.com

Subject Index

337